AF556737

Organometallic Reaction Mechanisms

Of the Nontransition Elements

ORGANOMETALLIC CHEMISTRY

A Series of Monographs

BRIAN G. RAMSEY: Electronic Transitions in Organometalloids, 1969.
R. C. POLLER: The Chemistry of Organotin Compounds, 1970.
RUSSELL N. GRIMES: Carboranes, 1970.
PETER M. MAITLIS: The Organic Chemistry of Palladium, Volume I, Volume II – 1971.
DONALD S. MATTESON: Organometallic Reaction Mechanisms of the Nontransition Elements, 1974.

In preparation

P. S. BRATERMAN: Metal Carbonyl Spectra
RICHARD F. HECK: Organotransition Metal Chemistry: A Mechanistic Approach
P. W. JOLLY AND G. WILKE: The Organic Chemistry of Nickel, Volume I, Organonickel Complexes
P. C. WAILES, R. S. P. COUTTS, AND H. WEIGOLD: Organometallic Chemistry of Titanium, Zirconium, and Hafnium

Organometallic Reaction Mechanisms

Of the Nontransition Elements

Donald S. Matteson
Department of Chemistry
Washington State University
Pullman, Washington

ACADEMIC PRESS New York and London 1974
A Subsidiary of Harcourt Brace Jovanovich, Publishers

ACADEMIC PRESS, INC.
111 Fifth Avenue, New York, New York 10003

United Kingdom Edition published by
ACADEMIC PRESS, INC. (LONDON) LTD.
24/28 Oval Road, London NW1

Library of Congress Cataloging in Publication Data

Matteson, Donald Stephen, Date
Organometallic reaction mechanisms of the nontransition elements.

(Organometallic chemistry)
Includes bibliographical references.
1. Organometallic compounds. 2. Chemical reaction, Conditions and laws of. I. Title.
QD411.M39 547'.05 73-795
ISBN 0-12-481150-7

PRINTED IN THE UNITED STATES OF AMERICA

TO MARIANNA
from her Mad Chemist

Contents

Chapter 4 **Electrophilic Displacement Involving Neighboring Sites**

Chapter 5 **Polar 1, 2-Additions and Eliminations**

Chapter 6 **Carbene Transfer Agents, Carbenes, and Related Topics**

Chapter 7 **Free-Radical and Photochemical Reactions**

Chapter 8 **Free Metals and Metal Anions**

Preface

The field of organometallic reaction mechanisms has grown very rapidly in the last few years and has now reached a stage of maturity where a critical review can be useful to research workers and graduate students. It would be presumptuous of anyone to believe he could write a book covering such a diverse and rapidly expanding topic, but I was enough of a Boron chauvinist to assume that most of what I needed to learn would be analogous to what I knew already. A note of realism was added with the expansion of the title by the disclaimer, "of the nontransition elements."

Readers may notice a more classical sort of chauvinism in that most of the references are to the *Journal of the American Chemical Society*. The *Journal of Organometallic Chemistry* has also been covered fairly thoroughly, but the rest of the literature has not been searched in any systematic way. The historical development of a topic has not been included unless I happened to be familiar with it and to consider the ideas involved to be of some current interest. There is no magic way that a mere human can search all the literature while he is busy writing a book on a rapidly expanding topic and get the book done before it is obsolete.

What I have tried to do is to provide a reasonably critical review of selected significant developments in the field and to outline a self-consistent set of interpretations of organometallic reaction mechanisms. Well-established mechanisms have been emphasized, with more speculative mechanisms introduced whenever it seemed likely that these might stimulate further research. A special effort has been made to keep descriptions of proposed transition states consistent with quantum mechanical principles at the level of qualitative molecular orbital descriptions. Stable compounds have been used as models for transition state structures whenever analogies can be drawn.

Errors or questionable conclusions in the literature have been dissected, no doubt too enthusiastically to win the friendship of their authors, whenever it seemed that such analysis would have instructional value. It should be quickly added that much of the recent literature is of high quality. The occasional human error should not be cause for undue pessimism, and I have tried to strike a reasonable balance between faith and doubt. I would like to believe that I have been infallible in revealing what the correct mechanisms really are, but will instead settle for the hope that any questionable interpretations of mine will provoke a reader into doing the experiments needed to prove me wrong.

I thank the National Science Foundation for support of my research on reaction mechanisms, my graduate students and postdoctorals for continuing the production of boron chemistry while I have been writing, numerous colleagues for contributing preprints, reprints, and commentary on their (and their rivals') work, and my typist, Mrs. Joanna C. Farnsworth, for a succinct statement of a cautionary principle useful in evaluating scientific work and in reading what is to follow: "Faith may lead but cannot guide."

DONALD S. MATTESON

CHAPTER 1

Bonding, Structure, and Potential Energy Surfaces

I. Introduction

An organometallic reaction mechanism involves a carbon–metal bond, either directly or in a neighboring-group effect. The coverage in this book is restricted to main-group metals and metalloids, including boron and silicon but not phosphorus. Emphasis will be placed on mechanisms which have been well defined by experimental evidence.

A reaction mechanism is an ordered sequence of molecular structures. Points of interest include the starting materials, any metastable intermediates, the transition state, and the products. The first and the last of these are the easiest to determine experimentally, but the journals are littered with the ruins of mechanism studies in which the products were considered too obvious to be worth proving. (Given an even chance, a human will guess wrong more than half the time.) Human error aside, the most challenging structure problems are presented by transition states and short-lived intermediates, and the main objective of any mechanism study is to establish good approximate descriptions of these structures.

Descriptions of molecular structure and bonding can be valid only if they are consistent with quantum mechanics. The symbols used by the organic chemist to represent stable molecules conform to quantum mechanical symmetry rules without conscious effort on his part. However, transition states and reactive intermediates are often not adequately described by conventional electron-pair bond symbolism. Resonance structures can be used, but are likely to turn out meaningless if the user is not somewhat sophisticated in understanding the underlying quantum mechanical rules. Simple molecular orbital descriptions are often the most useful aid to qualitative understanding of transition state structures. It is generally possible to isolate the delocalized bonding system of the forming and breaking bonds from the relatively undisturbed nonreacting portions of the molecule by means of bond symmetry rules, just as the familiar delocalized π bonds of benzene are separated from the σ bonds in the usual molecular orbital treatment, though of course such factoring is always an approximation. Most of the mechanisms described in this book will be done with the aid of dashed lines, which merely indicate some degree of electron delocalization and bonding throughout the region so marked. In this chapter some effort will be made to relate the various sorts of dashed-line symbolism used to the underlying quantum mechanical symmetry rules.

The best clues to transition state structures are provided by stable structures. Stable representatives of the types of bonding found in transition states and unstable intermediates can often be found. Accordingly, bonding in theoretically important types of organometallic compounds is discussed in this chapter.

Complete description of a reaction mechanism would involve mapping a potential energy surface. In general, it is not possible to map more than a few points on the surface by experimental methods. However, it is not at all difficult to concoct hypothetical reaction mechanisms which cannot be related to any sort of potential energy surface, and many chemists have therefore committed such absurdities in the literature. Accordingly, this chapter will conclude with a brief discussion of potential surfaces and their relation to mechanisms.

II. Carbon–Metal Bonds

A. Typical Structures

Carbon–metal electron-pair bonds are typical covalent bonds. If the metal is an electropositive one, the electron density naturally is higher on carbon than on the metal, but most commonly encountered organometallic compounds are not salts of carbanions and cannot ionize to carbanions under any known conditions. Some exceptions are found in the alkali metal series.

The structure of the Grignard reagent is a good place to begin. Rundle and co-workers have reported the X-ray crystal structures of phenylmagnesium bromide dietherate (Fig. 1-1)[1] and ethylmagnesium bromide dietherate.[2] The X-ray studies show that the magnesium atom is tetracoordinate with distorted tetrahedral geometry and a covalent carbon–

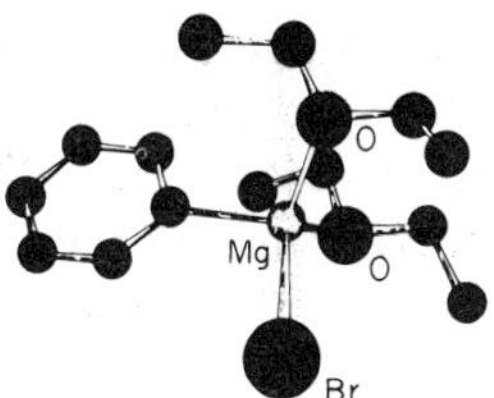

Fig. 1-1. The structure of phenylmagnesium bromide dietherate. [From G. Stucky and R. E. Rundle, *J. Amer. Chem. Soc.*, **86**, 4828 (1964).]

magnesium bond. For example, in ethylmagnesium bromide the Mg—C distance is 2.15 Å, the expected covalent bond length. The Mg—Br distance is 2.48 Å and the Mg—O distances are 2.03–2.05 Å.[2] The Br—Mg—C angle is 125° and the Mg—C—C angle is only slightly distorted from tetrahedral at 114.6°.

Although ordinary alkyllithium compounds are covalent tetramers (Section III,F), coordination of the lithium ion with a suitable amine can lead to an ionic salt of a carbanion. An example is the compound formed from benzyllithium and triethylenediamine {1,4-diazabicyclo[2.2.2]octane, $N(CH_2CH_2)_3N$}, which has been shown by X-ray crystallography to consist of a polymeric chain of the amine and lithium cations paired with benzyl anions (Fig. 1-2).[3] The lithium is not bound to any one carbon, but occupies

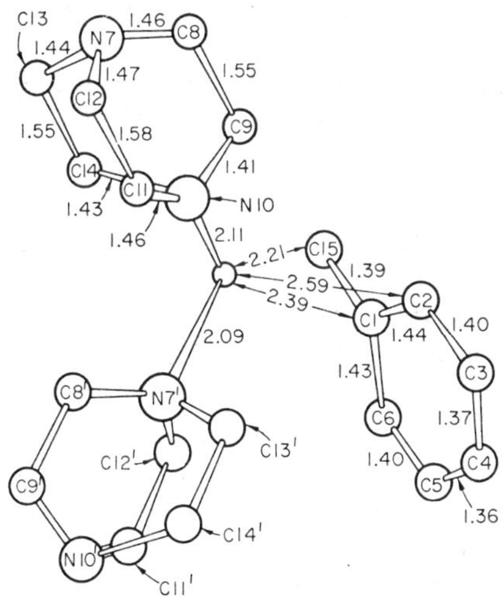

Fig. 1-2. Molecular structure of benzyllithium–triethylenediamine complex viewed down the *b* axis. [From S. P. Patterman, I. L. Karle, and G. D. Stucky, *J. Amer. Chem. Soc.* **92**, 1155 (1970).]

a position similar to a π-bonded ligand above the plane of the benzene ring and between the benzyl and adjacent ring carbons.

Numerous other X-ray structures of organometallic compounds can be found in the literature, but there seems no need to provide further illustrations of simple covalent and ionic carbon–metal bonding.

B. Some Thermodynamic Relationships

The chemistry of the carbon–metal linkage is generally dominated by the tendency of the metal atoms to undergo oxidation. It is characteristic that hydrolysis of the C—M bond yields C—H and M—OH linkages, not C—OH and M—H. With the more electropositive metals, hydrolysis and related reactions are very exothermic. For example, B—C bond energies in trialkylboranes average ~83–88 kcal/mole and the B—O bonds in boric acid, 133 kcal/mole.[4] The penalty for converting an O—H to a C—H bond is only 11 kcal/mole,[5] leaving the hydrolysis exothermic by 34–39 kcal/mole per B—C bond. It should be emphasized that this result does not reflect any weakness in the B—C bond itself, which appears to be 5–10 kcal/mole stronger than C—C bonds in alkanes.[5] Hydrolysis of the B—C bond in the opposite direction to B—H (as diborane) and C—OH is endothermic by ~15–20 kcal/mole. The difference in energy between these two directions of hydrolysis resides almost entirely in the difference between the B—O bond (133 kcal/mole) and the C—O bond (78.3 kcal/mole). The C—H bond (98.2 kcal/mole) is only ~4 kcal/mole stronger than the average of the B—H bonds in B_2H_6.

One important consequence of this general thermodynamic trend is that most α-metallocarbonyl compounds, M—CH_2COR, are not isolable because they rearrange rapidly and irreversibly to enolates, CH_2=C(R)OM. For example, no boron compound of the general formula R_2BCH_2COR' has been isolated and characterized. Another example is that the Reformatsky reaction intermediate, often thought of as $BrZnCH_2CO_2Et$, is actually the enolate isomer, CH_2=C(OEt)OZnBr.[6] Thus, the large body of information on enolate condensations is clearly outside the scope of this book. Not surprisingly, silicon–carbon bonds are kinetically inert enough that Ph_3Si—CH_2CO_2H can be made, but it isomerizes to Ph_3Si—$OCOCH_3$ at 200°C in accord with the general thermodynamic trend.[7] Ketones of the type R_3SiCH_2COPh have been isomerized to enolates CH_2=C(Ph)O—SiR_3 at 110°C.[8] Where two metals compete for one oxygen, the rule is that the more electropositive metal gets the oxygen, as illustrated by the isomerization of PhCH(Li)$OSiEt_3$ to PhCH($SiEt_3$)OLi.[9] The large number of synthetic preparations of organometallic compounds from less electropositive metal alkoxides or halides with Grignard or lithium reagents is another illustration.

The more noble metals do not always follow the rule that the C—M bond hydrolyzes to C—H and M—OH. Acid cleavage of organomercurials to hydrocarbon and mercury(II) halide occurs (Chapter 3), but there is also a solvolysis of RHg^+ to R^+ and Hg^0, which, of course, ultimately yields a C—OH linkage.[10] Mercury also forms a variety of α-mercuri carbonyl compounds. An example is 1,3-bis(acetoxymercuri)acetone.[11]

C. Hybridization and Orbital Symmetry

Simplified quantum mechanical descriptions of chemical bonding all rest on symmetry properties of wave functions. The symmetry is represented, at least implicitly, by any structural formula, whether the chemist intends it or not. The power of orbital symmetry arguments in predicting the course of chemical reactions has been illustrated recently by the Woodward-Hoffman rules for the stereochemistry of ring closures.[12] Another familiar example is the Hückel rule, which predicts stability for $(4n + 2)$ electrons in a cyclic system where the atomic orbitals are symmetric with respect to progression around the ring.[13,14]

Hybridized orbitals are a necessity for any localized description of bonding. Hybridization is not something that atoms do or have done to them. It is purely a mental process gone through by the chemist, who wants to group atomic orbitals according to their symmetry properties so he can talk about one localized bond and ignore the rest. Hybridization does not change the shape of the electron distribution in any atom, not only because the atom fails to experience hybridization, but because the mathematical procedure used in deriving hybridized orbitals does not change the electron density distribution around the atom in any way, provided that all the orbitals considered hybridized are equally occupied. However, this last condition is not normally exactly true, and hybridized orbital descriptions are therefore approximations.

If the preceding paragraph conflicts with any of your ideas about hybridization, the problem perhaps lies in the tangled descriptions found in the currently fashionable elementary textbooks. One sees pictures of very knobby-looking tetrahedral carbon atoms which had to go through some magic electron-promotion process to get that way so they could form four strong bonds. The picture is wrong. The actual shape calculated for the carbon atom in methane, using the $2s$ and $2p$ orbitals as the basis set, is close to spherical whether the orbitals are assumed to be hybridized or unhybridized. The only deviation from perfect spherical symmetry with the s and p basis set is the distortion resulting from orbital overlap. More will be said about the shapes of atoms in Section II,D, but first a look at how hybridization helps simplify wave functions is in order.

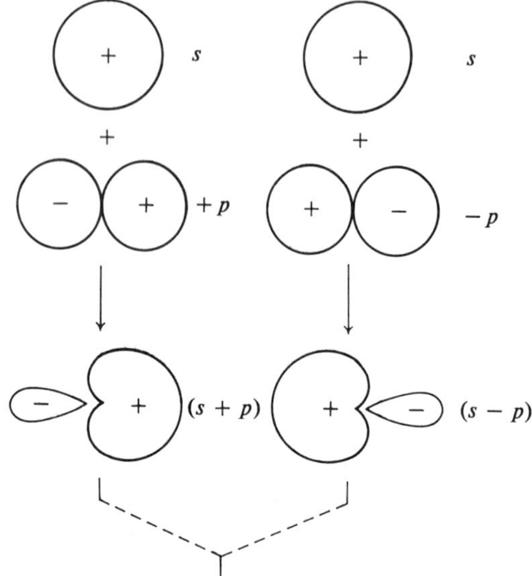

Omitting negative lobes of orbitals, $Hg(CH_3)_2$ is pictured as

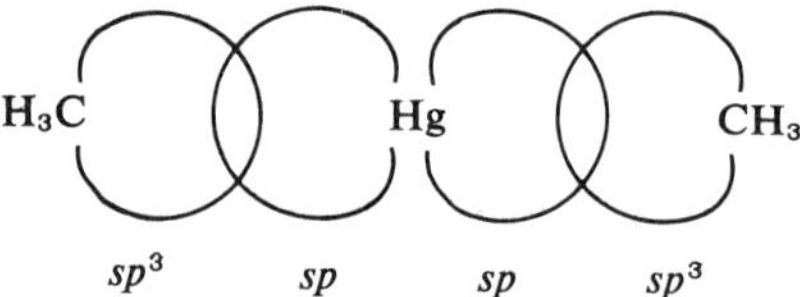

Each overlapping pair of hybrid orbitals combines to form a bond orbital

Fig. 1-3. Hybridized orbital picture of dimethylmercury.

Dimethylmercury has a simple linear C—Hg—C arrangement[15] which is easily described with or without hybridization of the mercury 6*s* and 6*p* orbitals. The hybridized orbital, localized bond picture is probably most familiar. The 6*s* and 6*p* wave functions can be combined in two different ways, $(s + p)$ and $(s - p)$ (Fig. 1-3), to form two orthogonal (noninteracting) *sp* hybrid orbitals pointed 180° apart. Each of these can interact with an sp^3 hybrid orbital of a CH_3 group, yielding two localized carbon–metal bonding orbitals.

Without hybridization, molecular orbitals[13,14] must be used. To avoid undue complexity, the illustration here will treat the methyl groups as contributing a hybridized sp^3 orbital apiece and the mercury atom as contributing an unhybridized *s* and an unhybridized *p* orbital. If the *s* orbital alone were used, the resulting molecular orbitals would be (A), (C), and (E) in Fig. 1-4. The mercury *s* orbital interacts only when the symmetry pattern of the methyl group orbitals is (+ +), yielding the bonding orbital (A), (+ + +), and the antibonding orbital (E), (+ − +). The methyl group orbital symmetry pattern (+ −) does not interact with the mercury *s* orbital, leaving the nonbonding molecular orbital (C), (+ 0 −). If only the mercury *s* orbital were available, then two electron pairs would occupy the bonding molecular orbital (A) and the nonbonding molecular orbital (C) to yield a relatively weak net bonding interaction. However, the mercury *p* orbital interacts with (C) and splits it into two molecular orbitals, (B) (+ +− −) (bonding) and (F) (+ −+ −) (antibonding). Thus, the second electron pair goes into orbital (B) and the nonbonding orbital (C) disappears from the scheme. Likewise, the nonbonding orbital (D) (+ +− +) drops out of the bonding scheme in favor of (A) and (E).

If the *s* and *p* orbitals of the mercury atom were equally occupied, the hybridized and unhybridized orbital descriptions would turn out to be equivalent, yielding the same total wave function. It makes no difference whether the component orbitals are added together before or after construction of the wave function. However, the *s* orbital is at a lower energy level than the *p* and is more highly populated, that is, makes a greater contribution to the total electron density. This is also true of the carbon 2*s* and 2*p* orbitals

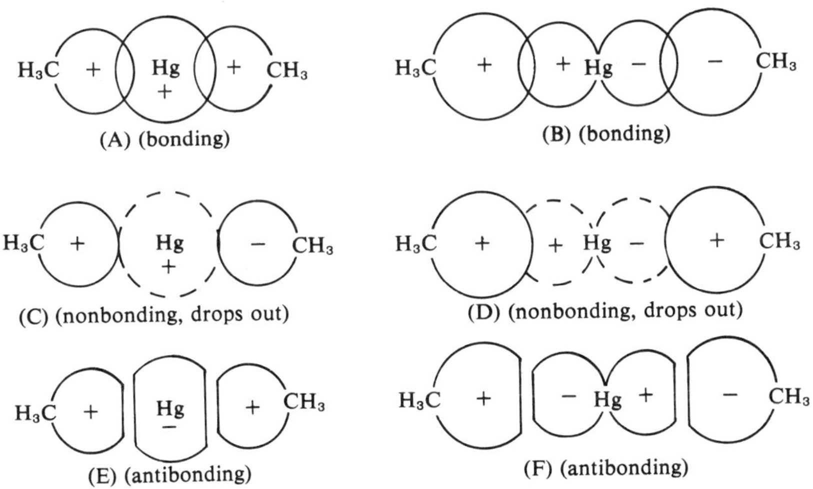

Fig. 1-4. Symmetries of molecular orbitals for linear $H_3C—Hg—CH_3$. Orbitals (A) and (B) are bonding and occupied; orbitals (C) and (D) drop out of the bonding scheme because the Hg orbital is orthogonal to the overall symmetry; orbitals (E) and (F) are antibonding and vacant.

in the methyl groups, which would not be considered hybridized in a proper molecular orbital treatment of $Hg(CH_3)_2$. Thus, the use of unhybridized atomic orbitals in a molecular orbital treatment is capable of yielding more accurate total wave functions, and therefore they are used in extended Hückel calculations[16] and more accurate calculations.[17]

But which is "right," localized bonds or delocalized bonds in saturated molecules? The answer is that there is no such thing as a localized electron pair,[18] except in H_2. All the electrons move in the electric field of the nuclei, all the electrons are indistinguishable, and all lack any memory that could keep them in a particular energy level. Only the total wave function for a molecule has physical reality. Dewar and Gleicher emphasized this point in arguing what might seem a contradictory position, that the bonds in conjugated acyclic polyenes can be considered "localized."[18] Insofar as the properties of any molecule are additive functions of localized bond properties, the bonds may be considered to be localized. Gross effects of electron delocalization generally appear only in aromatic systems, multicenter electron-deficient systems, electronically excited states, and probably in most transition states where bonds are stretched or distorted.

As an example, it may be pointed out that the bonding in $Hg(CH_3)_2^+$, which might be formed by removal of an electron from $Hg(CH_3)_2$ in a mass spectrometer, cannot be considered localized because removal of an electron from any one bond would violate symmetry requirements. If the electron is taken from molecular orbital (B), Fig. 1-4, leaving that level half-filled, the symmetry condition is fulfilled. For a complete description of $Hg(CH_3)_2^+$, loss of electron density from the other carbon and hydrogen orbitals would also have to be considered, but these are largely orthogonal to molecular orbital (B) and the approximation which takes all the electron density out of (B) is probably justified. Such factoring of wave functions into localized and delocalized portions is common practice in molecular orbital theory, as for example in calculating the π-bond systems in aromatic molecules,[14] and is justified whenever the localized and delocalized bonds are orthogonal. Such factoring in some form, if only a distinction between solid lines and broken lines, is absolutely essential to any sort of description of the bonding in transition states.

D. The Shapes of Atoms

It was asserted in Section II,C that closed-shell atoms are nearly spherical. The basis for this assertion and some of its chemical consequences are discussed here.

Unsöld's theorem requires that all rare gas atoms have perfect spherical symmetry.[19] The angular dependencies of p and higher orbitals are sine and cosine functions. The total set is arranged in space so that the sum of the squares, and hence the electron density, does not vary with angle, in accord with the trigonometric identity $\sin^2\theta + \cos^2\theta \equiv 1$. The tendency of atoms to acquire closed valence shells is a tendency toward spherical symmetry, which of course matches the spherical electric field of the nucleus.

Hybridization does not change the spherical symmetry. For example, the complete set of sp^3 hybrid orbitals consists of $\frac{1}{2}(s + p_x + p_y + p_z)$, $\frac{1}{2}(s - p_x - p_y + p_z)$, $\frac{1}{2}(s - p_x + p_y - p_z)$, and $\frac{1}{2}(s + p_x - p_y - p_z)$. The sum of the squares of these four sp^3 functions is just $(s)^2 + (p_x)^2 + (p_y)^2 + (p_z)^2$, which is spherically symmetrical. The cross terms $(s)\cdot(p_x)$, etc., are all cancelled by equal negative terms.

The two distortions from spherical symmetry in chemically bound atoms arise from orbital overlap integrals, which may include 20–30% of the valence electron population, and from unequal populations of p_x, p_y, etc., orbitals when the atom is not in a fully symmetric environment. Experimentally, precise X-ray diffraction measurements confirm the basically spherical shape of atoms and reveal the distortions due to chemical bonding, as, for example, in Stewart's experimental and spherical-atom-model electron density maps of uracil.[20]

The narrow bond angles and high coordination numbers characteristic of carboranes, free metals, and electron-deficient three-center-bonded systems in general (see Section III) can be understood on the basis of the spherical shape of completed *p*-orbital sets. Contrary to common folklore, the amount of space in which orbitals overlap does not vary with bond angle. What does vary is whether the overlap leads to attraction or repulsion (positive or negative overlap integrals), as required by the Pauli exclusion principle. As long as there is plenty of extra space available, it makes little difference which *p* orbital or what combination is involved in chemical bond formation. Carbon can have coordination numbers up to 6, boron up to 7, in carboranes (Section III, A). There is no ring strain in the usual sense in such systems, though atoms do tend to be arranged with near neighbors on all sides, so that no gross bare spots are exposed and repulsions between neighbors are minimized. In contrast, cyclopropane has too many electrons to fit inside the ring, where orbital interaction is strong, and the Pauli principle requires the maximum overlap integrals to be around the periphery where the potential energy is higher, resulting in ring strain.

E. Hückel Treatment of the Carbon–Metal Bond

For purposes of comparison with three-center bonds (Section III), it is useful to consider the results of a Hückel calculation on a generalized carbon–metal two-center bond. The numbers have only qualitative significance. The metal orbital may be chosen as a hybrid orthogonal to the other bonding orbitals on the metal atom. The Coulomb integral for the metal orbital is designated α_M, that for the carbon orbital is α_C, and the exchange integral between the two is β_{CM}. If $\alpha_M = \alpha_C + h\beta_{CM}$, the calculated energy levels are

$$E_1 = \tfrac{1}{2}[\alpha_C + \alpha_M + (\sqrt{4 + h^2})\beta_{CM}]$$

and

$$E_2 = \tfrac{1}{2}[\alpha_C + \alpha_M - (\sqrt{4 + h^2})\beta_{CM}]$$

The electron pair is assigned to the lower energy level E_1, leaving E_2 empty. The total energy of the bond is therefore

$$\begin{aligned} E &= \alpha_C + \alpha_M + (\sqrt{4 + h^2})\beta_{CM} \\ &= 2\alpha_C + (h + \sqrt{4 + h^2})\beta_{CM} \end{aligned}$$

The first expression provides direct comparison with the energy of a carbon radical α_C and a metal radical α_M, the second with the energy of a carbanion $2\alpha_C$ and a metal cation 0. The electron population in the carbon orbital is $2c_C{}^2 = 1 - h/(\sqrt{4 + h^2})$ (and is >1 because h is negative when the metal is more electropositive than carbon) and that on the metal is $2c_M{}^2 = 1 + h/(\sqrt{4 + h^2})$ (which is <1). The c's are the coefficients of the atomic orbitals which yield the minimum energy for the molecular orbital.

To illustrate with actual numbers, if $h = -1$, appropriate for boron, then

$$\begin{aligned} E &= \alpha_C + \alpha_M + 2.24\beta_{CM} \\ &= 2\alpha_C + 1.24\beta_{CM} \end{aligned}$$

the charge on the carbon is $-(2c_C{}^2 - 1) = -0.45$, and the charge on the metal is $+0.45$. If $h = -3$, a reasonable guess for lithium, then

$$\begin{aligned} E &= \alpha_C + \alpha_M + 3.61\beta_{CM} \\ &= 2\alpha_C + 0.61\beta_{CM} \end{aligned}$$

and the electric charges are ± 0.83. These charges are grossly overestimated in typical Hückel calculations, but even so, no reasonable parameter for h can yield a completely ionic bond, which would have $E = 2\alpha_C$ and charges of ± 1.0. Thus, simple theory agrees with the experimental results already noted, which require that carbon–metal bonds be covalent to some degree. Exceptions are known only where the metal ion is strongly complexed by another nucleophile.

III. Three-Center Bonds

A. General Considerations

Cyclic three-center, two-electron bonds, and related multicenter bonds are common in organometallic compounds and in the transition states for some of their reactions. This is hardly surprising, inasmuch as multicenter bonding is characteristic of free metals. An example of three-center bonding is provided by trimethylaluminum dimer (**1**).[21]

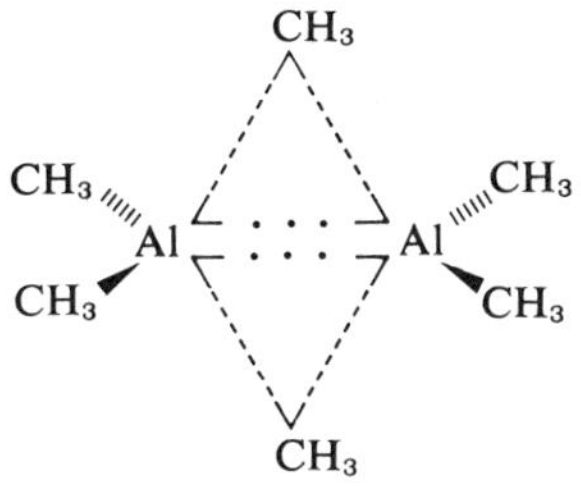

CH_3

CH_3 Al Al CH_3

CH_3 CH_3

CH_3

1

Strangely, some chemists have been reluctant to concede that carbon, too, can participate in three-center bonding. Perhaps it is because at first glance pentacoordination might seem to violate the sacred tetracovalence of carbon, but there is no violation when the quantum mechanical details are considered. X-Ray structure determinations have established the existence of pentacoordinate carbon in trimethylaluminum dimer[21] and medium-size carboranes,[22] and hexacoordinate carbon in a number of icosahedral carborane derivatives.[23] Three-center bonding is a fundamental structural feature of boron hydrides and carboranes, and the multicenter bonds in the larger borane cages can be described by combinations of resonance structures written with three-center bonds.[24]

Examples of the reluctance of chemists to accept three-center bonding include a naive challenge to the trimethylaluminum dimer structure,[25] which will be disposed of later (Section III,C). The controversy over the structure of the norbornyl cation is too famous to need review here, except to point out that Olah and co-workers' recent nmr, Raman, and ESCA spectra of solutions of this species strongly support the existence of an all-carbon three-center bond,[26] as originally advocated by Winstein.

B. Molecular Orbital Descriptions

The molecular orbital description of three-center bonding is straightforward. If all three atomic orbitals are equivalent, as in the π-bonding of the cyclopropenium cation,[27] the usual Hückel calculation yields a bonding energy level $(\alpha + 2\beta)$ and a degenerate pair of antibonding orbitals $(\alpha - \beta)$. The single electron pair goes into the bonding orbital, which distributes the electron density equally to all three carbon atoms, and the antibonding orbitals are empty. Note that the cyclic three-center bonding energy level ($\alpha + 2\beta$) is significantly lower than either the ethylenic two-center orbital ($\alpha + \beta$) or the noncyclic allylic three-center orbital ($\alpha + 1.414\beta$).

Typical examples of organometallic three-center bonds include the CAl_2 bridge bonds in trimethylaluminum dimer and the HB_2 bridges in diborane. The atomic orbitals used to construct these bonds may be generally considered to be sp^3 hybrids, except that the $1s$ orbital of hydrogen is used. Four parameters are needed to describe the CAl_2 three-center bond: α_C, α_{Al}, β_{CAl}, and β_{AlAl}. If $\alpha_{Al} = \alpha_C + h\beta_{CAl}$ and $\beta_{AlAl} = k\beta_{CAl}$, the bonding energy level is

$$E_1 = \alpha_C + \tfrac{1}{2}[h + k\sqrt{(h + k)^2 + 8}]\beta_{CAl}$$
$$= \tfrac{1}{2}\{\alpha_C + \alpha_{Al} + [k + \sqrt{(h + k)^2 + 8}]\beta_{CAl}\}$$

The antibonding orbitals are

$$E_2 = \alpha_C + \tfrac{1}{2}[h + k - \sqrt{(h + k)^2 + 8}]\beta_{CAl}$$

and

$$E_3 = \alpha_{Al} - \beta_{AlAl}$$

Assuming an electron pair in the bonding orbital, the electron density on carbon is $1 - (h + k)/\sqrt{(h + k)^2 + 8}$. Appropriate parameters for aluminum might be $h = -2$ and $k = 1$, which would make $E_1 = \alpha_C + \beta_{CAl}$ and the charge density on carbon -0.33, on aluminum $+0.17$. Increasing k always lowers the energy E_1, but as the electronegativity difference increases and h becomes more negative, the value of k has less effect on the result. For example, if $h = -2$ and $k = 0$, as in a linear Al—C—Al three-center bond, $E_1 = \alpha_C + 0.73\beta_{CAl}$. It is readily apparent that Al—Al bonding can do less to lower the energy than does the third C—C bond in cyclopropenium ion, where $h = 0$ and $k = 1$, as contrasted with allyl cation, $h = 0$ and $k = 0$. An obvious reason for this is the smaller value for the coefficient of the Al orbital than for the C orbital, which makes the C—Al bond order $c_C c_{Al}$ considerably larger than the Al—Al bond order c_{Al}^2. Even so, there is reason to consider the Al—Al bonding a significant effect.

Orbitals do not have sharp cutoff points, and it is impossible to bring two aluminum atoms close enough to bond with the same carbon sp^3 orbital without having the aluminum orbitals overlap with each other. According to the theoretical treatment just presented, as well as more advanced calculations, the result of that overlap is bonding. In accord with the theory, the Al—Al distance in trimethylaluminum dimer is relatively short, 2.60 Å.[21]

The orbital overlap may be pictured as in structure **2**, where the loops represent sp^3 hybrid orbitals on aluminum and carbon. It should be remembered that sp^3 orbitals are much fatter in the dimensions perpendicular to their bonding axis than can be shown easily in an uncluttered drawing, including this one.

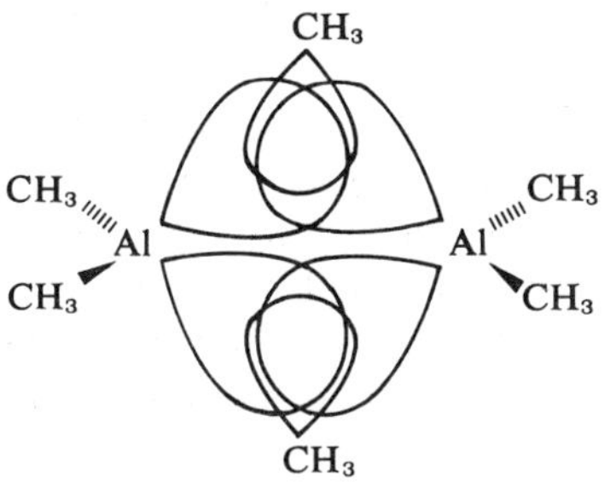

2

From the terminal C—Al—C bond angles of 122°[21] it is at least equally appropriate to factor the aluminum orbitals into sp^2 hybrids and a p orbital. Then the bonding with the bridging methyl groups looks like two superimposed four-center electron-pair bonds as shown for the sp^2 set **3** and the p set **4**. However, it will turn out that these can be factored by symmetry into two three-center bonds. The symmetric four-center orbital in **3** contains

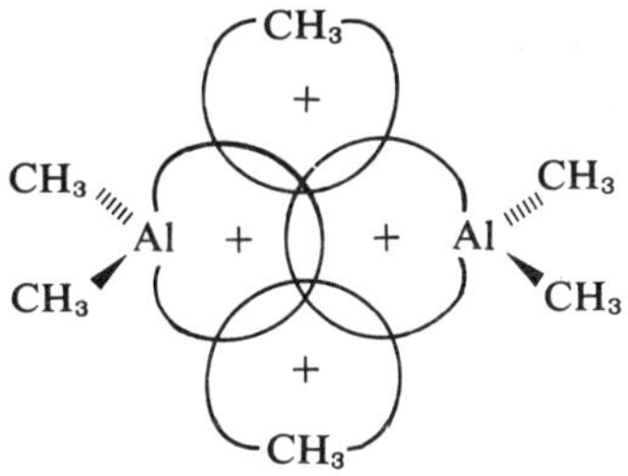

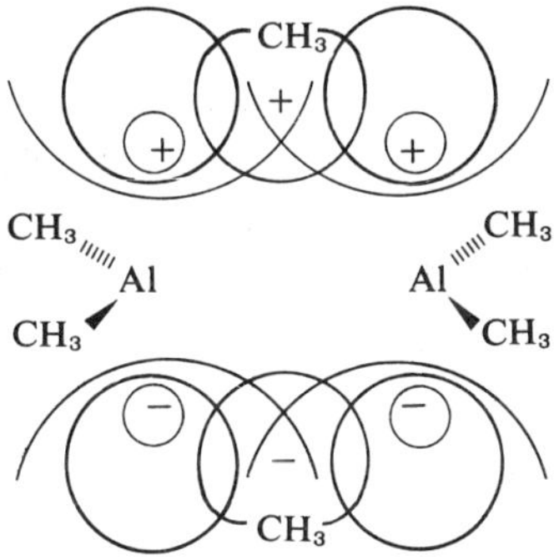

3 **4**

a bonding term between the two bridge methyl groups, whose orbitals do overlap slightly even though the drawing does not show it. However, the antisymmetric orbital in **4** contains an antibonding term between these same methyl groups. The sum of the two occupied orbitals **3** and **4** is nonbonding between the methyl groups. All Al—Al and C—Al interactions are bonding. Thus, two semilocalized three-center CAl_2 bonds are the ultimate result, whether greater delocalization is assumed at the start of the calculation or not. This result is in accord with the principles of hybridization discussed in Section II,C.

It is important to note that orbital symmetries do not allow any further factoring of the cyclic three-center bonds into two-center bonds or even linear three-center bonds. The metal–metal bonding might be weak, but it cannot be zero. A satisfactory universal symbol is needed to represent three-center bonds. The dashed-line triangle (**5**) seems satisfactory, but there will be legitimate objections to this symbol in cases where one of the bond orders is much smaller than the other two. Perhaps making one side of the triangle a dotted line (**6**) or partially open (**7**) will overcome these objections, without committing the quantum mechanical *faux pas* of implying that the low bond order is truly zero (**8**). Lipscomb has used symbols such as **9** and **10** for three-center bonds,[24] their only disadvantage being that an organic chemist often uses such line intersections to represent carbon atoms.

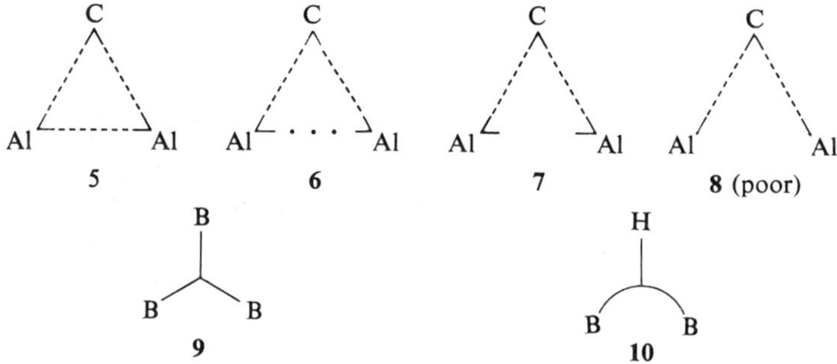

Advanced molecular orbital calculations have been carried out by the CNDO (complete neglect of differential overlap) method on trimethylaluminum dimer.[28] Bond orders in terms of the unhybridized atomic orbitals are summarized for the Al—Al bonding in Table 1-1 and the C—Al bonding in Table 1-2.

TABLE 1-1

BOND ORDERS[a] FOR THE Al—Al BOND IN $Al_2(CH_3)_6$ BY THE CNDO METHOD[b]

	$3s$	$3p_x$	$3p_y$	$3p_z$	$3d_{xz}$	$3d_{yz}$	$3d_{xy}$	$3d_{z^2}$	$3d_{x^2-y^2}$
$3s$	0.150								
$3p_x$	—	0.542							
$3p_y$	0.198	—	0.464						
$3p_z$	—	—	—	0.065					
$3d_{xz}$	—	—	—	—	—				
$3d_{yz}$	—	—	—	0.199	—	0.033			
$3d_{xy}$	—	0.106	—	—	—	—	0.021		
$3d_{z^2}$	0.040	—	0.070	—	—	—	—	0.023	
$3d_{x^2-y^2}$	0.030	—	0.090	—	—	—	—	0.014	0.017

[a] Bond orders less than 0.01 are omitted.

[b] From K. A. Levison and P. G. Perkins, *Discuss. Faraday Soc.* **47**, 183 (1969).

TABLE 1-2

BOND ORDERS FOR THE Al—C BRIDGE BOND IN $Al_2(CH_3)_6$ BY THE CNDO METHOD[a]

Al	$3s$	$3p_x$	$3p_y$	$3p_z$	$3d_{xz}$	$3d_{yz}$	$3d_{xy}$	$3d_{z^2}$	$3d_{x^2-y^2}$
C									
$2s$	0.130	0.246	0.195	—	—	—	0.080	0.049	0.029
$2p_x$	0.211	0.270	0.289	—	—	—	0.045	0.036	0.061
$2p_y$	0.082	0.038	0.022	—	—	—	0.021	—	0.017
$2p_z$	—	—	—	0.070	0.029	—	—	0.014	—

[a] From K. A. Levison and P. G. Perkins, *Discuss. Faraday Soc.* **47**, 183 (1969).

It turns out that $p\pi$–$d\pi$ bonding (**11**) is a considerable part of the Al—Al bonding, which is something one would hesitate to predict from simple molecular orbital descriptions. Another surprising result is that the CNDO calculations indicate considerable negative charge on the aluminum atoms,[28] contrary to expectation unless the Al—Al bonding is fairly strong ($k > -h$ in the simple calculations). The p orbitals are more involved in the bonding than the simple hybridization picture suggests. In the tabulated results, p_y is chosen along the Al—Al axis, p_x is in the plane of the bridging methyl groups, and p_z is in the plane of the terminal methyl groups.

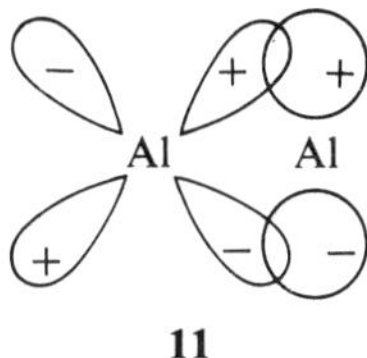

11

Hydrogen-bridged systems are simpler to compute than methyl-bridged systems and were studied earlier. For example, optimized minimum-basis-set SCF (self-consistent field) calculations on diborane (**12**) yielded bond populations of 0.392 for the bridge B—H, 0.288 for the B—B, and 0.826 for the terminal B—H bonds.[29] Palke and Lipscomb concluded that the similarity of the first two figures "... reinforces the notion that there is a moderate boron–boron bond in diborane."

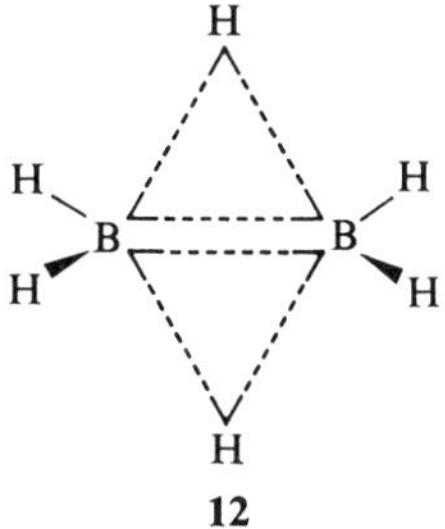

12

C. Structures of Compounds Containing Three-Center Bonds

The relative energies of bridge-bonded dimers and the dissociated monomers are governed by a complex set of factors, as can be appreciated from a glance at Tables 1-1 and 1-2. Trimethylboron does not dimerize, perhaps because the monomer is stabilized by hyperconjugation and its dimer destabilized by steric crowding. Equilibrium favors the monomer of trimethylgallium, perhaps because the dimer is slightly destabilized by inner-shell repulsion or other small negative influences on the strength of the gallium–gallium bond,[21] but the difference between gallium and aluminum compounds is small and vinyldimethylgallium is dimeric with bridging vinyl groups.[30] Dimethylberyllium[31] and dimethylmagnesium[32] (**13**) are polymers linked by methyl bridges.

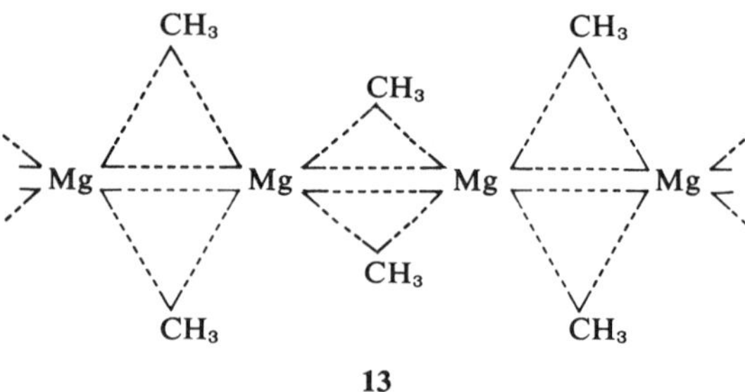

13

Methyl bridges between different kinds of metal atoms have also been found. A recent X-ray study has been reported on octamethyldialuminum-monomagnesium, $Me_2Al(Me)_2Mg(Me)_2AlMe_2$ (Fig. 1-5).[33]

A compound of special interest as a model for transition states in electrophilic displacement at saturated carbon is μ-diphenylamino-μ-methyltetra-

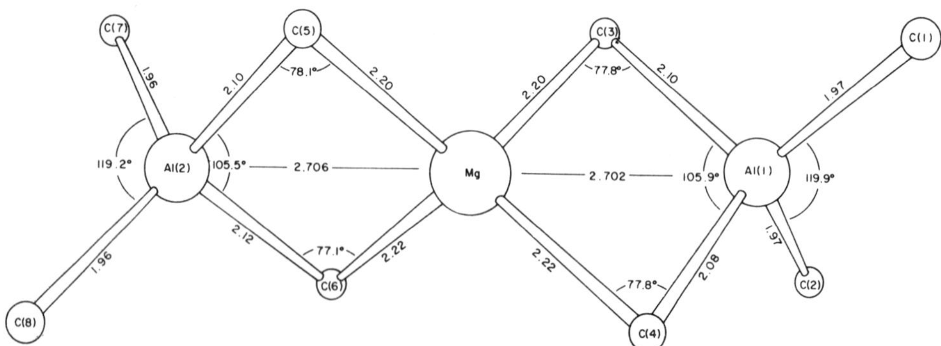

Fig. 1-5. The structure of $Me_2Al(Me)_2Mg(Me)_2AlMe_2$. [From J. L. Atwood and G. D. Stucky, *J. Amer. Chem. Soc.* **91**, 2542 (1969).]

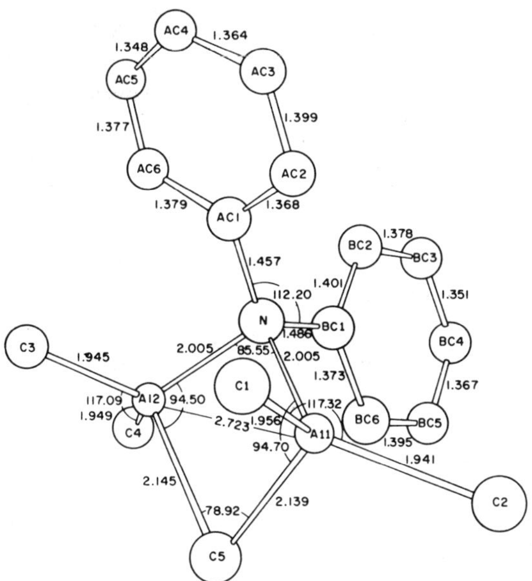

Fig. 1-6. The structure of $Me_5Al_2NPh_2$. [From V. R. Magnuson and G. D. Stucky, *J. Amer. Chem. Soc.* **91,** 2548 (1969).]

methyldialuminum (**14**) (Fig. 1-6).[34] Only the methyl bridge involves a three-center bond. Two electron pairs must be assigned to the nitrogen bridge.

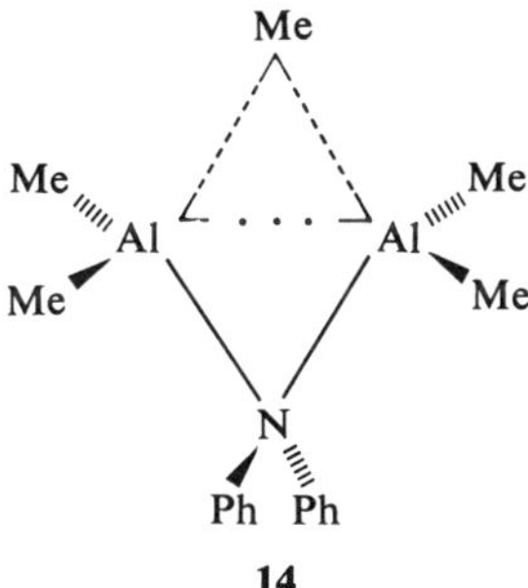

14

The simple way to do this is to factor the nitrogen orbitals into an sp^3 set and write two localized N—Al bonds. As usual in such situations, the molecular orbital approach yields the same net answer a harder way. One electron pair can be assigned to a totally symmetric orbital (**15**), but the second pair must be assigned to an antisymmetric orbital which has a node across the nitrogen and between the aluminum atoms (**16**). The nitrogen p orbital is included in the antisymmetric set and makes the N—Al interaction

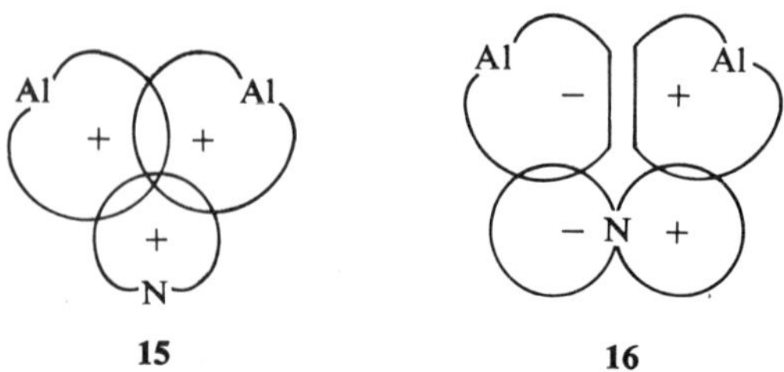

15 **16**

bonding. The sum of these two orbitals is nonbonding between the two aluminum atoms, bonding between aluminum and nitrogen, and is equivalent to localized Al—N bonds. Hybridization was used to simplify **15** and **16**, but by similar and more complex analysis it is possible to start with unhybridized atomic orbitals and still come out with a net sum of two localized N—Al bonds.

Extended Hückel calculations on the model compounds $H_2Al(Me)_2AlH_2$, $H_2Al(Me)(NH_2)AlH_2$, and $H_2Al(NH_2)_2AlH_2$ yielded the Al—Al bond orders 0.23, 0.04, and -0.13, respectively.[34] If the negative Al—Al bond order in the Al—N—Al bridge is constant at -0.065 per bridge, the positive Al—Al bond order in the Al—C—Al bridge is constant at $+0.11$ (± 0.01) per bridge.

It was mentioned in Section III,A that the three-center bond structure of trimethylaluminum dimer has been challenged. Nyburg and co-workers carried out a further refinement of Vranka and Amma's data and found some peaks which might be hydrogen atoms,[25] but their locations are most peculiar. The proposed structure (**17**) contains C—H—Al bridge bonds

H
C—H
H
Me Al Al Me
Me H Me
H C
H

17

(proposed but not reasonable)

which have exactly the same C—Al distance as the unbridged C—Al bonds, and the methyl groups are twisted at a strange angle, with their H—C—H bond angles being close to 90°.

Vranka and Amma had noticed the same peaks[21] but assumed they were the result of inaccuracies in their data. Structure **17** is irrational from any theoretical point of view and there is no justification for taking it seriously, as has been pointed out emphatically by Cotton.[35] Synthetic chemists who may have assumed that X-ray crystallographers have a leakproof pipeline to the Truth may be amused by the 1926 finding that yes, the central carbon atom in $C(CH_2OH)_4$, strange as it seemed, was unquestionably planar,[36] or the more recent mistaking of octaphenylcyclooctatetraene for octaphenylcubane[37] and the assignment of a strange basketlike structure to icosahedral carborane.[38] In crystallography as in any science, "If you give anything a chance to go wrong, it will."

Even though structure **17** is neither required by the facts nor consistent with theory, there is nothing quite like good experimental data for proving a point. Dewar and Patterson found structure **17** to be incompatible with NQR measurements.[39] Oliver and co-workers have recently determined the crystal structure of tricyclopropylaluminum dimer and have found that the bridging cyclopropyl groups are oriented in such a way that the hydrogen cannot possibly be involved in the bridging[40] (Fig. 1-7). The Al—Al distance is 2.618 Å and the average bridge Al—C distance is 2.087 Å, ranging from 2.06 to 2.10 Å. The terminal Al—C distance averages 1.93 Å, range 1.90–1.95 Å. The cyclopropyl groups are oriented *cis*.

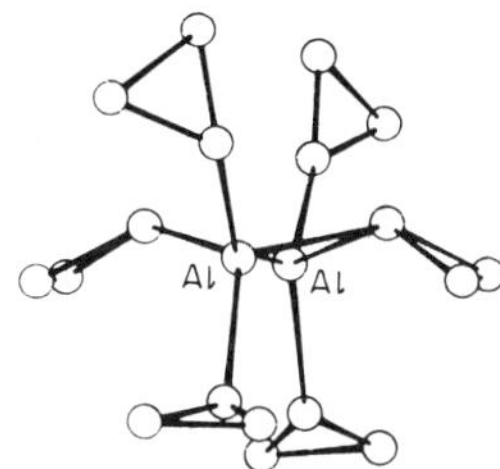

Fig. 1-7. The structure of tricyclopropylaluminum dimer. [From J. W. Moore, D. A. Sanders, P. A. Scherr, M. D. Glick, and J. P. Oliver, *J. Amer. Chem. Soc.* **93,** 1036 (1971).]

D. Olefin π Complexes

A somewhat different type of three-center bond occurs in metal–olefin π complexes. Nontransition metals generally do not form stable π complexes with olefins, though such complexes are probable intermediates in some kinds of additions to double bonds (Chapter 5). An example of a stable silver(I)–olefin complex has been characterized by X-ray diffraction.[41] The principal bonding in the silver(I)–olefin complex is presumably a three-center bond (**18**) formed from the symmetric ethylenic π orbital and the *s* or *sp* hybrid orbital of the silver, with the silver acting as an electron acceptor. However, there is also an antisymmetric component (**19**) formed from the filled silver *d* orbital and the vacant ethylenic π orbital. The antisymmetric component weakens the carbon–carbon π bond, but because this π bond

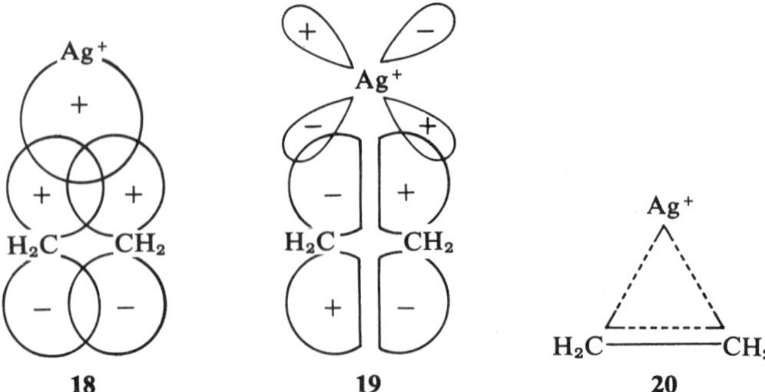

was very strong to begin with and probably remains stronger than the C—Ag bond, it seems appropriate to retain the usual three-center bond symbol (**20**). However, it should be kept in mind that this symbolism is to some degree inadequate, and two electron pairs are involved to some extent in the bonding.

It may seem outside the scope of this book to include a silver(I) complex here, but the distinction between transition and nontransition metals is blurred. Both silver(I) and mercury(II) are d^{10} ions and bond to olefins in the same way, and the former is illustrated here because its complexes are stable and isolable, while the latter's are not. Aluminum and other metals which lack filled *d* orbitals appear to form unstable π complexes containing only the symmetric (**18**) type of three-center bonding.

E. A Linear Methyl Bridge

Electrophilic displacement with inversion (Chapter 2, Section IV,C and Chapter 3, Sections III,C–F) requires that the reaction coordinate pass through a point where the carbon atom becomes trigonal bipyramidal and contributes a *p* orbital to an antisymmetric molecular orbital linking the electrophile E through the carbon to the electrofuge M (**21a**). No stable model for such a structure is available, but some degree of approach is provided by the linear B—CH_3—Li linkages in the crystal of lithium tetramethylboron (**21b**).[42] The X-ray results obtained by Stucky's group[42a] have

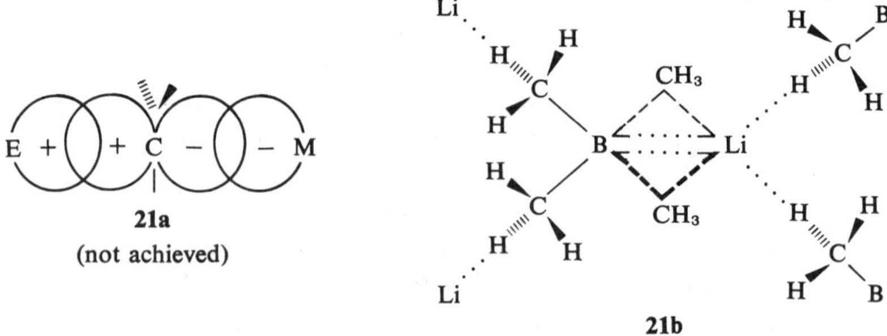

21a
(not achieved)

21b

been somewhat revised by recent neutron diffraction studies, which have provided precise locations for the hydrogen atoms.[42b]

The carbon atoms in the linear B—CH_3—Li linkages of **21b** are roughly tetrahedral, which may represent some flattening compared with the narrower H—C—H angle expected for carbon having considerable anionic character and found in the cyclic bridge methyl groups of **21b**. The interatomic distances suggest strong interaction of the lithium with both hydrogen and carbon of the linear B—CH_3—Li links. The failure of the carbon to approach planarity is in accord with the general favoring of retention over inversion in electrophilic displacements.

F. Lithium and Four-Center Bridge Bonds

Organolithium compounds tend to form cluster structures held together by two-electron four-center bridge bonds, which link each alkyl group to three lithium atoms. Weiss and Lucken showed by X-ray studies that methyllithium exists as a tetramer (**22**) in the crystalline state.[43] The tetrameric species is

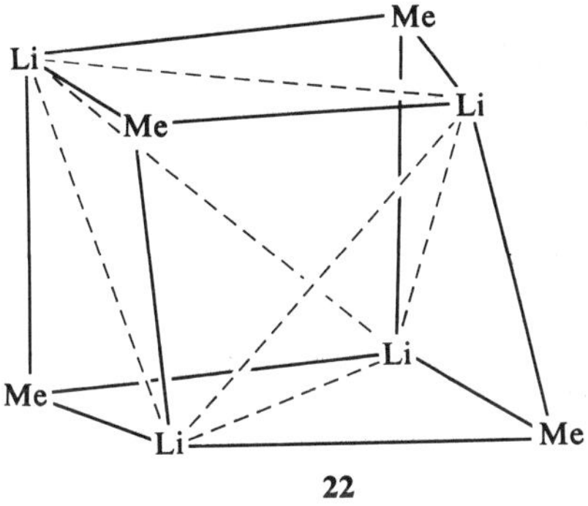

22

also favored in ether solution, and it has been shown by ^{7}Li nmr on 57% ^{13}C methyllithium that each lithium atom has three neighboring carbons, ^{7}Li—^{13}C coupling constant ~16 Hz.[44] A hexamer, probably octahedral, is the predominant species in hydrocarbon solvents.[45]

The coupling constant between ^{6}Li and ^{7}Li in methyllithium tetramer is less than 0.3 Hz.[46] Brown *et al.* interpreted this result as indicating that the Li—Li bond order is very small. However, this conclusion is questionable. The lithium atoms should also be coupled rather strongly through the Li—C—Li linkage, and it seems unlikely that the coupling constant would come out to be zero unless there were some chance factor involved, such as cancellation of two terms of opposite sign. It might further be pointed out that even a weak Li—Li bond would have gross chemical effects. For example, if there were only 1 kcal/mole per bond, this would lower the energy of the tetramer by 6 kcal/mole and increase the equilibrium constant for its formation from monomer by a factor of 10^4 at 25°C.

Cowley and White calculated strong Li—Li interactions in methyllithium tetramer, using the CNDO–SCF method.[47] They also found that the tetramer has a good deal more covalent character than the monomer. (This conclusion agrees with the trend already described in simple Hückel calculations, which suggest less negative charge on carbon with increasing metal–metal bonding in three-center bonds.) Overlap populations and atomic charges are summarized in Table 1-3.

TABLE 1-3

RESULTS OF CNDO–SCF CALCULATIONS ON MeLi AND $(MeLi)_4$[a]

Compound	Bond	Overlap population
CH_3Li	Li—C	0.482
	C—H	0.654
$(CH_3Li)_4$	Li—Li	0.424
	Li—C	0.333
	C—H	0.629
Compound	**Atom**	**Charge**
CH_3Li	Li	+0.53
	C	−0.39
	H	−0.05
$(CH_3Li)_4$	Li	+0.04
	C	−0.26
	H	+0.07

[a] The data in this table are from A. L. Cowley and W. D. White, *J. Amer. Chem. Soc.* **91**, 34 (1969).

IV. Potential Energy Surfaces

A. ELECTRONIC WAVE FUNCTIONS AND NUCLEAR POSITIONS

Reaction mechanisms are fundamentally determined by relationships between molecular structure and energy. The theoretical basis for relating structure and energy is the familiar (but hardly self-explanatory) Schrödinger equation, $H\psi = E\psi$. The brief review which follows is intended to provide qualitative insight into the meaning of the Schrödinger equation to aid in the interpretation of structures and mechanisms. The mathematical formalism and details may be found elsewhere.[13,14,19,48–50]

The Schrödinger equation treats a molecule as a collection of particles having wavelike properties. The Hamiltonian operator H consists of expressions for the kinetic and potential energy of the systems, which naturally must sum to the total energy E. The potential energy follows the classical laws of electrostatics, but the kinetic energy term must be written according to the postulates of quantum mechanics and has meaning only as an operation on the wave function ψ. The kinetic energy of a particle is inversely proportional to its mass, in order to conserve momentum about the center of mass of the total system, which means that 99.95% (in the case of hydrogen) or more of the internal kinetic energy of any molecule is carried by the electrons. Thus, the electronic wave function ψ and the energy E can be computed as if the nuclei were motionless. This approach is known as the Born–Oppenheimer approximation, and it introduces negligible error for ordinary chemical purposes.

A set of selected nuclear positions together with the corresponding electron probability density function $|\psi|^2$ constitutes the molecular structure of the instant, an "instant" being a time interval too short for appreciable nuclear motion but long enough for much electronic motion in this case. As the nuclei shift their positions, $|\psi|^2$ and E change continuously. It is in principle possible to map E as a function of the positions of the atomic nuclei throughout any region of interest. Chemical reactions are accordingly treated in terms of the movement and regrouping of a set of atomic nuclei. In general, a multidimensional surface would be necessary for the map of E versus the nuclear coordinates, but visualization is easier in three dimensions. Daydreamers may imagine a system of mountain valleys and passes, or sober scientists may look at the hypothetical contour map shown in Fig. 1-8. A cross section along the path of minimum energy is the conventional plot of energy versus reaction coordinate illustrated in Fig. 1-9.

Accurate calculations of potential energy surfaces are beyond present computer technology, but reasonably good approximations to scattered points on a surface can be made for simple systems. An example is shown in Fig. 1-10, with a reaction coordinate plot in Fig. 1-11.[51] The reaction is the displacement of one hydride from methane by another hydride, via transition

$H^- + H_3C{-}H \rightleftharpoons H{\cdots}CH_3{\cdots}H \rightleftharpoons H{-}CH_3 + H^-$

23

state **23**. The contours are interpolated from individual points computed by an *ab initio* self-consistent field method. The surface is kept to three dimensions by continuous adjustment of the three nonreacting hydrogens to their optimum positions at all points.

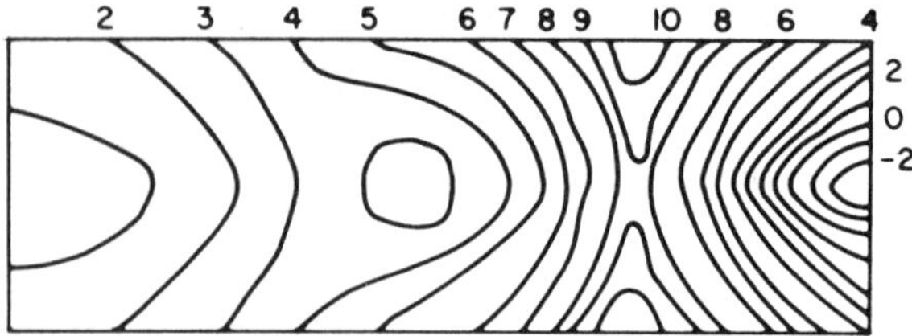

Fig. 1-8. Hypothetical contour map of a potential energy surface for a reaction having a short-lived intermediate.

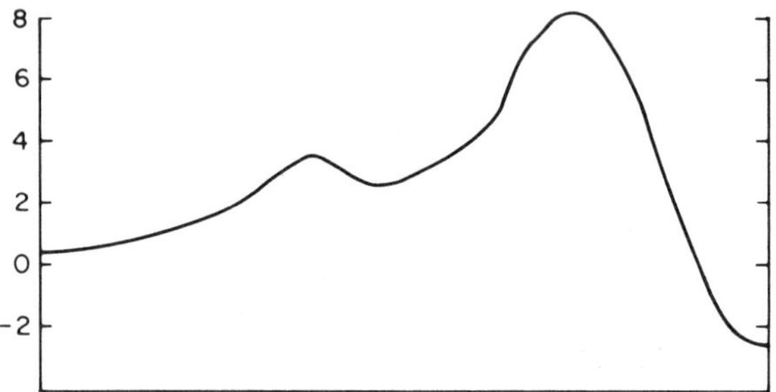

Fig. 1-9. Cross section of minimum energy through Fig. 1-8, the conventional plot of energy versus reaction coordinate.

B. Qualitative Description of Reaction Mechanisms

To relate electronic potential energy surfaces to reaction mechanisms, it is necessary to consider the kinetic energy of molecules. Vibrational excitation is the major factor which leads to the regrouping of chemical bonds. In principle, the electronic potential energy surface can be combined with the kinetic energy of the nuclei in the Schrödinger equation to yield a set of allowed vibrational energy levels. Just as the electrons are not allowed to have zero kinetic energy, the nuclei must also be continuously in motion, and the lowest allowed vibrational energy is a finite distance above the bottom of any potential energy well. This zero-point vibrational motion may be said to be a consequence of the Heisenberg uncertainty principle, which requires that a particle confined to a small location Δx must have an indeterminate momentum Δp such that $\Delta x \Delta p = h$ (Planck's constant), and that for measurements of energy and time, $\Delta E \Delta t = h$. The Schrödinger equation expresses kinetic energy in a manner consistent with the uncertainty principle.

At room temperature most vibrational modes are in the ground state and the thermal energy is distributed among the closely spaced rotational levels and near continuum of translational levels. For an idea of the magnitudes of the energies involved, the translational energy of a perfect gas is 1.5 *RT*, 0.9 kcal/mole at 300°K, and the rotational energy of three-dimensional

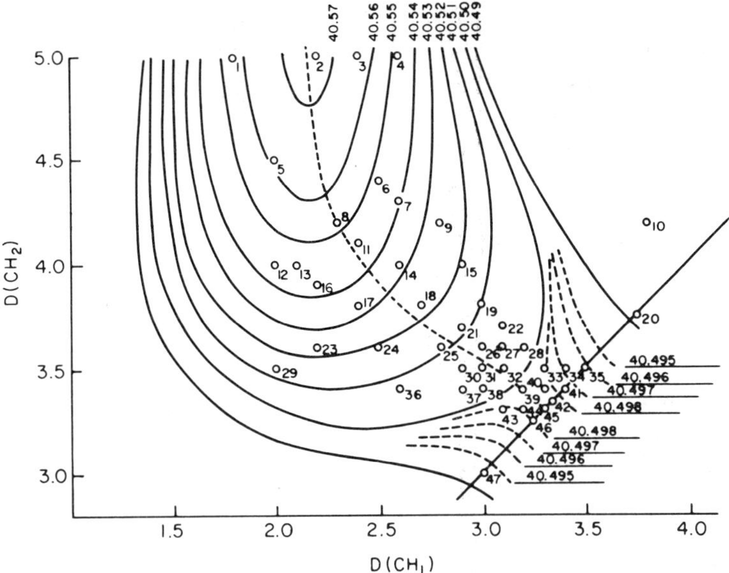

Fig. 1-10. Calculated portion of potential energy surface for the reaction of H^- with CH_4. The transition state is at the lower right and the completed plot would be symmetrical about the line $D(CH_2) = D(CH_1)$, which gives the plot a 90° turn. The distances $D(CH_1)$ and $D(CH_2)$ between carbon and the first and second hydrogen atoms are plotted in atomic units, which equal the Bohr radius of the H atom, 0.529 Å. The energy unit is 627.1 kcal/mole, twice the dissociation energy of the H atom. [From C. D. Ritchie and G. A. Chappell, *J. Amer. Chem. Soc.*, **92**, 1819 (1970).]

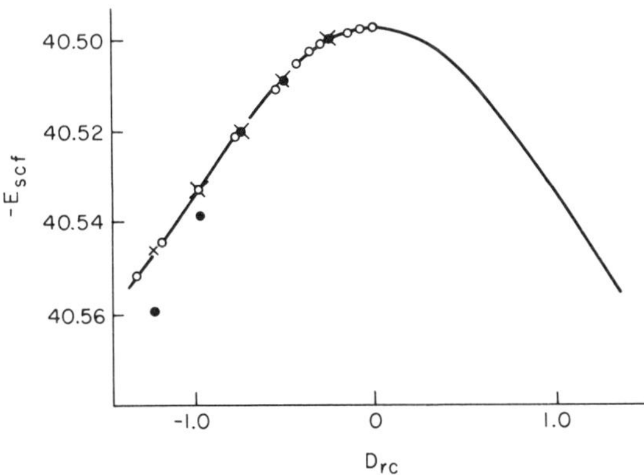

Fig. 1-11. Reaction coordinate diagram based on Fig. 1-10. [From C. D. Ritchie and G. A. Chappell, *J. Amer. Chem. Soc.*, **92**, 1819 (1970).]

molecules is similar in magnitude. Excitation of a 1000 cm^{-1} vibrational transition amounts to 2.86 kcal/mole, and three to ten times this amount is a typical activation energy for a reaction having a rate in the measurable range.

Qualitatively, a molecule may be pictured as moving randomly in an energy plane above the bottom of a potential energy well. As the atomic nuclei move, the electronic wave function readjusts continuously to maintain its optimum value at each point, which is the value corresponding to the potential energy surface below. This picture is thrown out of focus by the uncertainty principle, which does not allow accurate instantaneous measurement of the kinetic and potential energy, but the statistical average values of these terms are definite. The total energy of a molecule must remain constant (conservation of energy) except on collision with another molecule. If a fortuitous collision lifts the molecule or colliding pair to a high enough energy altitude, and if the nuclei happen to move in the right direction, the molecule flies over a relatively low saddle-shaped region in the potential energy wall separating one well from the next. Loss of kinetic energy in another collision drops the molecule into the potential well below and completes the reaction.

The foregoing description applies strictly to the gas phase, but there is no reason to expect any fundamental changes in the liquid phase, except that the size of molecular aggregate carrying the activation energy becomes hard to define.

C. Some Popular Misconceptions

The qualitative picture just described may be clarified by pointing out some popular notions with which it is inconsistent. First, the use of curved arrows (**24**) to show the movement of electron pairs is misleading. The electrons

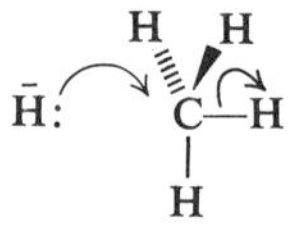

$$\bar{H}: \quad CH_4$$

24

Meaningless as a mechanism

cannot go anywhere without the nuclei, and the nuclei cannot progress without the electron distribution continuously adjusting to the optimum $|\psi|^2$ of the moment, within a range of statistical deviations required by the uncertainty principle. The electrons certainly do not push the nuclei over any electronic energy barrier. Such a thought is almost too illogical to put into words, yet the picture seems to imply it. The curved arrows might help a human keep the electrons counted, but they have no real meaning not contained in the structural formulas of the reactants and products.

A second misleading popular notion is that of "driving force" for a chemical reaction. Thermal energy is the only driving force, except in photochemical reactions. It is possible to ask whether one ψ is associated with a higher or lower E than another ψ. Sometimes it is even possible to understand why. The notion ends there.

Finally, reaction coordinate diagrams have a misleading aspect. It is not correct to say that a molecule takes the path of lowest energy up to and over a potential barrier, because it is not correct to picture the molecule as crawling around on the potential surface at all. Only the sum of kinetic and potential energy is quantized. Molecular configurations are distributed statistically according to the vibrational ψ, which in the ground state favors the region over the center of the potential well but in excited states (reaction pathways) also includes regions of high probability near the sides. Excitation may occur from any region of the potential well to any of several vibrationally excited states having as much energy as the transition state, and there is then a finite probability that the energy will be redistributed into the right bond so that the molecule crosses the potential barrier before it loses energy again. These random paths converge into a small region as the transition state structure is passed.

The structure of a transition state may be defined as a saddle point on the potential energy surface. However, there is a statistical distribution of structures among the individual reacting molecules as they cross the saddle-shaped region, just as there is a statistical distribution of nuclear positions about the energy minimum in a stable vibrating molecule. In practical measurements of transition state structures there is always entropy involved, and attempts to define the exact saddle point on a potential surface usually face impossible obstacles.

D. Absolute Reaction Rate Theory

Quantitative interpretation of the foregoing discussion leads to absolute reaction rate theory.[19,49,50] At equilibrium, the number of molecules N_i in an energy level E_i is related to the number N_j having energy E_j by the Boltzmann distribution law, which is the most probable way to distribute a finite amount of energy among the available levels [Equation (1-1).] The

$$\frac{N_i}{N_j} = \frac{g_i}{g_j} e^{-(E_i - E_j)/kT} \tag{1-1}$$

number of different states having the same energy E_i is g_i, or the entropy of level E_i is $k \ln g_i$, and k is the Boltzmann constant, T the absolute temperature. Macroscopic thermodynamic states contain many energy levels populated according to the Boltzmann law. The sum of the populations of these energy levels is called a partition function, and it can be used to calculate macroscopic thermodynamic properties. The resemblance of the Boltzmann law to the relation between an equilibrium constant K and the standard enthalpy and entropy changes ΔH° and ΔS° [Eq. (1-2)] is readily apparent. Here K replaces N_i/N_j, the molar gas constant R replaces k, and

$$K = e^{\Delta S^\circ/R} \times e^{-\Delta H^\circ/RT} \tag{1-2}$$

ΔH° is the energy with a pressure–volume work term included so that the expression holds at constant pressure.

The time scale of chemical reactions is set by the lifetime and population of vibrationally excited states. Any macroscopic population will have a Boltzmann distribution of energies. Thus, a collection of transition state molecules will not all have identical energies matching the saddle point on the potential energy surface, but a distribution which falls off according to the Boltzmann law, $e^{-E/kT}$. This corresponds to an uncertainty of about kT in the energy of a given molecule. Applying the Heisenberg uncertainty principle, $\Delta E \Delta t = h$, the average lifetime Δt of the transition state has to be about h/kT. This turns out to be the answer obtained from a true calculation of the time required to move across a potential barrier.[19]

The concentration C^* of a transition state is related by the equilibrium constant K^* to the concentrations of the reactants A, B, etc. For example, in a second-order reaction, $C^* = K^*AB$. The transition state decomposes to products with a rate constant $k^* = kT/h$, which makes the rate constant k for the conversion of reactants to products equal to k^*K^*. Straightforward substitutions in Eq. (1-2) lead to the relation between k, ΔH^*, and ΔS^* [Eq. (1-3)]. Theoreticians also discuss the transmission coefficient κ, which

$$k = \frac{kT}{h} e^{\Delta S^*/R} \times e^{-\Delta H^*/RT} \tag{1-3}$$

represents the fraction of transition states actually going on to products rather than reverting to starting materials, but experimentalists out of necessity take κ to be unity, which buries this small discrepancy in ΔS^*.

Rates should be measured at enough different reactant concentrations to establish the rate law, which indicates the stoichiometry of the transition state (except for solvent incorporation or loss) and thus fulfills the role played by elemental analyses of stable compounds. To find ΔH^* and ΔS^* the temperature dependence of k must be measured. The slope of a plot of $\ln k$ versus $1/T$ is the Arrhenius activation energy E_a. For conversion of E_a to ΔH^*, the relationship $\Delta H^* = E_a - RT$ holds in solution, where there is negligible pressure–volume work. For calculations of ΔS^*, it is useful to note that $kT/h = 6.250 \times 10^{12}\ \text{sec}^{-1}$ at 300°K (and is directly proportional to temperature). The value of kT/h is the theoretical maximum for any possible rate constant at temperature T. To obtain ΔS^* directly from the Arrhenius equation, $\ln k = Ae^{-E_a/RT}$, the expression $\Delta S^* = 4.575(\log_{10} A - \log_{10} T) - 49.20$ may be used, where the time unit is the second and ΔS^* is in cal/degree mole (eu, entropy units). (If conversion to joules is desired, 1 cal = 4.184 joules.) For qualitative visualization, ΔS^* may be regarded as the difference between the widths of the potential well of the starting materials and the saddle region of the transition state.

Isotope effects depend solely on shifts in the vibrational energy levels. The heavier isotope leads to a lower zero-point energy, so that the starting material is sunk slightly deeper into its potential well, but the transition state is unbound in the direction of bond making or breaking and the zero-point energy is lost. Thus, the energy gap between starting material and transition state is greater when a bond to a heavier isotope is being broken or made. It should be noted that the potential energy surface is constant and independent of the isotopes employed within the limits of the Born–Oppenheimer approximation, which is very accurate for chemical purposes.

From the foregoing discussion, it should be apparent that the actual shapes of potential energy surfaces are not measurable experimentally, except near the bottom of potential wells where detailed spectroscopic analysis is possible. To describe reaction mechanisms, only a few numbers for special points are available, such as ΔH^*, ΔS^*, and perhaps equilibrium constants for formation of intermediates or products. The chemist must usually be satisfied to write the mechanism in terms of a few rough structural formulas corresponding to points of interest. Even so, in discussing reaction mechanisms it can be very useful to keep the concept of a potential energy surface in mind. There is nothing unusual about describing a complex curve with a small set of special numbers, for example, 94-61-92. (Girl watchers who have not yet gone metric read 37-24-36.) The problem for scientists is to make sure that enough measurements have been included to give a true picture of the situation. As a cautionary tale, the missing dimension in the example just cited is 193 cm (6 ft 4 in.), and the sailor's description is of his own shape six weeks after the shipwreck.

E. Symmetry and Potential Surfaces

Any element of symmetry in a molecular structure must lie on a minimum or maximum on the potential energy surface with respect to that symmetry element. This self-evident requirement is sometimes overlooked in the interpretation of reaction mechanisms, but it can be very useful. Symmetrical reaction coordinates occur only in isotopic substitutions, but reasoning about the replacement of one group by another chemically related group may be aided by considering the analogous isotopic replacement. If a proposed mechanism violates the principle of microscopic reversibility in the case of isotopic exchange, it is probably also incorrect for replacement by other chemically related groups.[52]

Bartlett and Trachtenberg once suggested that the principle of microscopic reversibility requires a symmetrical transition state or intermediate in the case of isotopic exchange.[52] This is probably true in the majority of cases, which would have a single trough of minimum energy connecting the reactants and products. The midpoint of that trough, which must be either a transition state or an intermediate, must have X and X* related by some symmetry element which makes them chemically equivalent. However, in some cases the symmetrical structure may be located on a ridge of high energy, with two symmetrically related troughs passing around to either side. An individual trough may be unsymmetrical lengthwise, but if X* replaces X by one mechanism through one trough, X* must replace X by the exact reverse of that mechanism through the other, and both troughs must be equally used because both transition states have identical energies and structures (except for the position of the isotopic label, which does not affect the potential energy surface). A corollary is that the rate law must be the same in both directions, the relationship between the stoichiometry of the transition state and the reactants being independent of the route to the transition state.

It is not always easy to decide whether there is a single symmetrical energy trough or two symmetrically related unsymmetrical troughs. However, symmetry is favored if all the electrons are in bonding orbitals with respect to the symmetry element, and nonbonding or antibonding electrons promote dissymmetry. Exchange reactions of electron-deficient organometallic compounds tend to proceed by way of symmetrical transition states or intermediates (see Chapter 2). Where the metal has enough ligands to fill its valence shell, it often turns out that no reasonable structure can be written for a symmetrical transition state or intermediate (Chapters 2 and 3).

Some prominent chemists have proposed mechanisms which violate the principle of microscopic reversibility (Chapter 3, Section II,B). Merely leaving out the complementary path required for a symmetrical potential energy surface is trivial, but wrong reasoning about rate laws and ligand bridging is involved in the examples cited.

Another important point about symmetry is that the total electron density must have identical symmetry with that of the group of nuclei. Chemists are now generally aware of this point, and gross errors such as indicating a physical conversion of one resonance structure to another without moving nuclei have fortunately become rare in recent years. Molecular orbital calculations on closed-shell structures automatically yield matching electronic and nuclear symmetries. A partially filled group of degenerate orbitals may appear to have lower symmetry than that of the group of nuclei, but this is illusory. The entire degenerate set has to be superimposed to describe the electron distribution, which then matches the molecular symmetry. Also, such structures tend to be energy maxima without finite lifetimes, which distort in ways suggested by one or another of the degenerate orbitals. A familiar example is singlet cyclobutadiene, which avoids the square in favor of the rectangular structures.[53]

In outlining a reaction mechanism, it is important to keep the whole potential energy surface in mind and to establish its gross features first. Symmetry is of primary importance, but there will be many cases where it is impossible to decide with any reasonable effort whether a given symmetry element lies at an energy minimum or at a minor energy maximum between two lower energy regions to either side. Such minor bumps are of no great chemical importance, but heated mechanistic arguments have been waged over just such issues, with the contenders seemingly oblivious to any definition in terms of potential surfaces, and as a result exaggerating the importance of slight changes in the electronic wave function that would accompany a few degrees of distortion of an angle or a little lengthening of a bond. It is not reasonable to expect a transition state to be characterized with the same detailed accuracy that can be achieved by X-ray diffraction on a crystalline compound, but this is not good grounds for saying that reaction mechanisms are "uncertain." Certainty in science is a relative thing. Synthetic chemists generally consider a structure to be proved when they know which atom is connected to which and what the stereochemistry is. Transition state structures generally require more effort, but many have been characterized sufficiently so that the approximate positions of the atoms are well established, and, in addition, energy and electron distribution relationships with the starting materials are often available. Plenty of errors have been made, as will be pointed out in the following chapters, but the prevailing atmosphere should be one of optimism that much is already known and much more can be learned about organometallic reaction mechanisms.

References

1. G. Stucky and R. E. Rundle, *J. Amer. Chem. Soc.* **85**, 1002 (1963); **86**, 4825 (1964).
2. L. J. Guggenberger and R. E. Rundle, *J. Amer. Chem. Soc.* **86**, 5344 (1964); **90**, 5375 (1968).
3. S. P. Patterman, I. L. Karle, and G. D. Stucky, *J. Amer. Chem. Soc.* **92**, 1150 (1970).
4. A. Finch and P. J. Gardner, *Progr. Boron Chem.* **3**, 177 (1970).
5. J. Hine, "Physical Organic Chemistry," 2nd ed., pp. 30–32. McGraw-Hill, New York, 1962.
6. W. R. Vaughan, S. C. Bernstein, and M. E. Lorber, *J. Org. Chem.* **30**, 1790 (1965).
7. A. G. Brook, D. G. Anderson, and J. M. Duff, *J. Amer. Chem. Soc.* **90**, 3876 (1968).

8. A. G. Brook, D. M. MacRae, and W. W. Limburg, *J. Amer. Chem. Soc.* **89**, 5493 (1967).
9. R. West, R. Lowe, H. F. Stewart, and A. Wright, *J. Amer. Chem. Soc.* **93**, 382 (1971).
10. F. R. Jensen and R. J. Ouellette, *J. Amer. Chem. Soc.* **83**, 4477, 4478 (1961); **85**, 363, 367 (1963).
11. W. L. Waters and E. F. Kiefer, *J. Amer. Chem. Soc.* **89**, 6261 (1967).
12. R. B. Woodward and R. Hoffmann, *Accounts Chem. Res.* **1**, 17 (1968); "The Conservation of Orbital Symmetry." Academic Press, New York, 1969.
13. A. Liberles, "Introduction to Molecular Orbital Theory." Holt, New York, 1966.
14. A. Streitwieser, Jr., "Molecular Orbital Theory for Organic Chemists." Wiley, New York, 1961.
15. D. Grdenic, *Quart. Rev.* **19**, 303 (1965).
16. R. Hoffmann, *J. Chem. Phys.* **39**, 1397 (1963).
17. J. A. Pople, *Accounts Chem. Res.* **3**, 217 (1970).
18. M. J. S. Dewar and G. J. Gleicher, *J. Amer. Chem. Soc.* **87**, 692 (1965).
19. K. S. Pitzer, "Quantum Chemistry," pp. 1–278, 384–393. Prentice-Hall, Englewood Cliffs, New Jersey, 1953.
20. R. F. Stewart, *J. Chem. Phys.* **48**, 4882 (1968).
21. R. G. Vranka and E. L. Amma, *J. Amer. Chem. Soc.* **89**, 3121 (1967).
22. H. V. Hart and W. N. Lipscomb, *J. Amer. Chem. Soc.* **89**, 4220 (1967).
23. (a) J. A. Potenza and W. N. Lipscomb, *Inorg. Chem.* **3**, 1673 (1964); (b) D. Voet and W. N. Lipscomb, *ibid.* **3**, 1679 (1964); (c) J. A. Potenza and W. N. Lipscomb, *ibid.* **5**, 1471, 1478, 1483 (1966).
24. W. N. Lipscomb, "Boron Hydrides." Benjamin, New York, 1963.
25. S. K. Byram, J. K. Fawcett, S. C. Nyburg, and R. J. O'Brien, *J. Chem. Soc. D* 16 (1970).
26. (a) G. A. Olah, A. M. White, J. R. DeMember, A. Commeyras, and C. Y. Lui, *J. Amer. Chem. Soc.* **92**, 4627 (1970); (b) G. A. Olah, G. D. Mateescu, and J. L. Riemenschneider, *ibid.* **94**, 2529 (1972).
27. (a) R. Breslow, J. T. Groves, and G. Ryan, *J. Amer. Chem. Soc.* **89**, 5048 (1967); (b) D. G. Farnum, G. Mehta, and R. G. Silberman, *ibid.* **89**, 5048 (1967).
28. K. A. Levison and P. G. Perkins, *Discuss. Faraday Soc.* **47**, 183 (1969).
29. W. E. Palke and W. N. Lipscomb, *J. Chem. Phys.* **45**, 3949 (1966).
30. H. D. Visser and J. P. Oliver, *J. Amer. Chem. Soc.* **90**, 3579 (1968).
31. A. I. Snow and R. E. Rundle, *Acta Cryst.* **4**, 348 (1951).
32. E. Weiss, *J. Organometal. Chem.* **2**, 314 (1964).
33. J. L. Atwood and G. D. Stucky, *J. Amer. Chem. Soc.* **91**, 2538 (1969).
34. V. R. Magnuson and G. D. Stucky, *J. Amer. Chem. Soc.* **91**, 2544 (1969).
35. F. A. Cotton, *Inorg. Chem.* **9**, 2804 (1970).
36. M. L. Huggins and S. B. Hendricks, *J. Amer. Chem. Soc.* **48**, 164 (1926).
37. (a) (Error) H. H. Freedman and D. R. Peterson, *J. Amer. Chem. Soc.* **84**, 2837 (1962); (b) (Correction) G. S. Pawley, W. N. Lipscomb, and H. H. Freedman, *ibid.* **86**, 4725 (1964).
38. (a) (Error) L. I. Zakharkin, V. I. Stanko, V. A. Brattsev, Yu. A. Chapovsky, and Yu. T. Struchkov, *Izv. Akad. Nauk SSSR, Ser. Khim.* 2069 (1963); (b) (Correction) Ref. 23a.
39. M. J. S. Dewar and D. B. Patterson, *J. Chem. Soc. D* 544 (1970).
40. J. W. Moore, D. A. Sanders, P. A. Scherr, M. D. Glick, and J. P. Oliver, *J. Amer. Chem. Soc.* **93**, 1036 (1971).

41. M. B. Hossain and D. van der Helm, *J. Amer. Chem. Soc.* **90**, 6607 (1968).
42. (a) D. Groves, W. Rhine, and G. D. Stucky, *J. Amer. Chem. Soc.* **93**, 1553 (1971); (b) G. D. Stucky, personal communication (1973).
43. E. Weiss and E. A. C. Lucken, *J. Organometal. Chem.* **2**, 197 (1964).
44. L. D. McKeever, R. Waack, M. A. Doran, and E. B. Baker, *J. Amer. Chem. Soc.* **90**, 3244 (1968).
45. H. L. Lewis and T. L. Brown, *J. Amer. Chem. Soc.* **92**, 4664 (1970).
46. T. L. Brown, L. M. Seitz, and B. Y. Kimura, *J. Amer. Chem. Soc.* **90**, 3245 (1968).
47. A. L. Cowley and W. D. White, *J. Amer. Chem. Soc.* **91**, 34 (1969).
48. C. A. Coulson, "Valence." Oxford Univ. Press (Clarendon), London and New York, 1952.
49. K. B. Wiberg, "Physical Organic Chemistry." Wiley, New York, 1964.
50. J. E. Leffler and E. Grunwald, "Rates and Equilibria of Organic Reactions," pp. 1–75. Wiley, New York, 1963.
51. C. D. Ritchie and G. A. Chappell, *J. Amer. Chem. Soc.* **92**, 1819 (1970).
52. P. D. Bartlett and E. N. Trachtenberg, *J. Amer. Chem. Soc.* **80**, 5808 (1958).
53. P. Reeves, T. Devon, and R. Pettit, *J. Amer. Chem. Soc.* **91**, 5890 (1969).

CHAPTER 2

Electrophilic Displacement: NMR Studies of Metal Exchange

I. Introduction

If there is one reaction uniquely characteristic of organometallic compounds, it is electrophilic displacement at carbon. The most active electrofuges (leaving electrophiles) are metal cations. Even though electrophilic displacement of the proton and other reluctantly electrofugic species from aromatic systems is familiar, it is only with organometallic substrates that electrophilic displacement can be generalized to all types of organic structures and made specific for one particular carbon atom. The practical importance of electrophilic displacements can be appreciated by considering the synthetic utility of the Grignard reagent.

The conventional approach at this point would be to trace the historical development of the topic and to present some speculative suggestions about the possible classes of electrophilic displacement mechanisms. These matters will be postponed until Chapter 3, because recent studies of exchanges of alkyl groups between metal alkyls provide some of the best defined and most easily interpreted experimental evidence available. An example of this type of reaction is the exchange of methyl groups between dimethylzinc (**1**) and dimethylcadmium, the rate of which can be measured easily with proton nmr spectra.[1] The synthetic chemist might doubt that this sort of nonreaction is

$$\underset{\mathbf{1}}{\text{Me—Zn—}\overset{*}{\text{Me}}} + \text{Me—Cd—Me} \rightleftharpoons \text{Me—Zn—Me} + \overset{*}{\text{Me}}\text{—Cd—Me}$$

where chemistry is really happening, but keep in mind the virtues of simplicity and symmetry as aids to the understanding of reaction mechanisms and potential energy surfaces.

II. Principles of NMR Line Broadening

The method of determining rates by nmr line broadening is so simple and useful that any chemist not already familiar with it will want to invest three minutes right now and learn the basic principles. The method works wherever a separate but somewhat broadened nmr peak can be observed for each equilibrating species of interest. The pseudo-first-order rate constant k for the disappearance of a particular species is found from the equation

$$k = \pi(w - w_0) \tag{2-1}$$

where w is the width of the nmr absorption peak measured in hertz (Hz, cycles/second) at half its maximum height under the reaction conditions, and w_0 is the width under conditions of no reaction, such as absence of a second reagent or (less accurately) too low a temperature.[2,3] The factor π arises in making the units cycles per second, and the line shape is assumed to be Lorentzian.

The fundamental measurement is actually of the average lifetime τ of the species yielding the particular broadened line. (Do not confuse this use of the symbol τ with the unrelated τ scale of chemical shifts.) The broadening is a consequence of the Heisenberg uncertainty principle, which requires an uncertainty in the energy of a species when its lifetime is short (see Chapter 1, Section IV,B). If the reaction is first order, τ is independent of concentration and $1/\tau = k$, the first-order rate constant. Otherwise, k is pseudo-first-order and the true rate constant is calculated from the rate law and the concentrations in the usual manner.

From Eq. (2-1) it is evident that proton nmr spectra can generally be used to measure k values in the range 10–100 $\sec^{-1}$ (τ values 10^{-1} to 10^{-2} second). Other nuclei with larger chemical shifts can extend the range to an order of magnitude faster rates. Whenever exchange phenomena are observed, the rate can be brought into the range where Eq. (2-1) is applicable by adjusting the temperature, so long as problems of insolubility or instability do not interfere.

At somewhat higher temperatures where broadened lines begin to overlap, line shapes and rates can be matched with the aid of a computer program.[3] For the special case of two lines which have coalesced into one because of rapid exchange, the lifetimes are related to line width by Eq. (2-2).[2,4]

$$(\tau_A + \tau_B) = \frac{(w - w_0)}{4\pi x_A{}^2 x_B{}^2(\nu_A - \nu_B)^2} \tag{2-2}$$

Here τ_A and τ_B are the average lifetimes of species A and B, x_A and x_B are the "mole fractions" of nuclei A and B involved in the observed line, and ν_A and ν_B are the chemical shifts in hertz of the lines arising from A and B under conditions of no reaction. This equation is also approximately correct in cases where lines A and B arise from a spin–spin splitting which is collapsed by an exchange reaction. To extract individual τ values from Eq. (2-2), the relation $x_A/\tau_A = x_B/\tau_B$ is used.[2] As usual, the pseudo-first-order rate constant $k_A = 1/\tau_A$. If the chemical shifts ν_A and ν_B are 100 Hz apart, τ values as short as 10^{-4} second yield measurable line broadening which can be treated by Eq. 2-2.

III. Group III

A. Monomer–Dimer Equilibria

The alkyls of the Group III metals provide some of the best-defined evidence regarding mechanisms of alkyl exchange, which is of course simultaneously the electrophilic displacement of one metal atom from an alkyl group by another metal atom. An example is the exchange of methyl groups between two molecules of trimethylaluminum monomer, which obviously requires the intermediacy of the dimer **2**. This example is

$$Me_2\overset{*}{Al}-\overset{*}{Me} + Me_2Al-Me \rightleftharpoons Me_2\overset{*}{Al}(\mu\text{-}\overset{*}{Me})(\mu\text{-}Me)AlMe_2 \rightleftharpoons Me_2\overset{*}{Al}-Me + \overset{*}{Me}MeAl-Me$$

2

deliberately chosen because it has its energy relationships backwards. The dimer **2** is not an evanescent transition state or short-lived intermediate but the stable state of the system, and its structure has been established by X-ray crystallography (see Chapter 1, Section III). At sufficiently high dilutions and temperatures, the monomer, $AlMe_3$, would become the relatively stable state. If one of the heavier Group III elements (Ga, In, or Tl) is put in place of aluminum, or isobutyl groups in place of methyl groups, the energies are shifted so that the monomer is stable and the dimer becomes the transition state or unstable intermediate for the exchange of alkyl groups between monomers under standard conditions (idealized 1 molar at 25°C). Plausible plots of enthalpy and free energy versus reaction coordinate are shown in Fig. 2-1.

The curves drawn in Fig. 2-1 are based on a considerable body of data. They probably resemble reality more closely than Picasso's Chicago sculpture

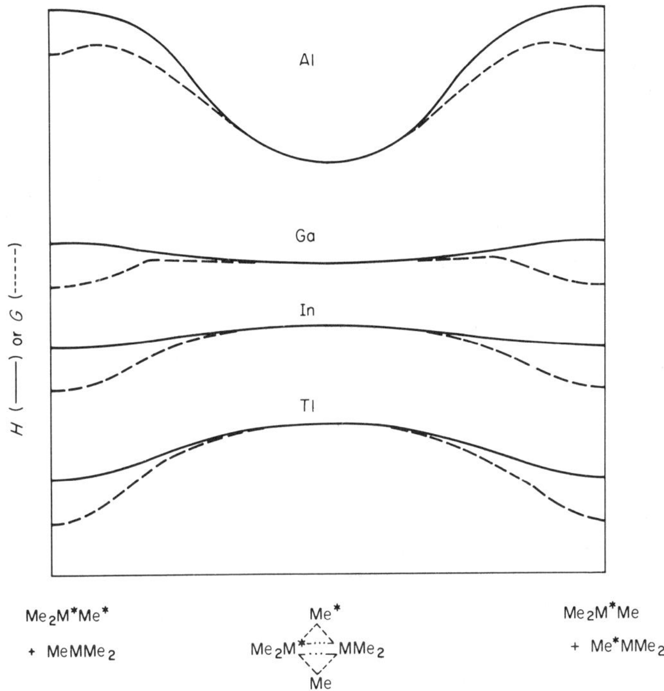

Fig. 2-1. Qualitative plots of enthalpy, *H* (——) and free energy, *G* (-----) versus reaction coordinate for methyl group exchange in the trimethyl compounds of Group III metals.

resembles a woman's head, though of course Picasso's distortions have nothing to do with ignorance. Starting with the curve based on the simplest data, broadening and collapse of the ^{203}Tl and ^{205}Tl satellites in the proton nmr spectrum of trimethylthallium ($TlMe_3$) indicate that the reaction is second order in $TlMe_3$ with a rate constant of 3430 ± 150 liters/mole second at 26°C in toluene-α-d_3 ($PhCD_3$).[5] From the temperature dependence, $\Delta H^* = 5.7$ kcal/mole and $\Delta S^* = -21$ cal/deg mole (eu).

The exchange of methyl groups between trimethylgallium ($GaMe_3$) and trimethylindium ($InMe_3$) is so fast at $-65°C$ in toluene that only a single peak appears in the proton nmr spectrum.[6] This result implies that the lifetime τ is $<10^{-2}$ second. Then if ΔS^* is the usual -20 eu for a bimolecular reaction and the concentrations are $\sim 0.3\ M$, the upper limit on ΔH^* is about 5 kcal/mole. It seems probable that trimethylgallium dimer (Ga_2Me_6) rests in a shallow potential energy well, inasmuch as vinyl bridging groups are sufficient to make dimethylvinylgallium dimer stable with respect to the monomer,[7] but the well is not deep enough to offset the entropy gain on dissociation of Ga_2Me_6 to 2 $GaMe_3$.

Triisobutylaluminum monomer and dimer provide a well-documented example of the sort of enthalpy and entropy relationship just described. From measurements of the heat of dilution in the hydrocarbon tetradecane, $\Delta H°$ for the dissociation of $Al_2[CH_2CH(CH_3)_2]_6$ to 2 $Al[CH_2CH(CH_3)_2]_3$ is 8.16 ± 0.12 kcal/mole and $\Delta S°$ is $+30.49 \pm 0.34$ eu.[8] These correspond to $\Delta G° = -0.9$ kcal/mole at 25°C and the dissociation constant $K = 4.8$. Accordingly, triisobutylaluminum is mostly monomeric under the usual conditions of observation but there is a substantial fraction of dimer in equilibrium in concentrated solutions. It is reasonable to expect that the free-energy curve has minima for the monomer and the symmetrically bridged dimer, with a maximum in between for a distorted dimer structure which lacks much of the entropy advantage of the monomer and lacks the energy advantage of the fully formed dimer. There are no kinetic data available to establish whether this maximum in G is associated with a maximum in H, though a slight rise in H similar to barriers to internal rotations seems likely.

Thermodynamic data are also available for the dissociation of trimethylaluminum dimer (Al_2Me_6) in the gas phase at 100°–155°C,[9] and for triethylaluminum dimer (Al_2Et_6) in tetradecane solution.[10] Hay, Hooper, and Robb have reviewed the literature critically and made revised interpretations of these results.[11] For the dissociation of Al_2Me_6 to 2 $AlMe_3$ in the gas phase, $\Delta H° = 20.2 \pm 1.0$ kcal/mole,[9] but when heats of vaporization are taken into account this should be revised to $\Delta H° = 16.3 \pm 1.5$ kcal/mole in the liquid phase or an inert solvent.[11] For Al_2Et_6 in solution, Smith found $\Delta H° = 16.93 \pm 0.23$ kcal/mole and $\Delta S° = 32.19 \pm 0.63$ eu,[10] but Hay and co-workers pointed out that the conditions of these experiments may have allowed some vaporization to occur, and they estimated $\Delta H°$ to be in the range 8–14 kcal/mole, with the most probable value near 12.5.[11] Thus, the heats of dissociation of Al_2Me_6, Al_2Et_6, and $Al_2[CH_2CH(CH_3)_2]_6$ show a self-consistent trend, decreasing with increasing size of the alkyl groups.

Kinetic data on the dissociation of Al_2Me_6 have been obtained by Ramey and co-workers,[12] Williams and T. L. Brown,[6] and Jeffrey and Mole.[13] The phenomena actually observed are proton nmr line broadenings resulting from methyl group exchanges between bridge and terminal positions of Al_2Me_6. It has been shown that the rate-determining step is dissociation of Al_2Me_6 to 2 $AlMe_3$. Evidence in favor of this conclusion has been obtained not only from data on the bridge–terminal exchange in Al_2Me_6 itself, to be discussed in the following paragraphs, but also from the exchange of methyl groups between Al_2Me_6 and $GaMe_3$, the kinetic complexities of which are analyzed in Section III,B.

The nmr spectrum of Al_2Me_6 in cyclopentane at $-60°C$ shows two separate peaks, the bridge methyl at $\delta 0.03$ and the terminal methyl upfield at $\delta -0.64$. As the temperature is increased these peaks broaden and then coalesce. The values of ΔH^* found are in the range 15–16 kcal/mole,[6,12,13] which is the same as the thermodynamic dissociation energy of Al_2Me_6 within experimental error. The temperature control equipment available for these experiments was not very accurate in an absolute sense, and an estimated ΔS^* value based on all three sources would be about $+15 \pm 4$ eu. If the solvent is changed to toluene, the rate becomes faster by a factor of 20,[13] and ΔH^* values are in the range 14.9–16.0 kcal/mole.[6] Thus, the transition state for bridge–terminal methyl exchange is very close in energy to a pair of dissociated $AlMe_3$ molecules, and therefore it is reasonable to postulate a close structural similarity between the transition state and 2 $AlMe_3$.

The transition state for dissociation of Al_2Me_6 may resemble the symmetrically distorted structure **3** if defined as a saddle point on the potential energy surface, though the unsymmetrical structure **4** probably does not differ much in energy, and statistics would favor the breaking of one C—Al linkage somewhat ahead of the other.

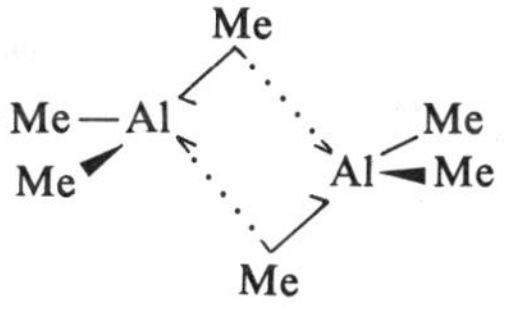

3 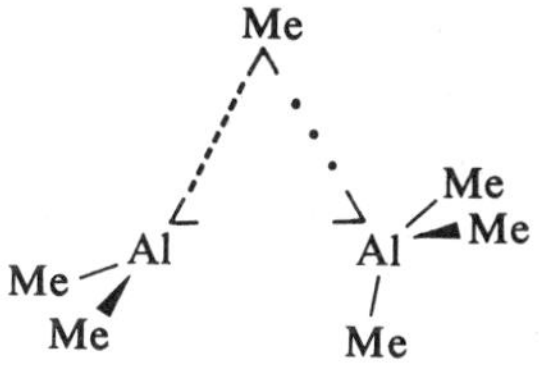 **4**

If **4** is not a transition state but an intermediate with some bonding remaining between the two $AlMe_3$ units, then rotation of one $AlMe_3$ group 120° with respect to the other and reclosure to Al_2Me_6 would interchange a bridge and a terminal methyl group, with the transition state coming at the eclipsed 60° angle of twist.[12] However, all available evidence indicates that dissociation of **3** or **4** is faster than internal rotation. Dissociation governs both the rate of bridge–terminal exchange and the rate of methyl exchange between Al_2Me_6 and $GaMe_3$, discussed in Section III,B. Single-bridged intermediates for bridge–terminal exchange are known where one bridging ligand is a nucleophile such as *t*-butoxide or diphenylamide. These have quite different characteristics and will be discussed in Section III,G.

The qualitative features of the potential energy surface for the bridge–terminal exchange reaction of Al_2Me_6 are illustrated in Fig. 2-2.

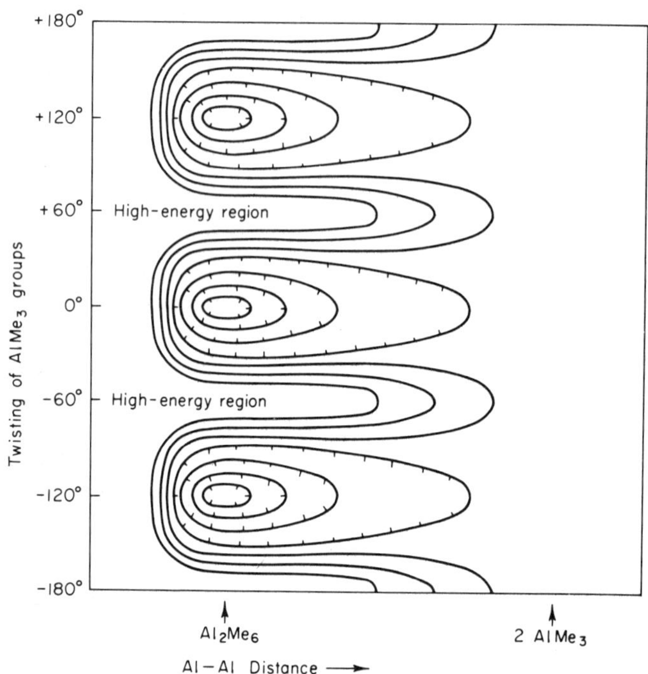

Fig. 2-2. Qualitative contour map of potential energy surface for bridge–terminal methyl exchange of Al_2Me_6. The two atom-position dimensions shown are supposed to represent Al—Al distance along the abscissa and the angle of twist of the two $AlMe_3$ groups along the ordinate, with the methyl group positions assumed to be otherwise optimized for the given parameters.

B. Kinetic Analysis of $GaMe_3$–Al_2Me_6 Methyl Group Exchange

Williams and Brown found that the bridge–terminal exchange of Al_2Me_6 and the methyl exchange between $GaMe_3$ and Al_2Me_6 have the same activation energy within experimental error.[6] The ΔH^* values found for $GaMe_3$ exchange with Al_2Me_6 are 15.4 ± 0.5 kcal/mole in cyclopentane, 16.0 ± 1 in toluene, and for $InMe_3$ exchange in toluene, 15.9 ± 1, which may be compared with the 14.9–16.0 values found for the bridge–terminal exchange of Al_2Me_6 described in the preceding section. A typical Arrhenius plot is shown in Fig. 2-3.

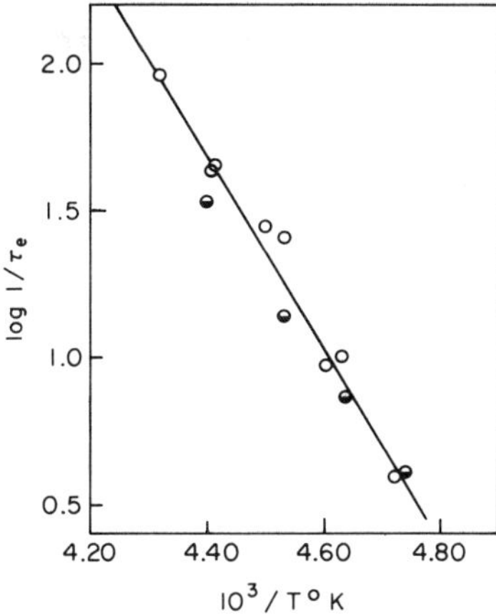

Fig. 2-3. Arrhenius plot of bridge–terminal methyl exchange rates for Al_2Me_6 in toluene: ◒, two-site exchange computer program data; ○, half-intensity line-width data. [From K. C. Williams and T. L. Brown, *J. Amer. Chem. Soc.* **88,** 5462 (1966).]

The rates of the bridge–terminal and $GaMe_3$–Al_2Me_6 exchange are also comparable, as may be seen from the relative line widths in the nmr curves illustrated in Fig. 2-4. For complete understanding of the rate relationships, a detailed kinetic analysis is necessary. Before launching into that, some initial errors in interpretation of the data will be examined for their instructive and cautionary value. In a field as complex and experimentally difficult as this one, even the best of investigators are unlikely to get everything right on the first try.

It was immediately clear that the methyl group lifetimes are about the same in toluene solutions of Al_2Me_6 and of $GaMe_3$–Al_2Me_6 mixtures.[6] However, one research group found exchange rates for Al_2Me_6 alone in cyclopentane[12] which were an order of magnitude greater than those found later for $GaMe_3$ in mixtures with Al_2Me_6 in cyclopentane.[6] It was postulated that the slower rate was caused by a solvent cage effect which retarded the escape of $AlMe_3$ to where it could react with $GaMe_3$ before recombining to Al_2Me_6.[6] Overlooked was the fact that the similar lifetimes of the $GaMe_3$ and Al_2Me_6 in the same cyclopentane solution (Fig. 2-4) preclude such an explanation.[13]

Jeffery and Mole found that the real reason for the discrepancy between the different sets of data was inaccurate temperature control, which is a serious problem with the control equipment generally available.[13] They obtained a self-consistent set of data for Al_2Me_6 alone in cyclopentane and with $GaMe_3$, which is summarized in Table 2-1.

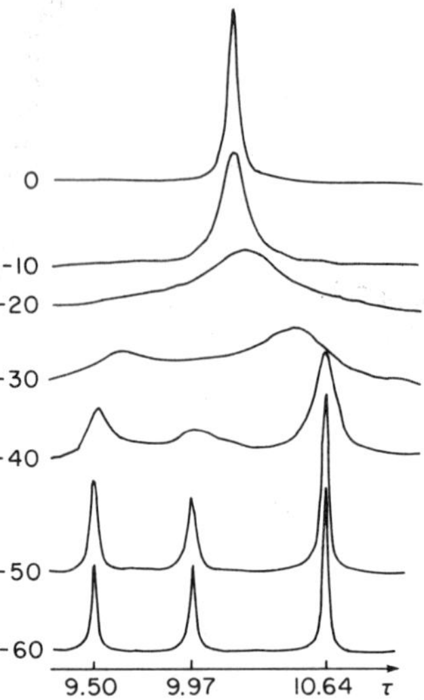

Fig. 2-4. Proton nmr spectra of an Al_2Me_6–$GaMe_3$ mixture in cyclopentane at various temperatures. The line at τ 9.50 is $GaMe_3$, that at 9.97 is Al_2Me_6 bridge methyl, and that at 10.64 is Al_2Me_6 terminal methyl. [From K. C. Williams and T. L. Brown, *J. Amer. Chem. Soc.* **88,** 5462 (1966).]

The data in Table 2-1 support the mechanism which was proposed,[6,13] with slight modification and simplification [Eqs. (2-3) and (2-4)]. Jeffery and Mole

$$Al_2Me_6 \underset{k_{-1}}{\overset{k_1}{\rightleftharpoons}} 2AlMe_3 \tag{2-3}$$

$$AlMe_3 + GaMe_3 \underset{k_{-2}}{\overset{k_2}{\rightleftharpoons}} Me_2Al \begin{smallmatrix} Me \\ \cdots \\ \cdots \\ Me \end{smallmatrix} GaMe_2 \tag{2-4}$$

found that their data fit the empirical rate law of Eq. (2-5), where τ_{Ga} represents the lifetime of a methyl group in a $GaMe_3$ site. The derivation of Eq.

$$\frac{1}{\tau_{Ga}} = \frac{2k_1[Al_2Me_6]}{[GaMe_3] + 0.3} \tag{2-5}$$

(2-5) had to begin with the dubious assumption that the 2 $AlMe_3$ remained caged by the cyclopentane solvent until either they recombined or reacted with $GaMe_3$. The term 0.3 in the denominator of Eq. (2-5) was then derived to be

TABLE 2-1

RECIPROCAL LIFETIMES OF METHYL GROUPS IN $GaMe_3$–Al_2Me_6 MIXTURES IN CYCLOPENTANE AT −47°C[a]

$[Al_2Me_6]$, *M*	$[GaMe_3]$, *M*	$1/\tau_{Ga}$, sec⁻	$1/\tau_{Al(t)}$ [b]	$1/\tau_{Al(b)}$ [b]
0.31	0.57	6.0	7.7	9.4
0.31	0.38	9.0	7.9	8.5
0.31	0.19	15.5	6.9	6.6
0.32	0.095	18 ± 5	6.9	7.4
0.32	0	none	3.5	6.0
0.20	0.38	3.9	7.0	8.3
0.10	0.19	2.7	5.7	7.0
0.05	0.095	1.5 ± 0.5	4.7	5.3

[a] Data from E. A. Jeffery and T. Mole, *Aust. J. Chem.* **22,** 1129 (1969).
[b] Al(t) refers to terminal methyls, Al(b) to bridge methyls of Al_2Me_6.

k_{-1}/k_2.[13] It is not likely that $AlMe_3$ would react faster with $GaMe_3$ than with another $AlMe_3$, and keeping the $AlMe_3$ trapped in a solvent cage decreases the plausibility of this result even further. The following derivation of the kinetic expression avoids any unusual assumptions, provides a general method for handling equilibrium kinetics, and yields results consistent with the experimental data, at least to a good first approximation.[14]

Equilibrium kinetics as measured by nmr must be analyzed in such a way that the expressions are in terms of quantities actually measured and that the equilibrium condition is an aid rather than an obstacle to calculation.[14] The concentrations of $AlMe_3$ and $AlGaMe_6$ are too small to observe, and one possible approach is to use the steady-state approximation that $d[AlMe_3]/dt = 0$ as a point of departure,[6] but there are two pitfalls. First, the equations must be rewritten in terms of the units actually measured, which are methyl groups (protons) and not molecules. If this is not done, the derived kinetic expressions will still have the right form but the wrong relation between the observed rates and calculated k values except in those cases where just one methyl group is transferred in every step of the reaction. The second pitfall is that the steady-state treatment applies to either the forward or the reverse path but will not work for both at once unless some way to distinguish between them symbolically is introduced. If all four of the processes represented in Eqs. (2-3) and (2-4) are put into the steady-state treatment without labeling the methyl groups with respect to their source, which is either Al_2Me_6 or $GaMe_3$, then attempted calculations lead to the dead end $0 = 0$ because there is no net change in the concentration of anything. The different approach which follows takes advantage of the equilibrium condition from the start.

In order to contribute to the observed line broadening, a methyl group exchange must be completed between two of the relatively long-lived sites, which are $GaMe_3$ (to be designated by the subscript Ga), terminal methyl of Al_2Me_6 [designated Al(t)], and a bridge methyl of Al_2Me_6 [designated Al(b)]. The subscript Al will be used to refer to both types of Al_2Me_6 sites collectively. An incomplete transfer, such as from $GaMe_3$ to $AlMe_3$ and back to $GaMe_3$, will not contribute to line broadening. This may be verified by substituting appropriate numbers into the Gutowsky-Holm Equation [Eq. (2-2)], which yields negligible broadening if the lifetime τ_{Ga} is short but τ_{AlMe_3} is much shorter, requiring that the proportion of $AlMe_3$, x_{AlMe_3}, be much smaller than x_{GaMe_3}.

The equilibrium condition supplies Eqs. (2-6) and (2-7). Equations (2-6)

$$k_1[Al_2Me_6] = k_{-1}[AlMe_3]^2 \tag{2-6}$$

$$k_2[AlMe_3][GaMe_3] = k_{-2}[AlGaMe_6] \tag{2-7}$$

and (2-7) can be solved to yield the concentrations of $AlMe_3$ and $AlGaMe_6$ in terms of the measurable concentrations of Al_2Me_6 and $GaMe_3$, Eqs. (2-8) and (2-9).

$$[AlMe_3] = \left[\frac{k_1}{k_{-1}}[Al_2Me_6]\right]^{1/2} \tag{2-8}$$

$$[AlGaMe_6] = \frac{k_2}{k_{-2}}[GaMe_3]\left[\frac{k_1}{k_{-1}}[Al_2Me_6]\right]^{1/2} \tag{2-9}$$

The key to solving the problem is finding the fraction Q of the methyl groups in the $AlMe_3$ which have come from the $GaMe_3$. Each dimerization of $AlMe_3$ transfers $6Q$ methyl groups from $GaMe_3$ to Al_2Me_6. The rate is given by Eq. (2-10). The rate of transfer of methyl groups in the opposite

$$-d[Me_{Ga}]/dt = 6Qk_{-1}[AlMe_3]^2 = 6Qk_1[Al_2Me_6] \tag{2-10}$$

direction, from Al_2Me_6 to $GaMe_3$, must be equal. This rate can be related to the dissociation of $AlGaMe_6$. Half of these dissociations result in transfer of one methyl group from $AlMe_3$ to $GaMe_3$. Since the fraction of methyl groups in the $AlMe_3$ which came originally from Al_2Me_6 is $(1 - Q)$, each dissociation of $AlGaMe_6$ transfers $\frac{1}{2}(1 - Q)$ methyl groups from Al_2Me_6 to $GaMe_3$ and the rate is given by Eq. (2-11). Equating (2-10) and (2-11) and

$$\begin{aligned} -d[Me_{Al}]/dt &= \tfrac{1}{2}(1 - Q)k_{-2}[AlGaMe_6] \\ &= \tfrac{1}{2}(1 - Q)k_2[GaMe_3]\left[\frac{k_1}{k_{-1}}[Al_2Me_6]\right]^{1/2} \end{aligned} \tag{2-11}$$

solving for Q yields Eq. (2-12).

$$Q = \frac{[\mathrm{GaMe_3}]}{[\mathrm{GaMe_3}] + \frac{12k_{-1}}{k_2}\left[\frac{k_1}{k_{-1}}[\mathrm{Al_2Me_6}]\right]^{1/2}} \tag{2-12}$$

The observed pseudo-first-order rate constant $1/\tau_{\mathrm{Ga}}$ is $-d[\mathrm{Me_{Ga}}]/dt$ from Eq. (2-10) divided by the number of methyl groups in $GaMe_3$ and its molarity, Eq. (2-13). If k_1 were rate-controlling, Q would approach unity and Eq. (2-13)

$$\frac{1}{\tau_{\mathrm{Ga}}} = \frac{-d[\mathrm{Me_{Ga}}]}{3[\mathrm{GaMe_3}]\,dt} = \frac{2Qk_1[\mathrm{Al_2Me_6}]}{[\mathrm{GaMe_3}]}$$

$$= \frac{2k_1[\mathrm{Al_2Me_6}]}{[\mathrm{GaMe_3}] + \frac{12k_{-1}}{k_2}\left[\frac{k_1}{k_{-1}}[\mathrm{Al_2Me_6}]\right]^{1/2}} \tag{2-13}$$

would approach first order in $[\mathrm{Al_2Me_6}]$ and zero order in $[\mathrm{GaMe_3}]$, which in this case appears in the denominator of the rate expression because $1/\tau_{\mathrm{Ga}}$ is pseudo-first-order in $[\mathrm{GaMe_3}]$. If k_2 were rate-controlling, Q would be small, the second term in the denominator of the final expression in Eq. (2-13) would be large, and Eq. (2-13) would approach half-order in $[\mathrm{Al_2Me_6}]$ and first order in $[\mathrm{GaMe_3}]$. Neither extreme prevails, and further discussion of the magnitudes of the k values will be taken up after examining the expressions for $1/\tau_{\mathrm{Al(t)}}$ and $1/\tau_{\mathrm{Al(b)}}$.

Each dimerization of $AlMe_3$ introduces four methyl groups into terminal sites, Al(t). Of these, the fraction which have come from $GaMe_3$ is Q, the fraction from bridge sites Al(b) is $\frac{1}{3}(1 - Q)$, and the fraction returning from Al(t) sites and not contributing to line broadening is $\frac{2}{3}(1 - Q)$. Since the numbers entering and departing the Al(t) sites are equal, the rate constant $1/\tau_{\mathrm{Al(t)}}$ is derived from the sum of the fractions coming from Ga and Al(b) sites [Eq. (2-14)]. A similar derivation of $1/\tau_{\mathrm{Al(b)}}$ yields Eq. (2-15).

$$\frac{1}{\tau_{\mathrm{Al(t)}}} = \frac{4[Q + \frac{1}{3}(1 - Q)]k_{-1}[\mathrm{AlMe_3}]^2}{4[\mathrm{Al_2Me_6}]} = k_1(\tfrac{1}{3} + \tfrac{2}{3}Q) \tag{2-14}$$

$$\frac{1}{\tau_{\mathrm{Al(b)}}} = k_1(\tfrac{2}{3} + \tfrac{1}{3}Q) \tag{2-15}$$

The derived rate expressions (2-13), (2-14), and (2-15) appear to fit the published kinetic data within experimental error. If k_1 is 9 $\mathrm{sec^{-1}}$ at -47°C and the expression $12(k_{-1}/k_2)(k_1/k_{-1})^{1/2}$ is 0.5, then the $1/\tau_{\mathrm{Ga}}$ values calculated from Eq. (2-13) match the experimental values in Table 2-1 with an average deviation of ± 2.2 $\mathrm{sec^{-1}}$, maximum deviation 3.6 $\mathrm{sec^{-1}}$. The $1/\tau_{\mathrm{Al}}$ values from Eqs. (2-14) and (2-15) match within 2.0 $\mathrm{sec^{-1}}$ in all cases. Since a 3 $\mathrm{sec^{-1}}$ error in $1/\tau$ corresponds to an increment of only 1 Hz in the half-intensity line width, this appears to be good agreement. No attempt has been made to find the truly optimum values for the rate constants.

The rate expressions also correlate the data published by Williams and Brown.[6] If k_1 is 7.5 sec^{-1} and [$GaMe_3$] is 0.3 M, which was not specified precisely, their $1/\tau_{Ga}$ values are within $\pm 1\%$ of the calculated values. Twofold dilution of one of the mixtures they used would reduce $1/\tau_{Ga}$ by only 15% according to Eq. (2-13), which appears to agree with their general statement that dilution did not affect the rates.

The results allow rough estimates of the values of k_{-1}, k_2, and k_{-2}, which are significant mainly for demonstrating self-consistency of the kinetic treatment. If k_1 is 9 sec^{-1} and $12(k_{-1}/k_2)(k_1/k_{-1})^{1/2}$ is 0.5, then $k_{-1} = 2 \times 10^{-4}k_2^2$. Thus, if both k's exceed 5×10^3, k_{-1} for the recombination of $2AlMe_3$ is greater than k_2 for reaction of $GaMe_3$ with $AlMe_3$, which is entirely consistent with the expected trends. Possible actual values might be $\sim 10^9$ for k_{-1}, assuming diffusion control with a barrier of about -15 eu, which would make k_2 about 2×10^6. A lower limit for k_{-2} can be estimated. If it is assumed that 0.006 M $AlGaMe_6$ would not be noticed in 0.6 M $GaMe_3$ and 0.3 M Al_2Me_6, then k_{-2} is at least 10^4 sec^{-1}, an estimate which depends on the relative but not the absolute values of k_{-1} and k_2^2. The magnitude of the formation constant k_2/k_{-2} for $AlGaMe_6$ may therefore be as high as 10^2, which seems reasonable.

The kinetic analysis just described can be shown to be equivalent to the steady state treatment. The fraction Q by definition equals the rate of delivery of methyl groups to $AlMe_3$ from $GaMe_3$ divided by the sum of the rates of delivery to $AlMe_3$ from $GaMe_3$ and Al_2Me_6. [Indeed, application of this principle gives an easier derivation of Eq. (2-12) than that presented, provided all pathways are taken into account without error.] Because of the equilibrium condition, the denominator of Q also equals the total rate of departure of methyl groups from $AlMe_3$. Then the identity $Q + (1 - Q) \equiv 1$ corresponds to the steady state condition that the sum of the rates of delivery of methyl groups to $AlMe_3$ from all sources equals the sum of the rates of departure. If the steady state treatment is applied without explicit derivation of Q, it is necessary to label all methyl groups of $AlMe_3$ with respect to their source, and to note that not only reactions of $2AlMe_3{}^{Al}$, but also reactions of $AlMe_3{}^{Al}$ with $AlMe_3{}^{Ga}$ return Me^{Al} groups to their source, Al_2Me_6. With proper inclusion of all mixed species $AlMe_2{}^{Al}Me^{Ga}$ and $AlMe^{Al}Me_2{}^{Ga}$ and proper statistical corrections for the methyl groups transferred per reaction, this computation becomes too complex for inclusion here, but the author has verified that it yields the same rate law.

It should be noted that this kinetic treatment applies only to lines separated by chemical shifts, not to spin multiplets, which will collapse from disconnection of a coupled pair of atoms for even the briefest of times.

Brown and Murrell have rechecked some of the data on $GaMe_3$–Al_2Me_6 methyl group exchange and compared their results with the foregoing kinetic analysis.[15] Their results indicate that the value of the expression $12(k_{-1}/k_2)(k_1/k_{-1})^{1/2}$ is about 0.4, but this value is subject to considerable uncertainty because the fit of the experimental data is not very sensitive to it. The value of k_1, the rate of dissociation of Al_2Me_6, at $-52°C$ is 5.3 sec^{-1} if calculated from $1/\tau_{Ga}$, 6.8 sec^{-1} from $1/\tau_{Al(t)}$, and 7.0 sec^{-1} from $1/\tau_{Al(b)}$, with standard deviations of 1.5–1.8 sec^{-1}. Thus, all the k_1 values are the same within experimental error, and there is no evidence that recombination of 2 $AlMe_3$ within the solvent cage competes significantly with diffusion apart, though the data would be equally consistent with about 20–30% cage recombination, which would broaden the Al_2Me_6 lines more than the $GaMe_3$ line. More positive evidence against cage recombination was obtained by running the reaction in a more viscous solvent, decalin, which should encourage cage recombination by retarding diffusion (see Chapter 7, Section II). There was no change in the relative line widths of the Al_2Me_6 and $GaMe_3$ on changing solvents.

Brown and Murrell tried exchange between Al_2Me_6 and $ZnMe_2$ in the hope of finding a reaction with a detectably different value of k_2, since $ZnMe_2$ was expected to react slower than $GaMe_3$. However, no difference could be detected. Part of the problem may be experimental, the chemical shifts of $ZnMe_2$ and Al_2Me_6 being only about 0.1 δ apart, and part of the failure may be due to the insensitivity of the results to the value of $12(k_{-1}/k_2)(k_1/k_{-1})^{1/2}$.[15]

Because of the inconsistency with the Al_2Me_6–$ZnMe_2$ exchange data, Jeffery and Mole have remained unconvinced that the foregoing kinetic treatment is adequate.[13b] Their most recent data fit Eq. (2-5), with the term in the denominator revised from 0.3 to 0.24, better than Eq. (2-13). To derive Eq. (2-5), they have assumed collision-controlled exchange of methyl groups between solvent-caged pairs $(AlMe_3)_2$ and cage-wall $GaMe_3$. However, they have not presented a detailed set of reversible equilibrium processes, and the author has been unsuccessful in attempts to derive Eq. 2-5 from any set of assumptions consistent with the requirements of collision-controlled rates and the principle of microscopic reversibility. If $(AlMe_3)_2$ remains caged throughout total methyl exchange with $GaMe_3$, then the experimentally evaluated term 0.24 in the denominator of revised Eq. (2-5) must represent $12k_{-1}/k_2$ moles/liter, hence $k_2 = 50k_{-1}$ M. However, allowing one or two $GaMe_3$ in the cage wall and considering the aggregate to be unimolecular, k_2 cannot exceed k_{-1} if both are collision-controlled. More likely, k_{-1} for recombination of caged $(AlMe_3)_2$ to Al_2Me_6 would be first order, 10^{11}–10^{12} sec^{-1}, while k_2 for the reaction of $(AlMe_3)_2$ with $GaMe_3$ should be treated as second order, upper limit $\sim 10^9$ liters/mole-sec, which is too small by a factor of 5000–50,000. If one assumes that a collision-controlled reaction $(AlMe_3)_2 + GaMe_3 = AlMe_3 + AlGaMe_6$ is followed by complete methyl exchange between $GaMe_3$ and $AlMe_3$ or $AlGaMe_6$, one cannot ignore the diffusion-controlled direct recombination of two uncaged $AlMe_3$, and any reasonable and self-consistent set of *k*'s will yield a term in the denominator of the rate law that is half-order in $[Al_2Me_6]$, that is, Eq. (2-13) and not Eq. (2-5). One is left with postulating some sort of systematic experimental error, perhaps nonideal line shapes influenced by appreciable though undetected amounts of $AlGaMe_6$. The skeptical reader is invited to try deriving equations himself, remembering the rule that every step must be written as a balanced equilibrium between explicitly assigned structures.

The kinetic analysis proposed by Jeffery and Mole[13] is consistent with the formation of a single-bridged intermediate, Me_2Al—Me—$AlMe_3$ (**4**), which could react with $GaMe_3$ or revert to Al_2Me_6, but this hypothesis seems unlikely on other grounds. Jeffery and Mole have obtained definitive evidence for another single-bridged intermediate, Me_2Al–(O-*t*-Bu)–$AlMe_3$ from Al_2Me_5(O-*t*-Bu), to be discussed in detail in Section III,G. The ΔS^* for formation of this single-bridged species is considerably less positive (by ~10 eu) than that for dissociation of Al_2Me_6, and the single-bridged species does not react measurably with $GaMe_3$. In contrast, Me_2Al–Me–$AlMe_3$ would have to have a higher numerical second-order k_2 for reaction with $GaMe_3$ than first-order k_{-1} for reclosure to Al_2Me_6 in order to yield the observed kinetics.

C. Structural Factors Affecting Bridging

It has already been noted in Section III,A that the stability of dimer compared to monomer is decreased by increasing size of the alkyl groups. For an understanding of the factors affecting electrophilic displacement at saturated carbon, the bridging ability of a particular alkyl group is more relevant than the total effect of all the terminal and bridging alkyl groups. Steric effects turn out to be somewhat greater for bridging groups than for terminal groups in hexaalkyldialuminum compounds.

From correlation of nmr chemical shifts with the composition of mixtures at room temperature, Yamamoto and Hayamizu concluded that the relative bridging abilities compared with methyl are 1/6 for ethyl, 1/7 for *n*-propyl, and 1/17 for isobutyl.[16] These figures represent the degree to which it is preferable to put the methyl group in the bridge position and the other group in a terminal position rather than vice versa. Methylpentaisobutylaluminum (**5**) is dimeric and has the lone methyl group in a bridge position.[17] The stability of **5** is consistent with the thermodynamic measurements, which show that triisobutylaluminum dimerizes to an appreciable extent.[8]

If steric hindrance were effective only in the bridge positions, the difference between the equilibrium constants for association of trimethylaluminum and of triisobutylaluminum should be about a factor of 17^2 or 289 in favor of

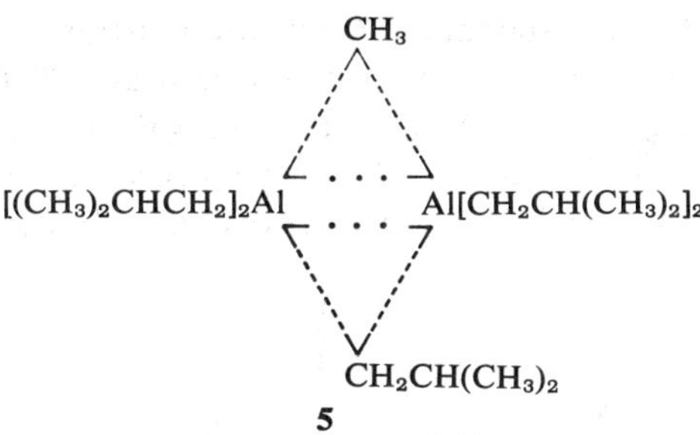

the former. From the thermodynamic data,[8,11] the actual factor is roughly 10^6, which indicates a substantial steric interaction involving the terminal alkyl groups in the dimer. The difference between the steric repulsions in Al_2Me_6 and $Al_2(iso\text{-}Bu)_6$ averages out to about 1.2 kcal per terminal group and 1.7 kcal per bridging group. Although these effects have been labeled "steric," there may also be a substantial electronic contribution, for which there is no means of estimating the magnitude or direction when the total effect is this small and the structures this complicated.

Cyclopropyl groups form stronger bridge bonds than methyl. Tricyclopropylaluminum dimer shows no exchange at room temperature,[18] and the heat of activation for bridge–terminal exchange is 20 ± 2 kcal/mole.[19] The X-ray structure of tricyclopropylaluminum dimer has been described in Chapter 1, Section III,C.

Vinyl groups accelerate methyl exchange in mixtures of vinyl and methyl thallium.[5] The rate enhancement is about a factor of 30 at 26°C for a 1 *M* solution of a 50–50 mixture. The vinyl groups themselves undergo faster exchange than the methyl groups, and transfer of the latter may be facilitated by intermediates having one vinyl and one methyl bridge.

Unsaturated and aromatic bridging groups are the most favored of all. The dimeric nature of dimethylvinylgallium has already been mentioned.[7] Trivinylgallium is also dimeric and shows only one kind of vinyl group at -90°C, indicating rapid bridge–terminal exchange with a dissociation energy <10 kcal/mole.[20] Dialkyl(phenylethynyl)aluminum,[21] -gallium, and -indium[22] compounds are dimers bridged through the phenylethynyl groups. The phenyl group in phenylpentamethyldialuminum occupies one of the bridge positions.[23]

Triphenylaluminum naturally exists as the dimer. The preliminary report of the X-ray structure indicates that the Al—Al distance is 2.7 Å (0.1 Å longer than in Al_2Me_6) and the bridging phenyl groups are perpendicular to the plane of the Al—C—Al three-center bond.[24] There are good theoretical reasons for this geometry. The *p* orbital of the bridging phenyl carbon overlaps with the set of aluminum orbitals that is antisymmetric along the Al—Al bond, strengthening the Al—C bonding at the expense of the Al—Al bond.

The σ-bonding (**6**) and π-bonding (**7**) sets are illustrated separately, and the actual orbital structure is a superposition of **6** and **7**. For simplicity, the aluminum orbitals are described as a set of sp^3 hybrids, but as noted in Chapter 1, Sections II,C and III,B, other correct choices of hybridized or unhybridized orbitals will yield the same overall symmetry.

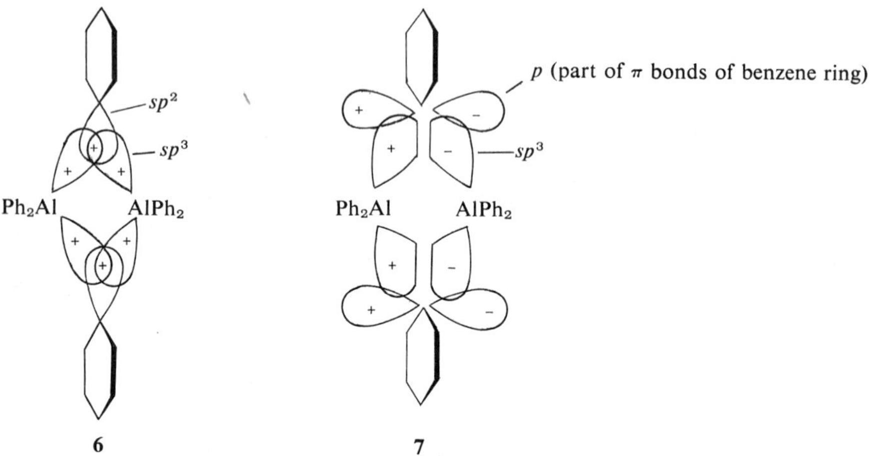

Since the π overlap with the antisymmetric set of aluminum orbitals cancels some of the Al—Al bonding, it can no longer be argued on symmetry grounds that there is net attraction between the aluminum atoms. One way to indicate this situation is to write the three-center bond symbol with one side open, as in structure **8**. If the aluminum atom were a strong enough

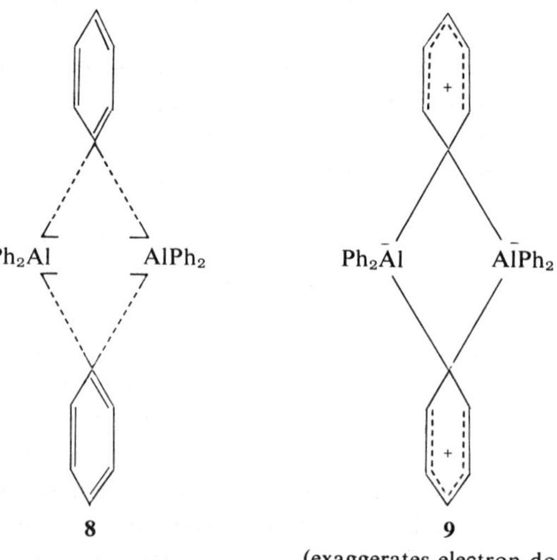

(exaggerates electron donation)

electrophile to disrupt the benzene ring aromaticity and seize an extra bonding electron pair, all vestiges of Al—Al bonding would be lost and structure **9** would become appropriate, but this undoubtedly greatly exaggerates the degree of electron donation from the rings to the aluminum atoms. Structures analogous to **9** become appropriate where the substituents are strong electrophiles, as in the intermediates or transition states for electrophilic aromatic substitution. That there is some degree of electron donation of this type is suggested by distortions in the bridging phenyl groups of Al_2Ph_6.[24]

Alkyl bridging groups also place a slight electron population in the Al—Al antibonding orbital set as a result of hyperconjugation, which was not mentioned in Chapter 1, Section III,B, but this is a relatively weak effect and, according to the CNDO calculations cited, does not counterbalance the Al—Al bonding. Cyclopropyl groups are somewhere in between alkyl and aryl groups in their ability to participate in this sort of π-bonding, hence the extra stability of tricyclopropylaluminum dimer.

D. Methyl Exchange between Al_2Me_6 and $Me_4Al_2(C{\equiv}C{-}Ph)_2$

The exchange of methyl groups between Al_2Me_6 and $Me_4Al_2(C{\equiv}C{-}Ph)_2$ involves direct reaction of the latter with $AlMe_3$ from dissociation of the former.[25] The reaction is half-order in Al_2Me_6 and first order in $Me_4Al_2(C{\equiv}C{-}Ph)_2$. Ham *et al.* suggested that the $MeAl_3$ may attack an

$$Al_2Me_6 \rightleftharpoons 2\,AlMe_3 \text{ (rapid equilibrium)}$$

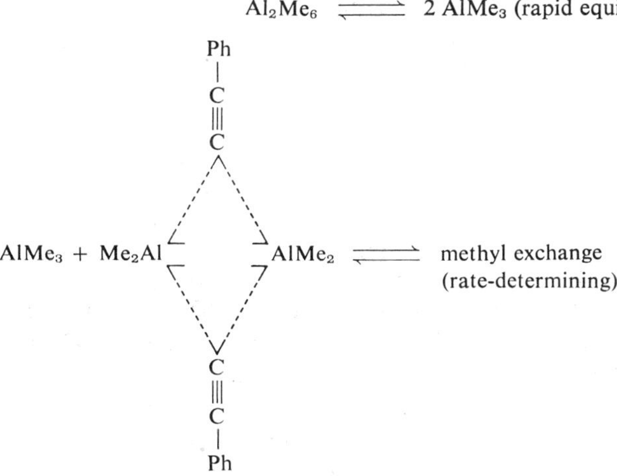

$AlMe_2$ group of $Me_4Al_2(C{\equiv}C{-}Ph)_2$ directly in the transition state **10**.[25] A possible alternative is attack of the electrophilic $AlMe_3$ at the nucleophilic triple bond followed by intramolecular rearrangement of the complex **11**.

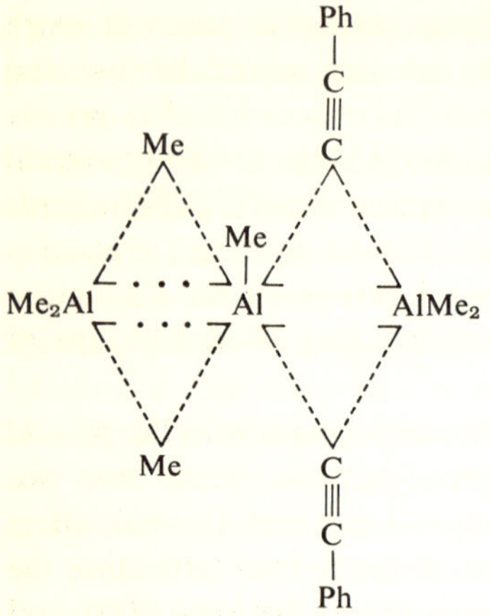

Ph
C≡C
Me Me
Me_2Al Al $AlMe_2$
Me
C≡C
Ph

10

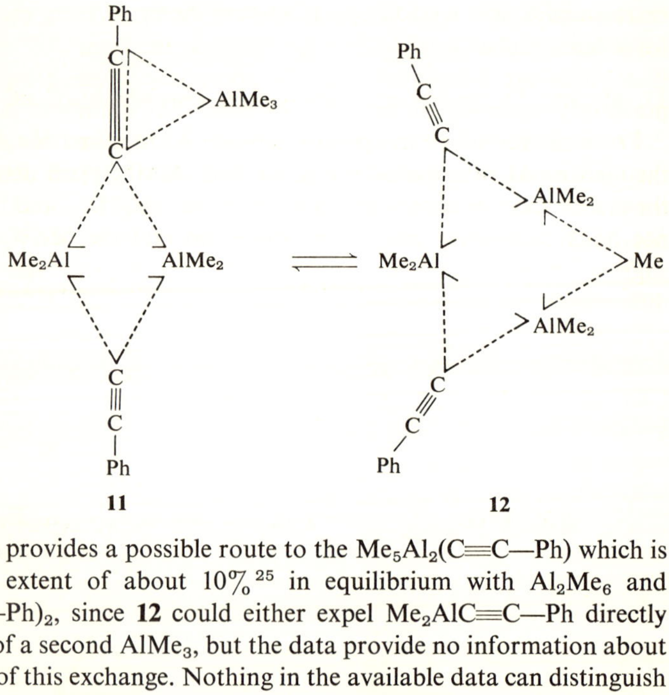

Ph
C≡C >$AlMe_3$
Me_2Al $AlMe_2$ ⇌ Me_2Al >$AlMe_2$ >Me >$AlMe_2$
C≡C
Ph

11 **12**

Intermediate **12** provides a possible route to the $Me_5Al_2(C{\equiv}C{-}Ph)$ which is present to the extent of about 10%[25] in equilibrium with Al_2Me_6 and $Me_4Al_2(C{\equiv}C{-}Ph)_2$, since **12** could either expel $Me_2AlC{\equiv}C{-}Ph$ directly or with the aid of a second $AlMe_3$, but the data provide no information about the mechanism of this exchange. Nothing in the available data can distinguish between the paths through structure **10** or through **11** and **12** or other structural variations which can be imagined. The path via **10** transfers only methyl groups between Al_2Me_6 and $Me_4Al_2(C{\equiv}C{-}Ph)_2$ while that through **11–12** exchanges aluminum atoms as well. Inasmuch as ^{27}Al nmr shows evidence for alkyl exchange between different AlR_3 compounds,[26] it might be possible to use this technique to find out whether methyl exchange is much faster than aluminum exchange (**10**) or the two rates are essentially the same (**11–12**).

The energy of activation of the Al_2Me_6–$Me_4Al_2(C{\equiv}C{-}Ph)_2$ methyl exchange is 14.1 kcal/mole and the second-order rate constant is 29 liters/mole second at 6°C ($\Delta G^* = 14.4$).[25] The calculated ΔS^* for the total process is -3 eu. It may seem strange that the energy of activation for this entire exchange is less than that for the first step, dissociation of Al_2Me_6, which requires ~16 kcal/mole.[6,12] The explanation is that only half a mole of Al_2Me_6 enters into the stoichiometry of the rate-determining step, and ΔH^* for $\frac{1}{2}Al_2Me_6 \rightarrow AlMe_3$ is only ~8 kcal. The equilibrium $\Delta H°$ for formation of 1 mole of $AlMe_3$ is also about 8 kcal, and the remaining ~5.5 kcal of the total 13.5 kcal of ΔH^* represent the heat of activation for the reaction of $AlMe_3$ with $Me_4Al_2(C{\equiv}C{-}Ph)_2$. The $\Delta S°$ for $\frac{1}{2}Al_2Me_6 \rightarrow AlMe_3$ is about $+17$ eu, which requires that ΔS^* in the rate-determining reaction of $AlMe_3$ with $Me_4Al_2(C{\equiv}C{-}Ph)_2$ be -20 eu, the expected figure for a bimolecular process, in order to yield the total ΔS^* of -3 eu. Thus, these data are consistent with the proposed mechanisms and with the previously discussed energy parameters for dissociation of Al_2Me_6.

From the foregoing discussion, it is apparent that the value of ΔH^* for the $GaMe_3$–Al_2Me_6 methyl exchange (Section III,B) will depend on whether the first or second step is rate-determining. Because the first step is almost but not quite completely rate-determining, the expected ΔH^* value would be slightly smaller than that for dissociation of Al_2Me_6, but in practice the difference is within experimental error.[15]

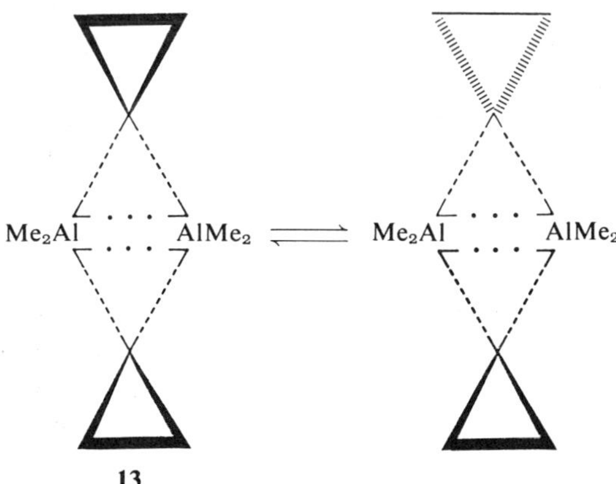

13

E. Rotation of Cyclopropyl Bridging Groups

The rotation of the bridging cyclopropyl groups in dicyclopropyltetramethyldialuminum (**13**) is restricted.[27] The terminal methyl groups are nonequivalent at $-70°C$, and Oliver and co-workers concluded that the bridging cyclopropyl groups are oriented *syn* as they are in crystalline hexacyclopropyldialuminum (see Chapter 1, Section III,C, Fig. 1-7). The rotation of the cyclopropyl groups has an activation energy of 11 kcal/mole,[27] far below the dissociation energy, which should be about 20 kcal/mole.[19]

F. Aluminum Trialkyl–Lewis Base Adducts

Lewis bases such as tertiary amines, ethers, or phosphines cleave Al_2Me_6 to 2 $AlMe_3 \cdot B$, where B is the Lewis base molecule. Brown and Murrell have studied reactions of $AlMe_3 \cdot B$ with $GaMe_3$ in cyclopentane where B was pyridine, 2,6-lutidine, or ether.[15] Conveniently measurable line broadenings ($1/\tau$ in the 5–15 sec^{-1} range) occurred at $+47.5°C$ with the pyridine adduct, $-17.5°C$ with the ether adduct, and $-40°C$ with the lutidine adduct. In the first two cases, the dependencies of line widths on concentrations indicate that dissociation of $AlMe_3 \cdot B$ to $AlMe_3 + B$ occurs as an equilibrium prior to the rate-determining step, which is reaction of $AlMe_3$ with $GaMe_3$. This conclusion is strengthened by the finding that the reaction of $AlMe_3$–pyridine with $ZnMe_2$ is much slower than with $GaMe_3$, requiring a temperature 25°C higher for the same amount of line broadening. The reaction of the 2,6-lutidine complex with $GaMe_3$ is kinetically more complex and requires a treatment analogous to that presented for the Al_2Me_6–$GaMe_3$ methyl exchange[14] discussed in Section III,B. Recombination of $AlMe_3$ and lutidine has a rate comparable to that for exchange between $AlMe_3$ and $GaMe_3$, the fraction Q of the methyl groups in the $AlMe_3$ which have come from $GaMe_3$ being about 0.55–0.8 in the concentration ranges studied, 0.1–0.5 *M*.[15] A similar situation was found for the exchange between $AlMe_3 \cdot PBu_3$ and $ZnMe_2$ at $+8°C$. The general usefulness of the author's kinetic analysis (Section III,B)[14] is thus demonstrated.

Exchange between $AlMe_3 \cdot NEt_3$ and Al_2Me_6 has dissociation of the complex to $AlMe_3$ and NEt_3 as the rate-determining step, and the NEt_3 reacts rapidly with $AlMe_3$ in equilibrium with Al_2Me_6.[15] However, $AlMe_3 \cdot SMe_2$ exchanges methyl groups with Al_2Me_6 at too high a rate at $-93°C$ to be attributed to dissociation of either the $AlMe_3 \cdot SMe_2$ or the Al_2Me_6, and the available kinetic data suggest a bimolecular process. This base-assisted type of exchange only occurs with ligands having two electron pairs which could coordinate with two aluminum atoms. Brown and Murrell suggested structure **14** as a possible transition state or intermediate for this base-assisted methyl exchange.[15]

Exchange of alkyl groups between molecules of trialkylaluminum–pyridine complex, $pyAlR_3$, proceeds very slowly. Mole found that several minutes at 90°C are required for the redistribution of methyl and ethyl groups between $pyAlMe_3$ and $pyAlEt_3$.[28] The reaction of 0.2 *M* $pyAlPh_3$ and $pyAlMe_3$ in the

$$Me_2Al(\mu\text{-}Me)_2AlMe_2 + Me_3AlSMe_2 \longrightarrow Me_2Al(\mu\text{-}Me)_3(AlMe_2)_2SMe_2$$

14

presence of 0.5–1 *M* pyridine in toluene does not involve a dissociative process analogous to the $pyAlMe_3$–$GaMe_3$ exchange, but proceeds by a bimolecular mechanism in a few hours at 40°C.[29] The rate is independent of excess pyridine at high pyridine concentrations. The stoichiometry of the transition state is ($pyAlMe_3$ + $pyAlPh_3$). The transition state has a larger volume than the starting materials, as shown by a 50% rate reduction when the reaction vessel was pressured with 2000 atm of nitrogen. Jeffery and Mole proposed a transition state structure consisting essentially of Al_2R_6 (where R is Me and Ph) solvated by two molecules of pyridine.[29] This seems reasonable in principle, even though many details are not defined by the available data.

Lewis bases can also coordinate with thallium compounds and slow exchange rates. Complexing of triphenylthallium by trimethylamine was found to reduce the phenyl exchange rate by a factor of 10^3 under the conditions chosen.[5] However, trimethylamine and trimethylthallium complex only weakly, and the exchange rate of the latter is not significantly affected.

G. Pentamethyldialuminum Alkoxides and Amides

This group of compounds is particularly important as a model for transition state structures in a common class of electrophilic displacements. The dimer structure having one alkyl bridge and one anionic ligand bridge is at an energy minimum for aluminum compounds but is at or near an energy maximum for compounds of most other common metals (see Chapter 3). The X-ray structure of a member of this class of compounds, $Me_5Al_2NPh_2$, has been determined (Chapter 1, Section III,C, Fig. 1-6). Magnuson and Stucky also found that the bridge and terminal methyl groups of $Me_5Al_2NPh_2$ are exchanged rapidly on the nmr time scale at 25°C and slowly at −5°C.[30]

Jeffery and Mole have carried out a detailed study of the bridge–terminal methyl exchange of *μ-t*-butoxypentamethyldialuminum (**15**) and related compounds.[31] The *t*-butoxy compound **15** is synthesized from Al_2Me_6 and a limited amount of acetone, and from the nmr data its structure must be analogous to that of $Me_5Al_2NPh_2$. Remarkably, Me_5Al_2O-*t*-Bu (**15**) does not disproportionate to the thermodynamically favored mixture of Al_2Me_6 and $Me_4Al_2(O\text{-}t\text{-}Bu)_2$ in less than several weeks at 110°C. Thus, dissociation to $AlMe_3$ and $Me_2AlO\text{-}t\text{-}Bu$ must be very slow. Bridge–terminal exchange is much faster. Exchange broadening shows that k for migration of the bridge methyl group to the terminal positions is 2.2 sec^{-1} at 36°C and 21 sec^{-1} at 59°C. The activation energy is 20 ± 3 kcal/mole and ΔS^* is +3.5 ± 7 eu, which is a reasonable value for a ring opening. The exchange mechanism evidently involves rotation of an $AlMe_3$ group in the opened intermediate (**16**).

Me* Me Me Al Al Me Me, ⇌ Me* Me Me—Al Al Me Me ⇌ O CMe₃ **15** O CMe₃ **16**

Me Me* Me Al Al Me Me O CMe₃ **15**

The compounds derived from Al_2Me_6 and acetophenone or benzophenone, $Me_5Al_2(OCMe_2Ph)$ and $Me_5Al_2(OCMePh_2)$, undergo bridge–terminal methyl exchange faster than $Me_5Al_2O\text{-}t\text{-}Bu$. A reasonable explanation is that phenyl groups in the alkoxide unit can complex with the trivalent aluminum atom in the intermediate analogous to **16**.[31] This idea is supported by the finding that in the Al_2Me_6–benzonitrile adduct (**17**) only half the terminal methyl groups exchange rapidly with the bridge position at 40°C, though all exchange at somewhat higher temperatures. The Me—C—Ph group is aligned parallel to the Al—Al bond axis in **17**, allowing the phenyl group to act as a π donor toward one of the aluminum atoms but not the other. Consequently, the energy of intermediate **18** is lower than that of **19**.

Methyl exchange between $Me_5Al_2O\text{-}t\text{-}Bu$ and $GaMe_3$ is too slow to observe by nmr even at 110°C.[31] Therefore, the ring-opened intermediate **16** is also

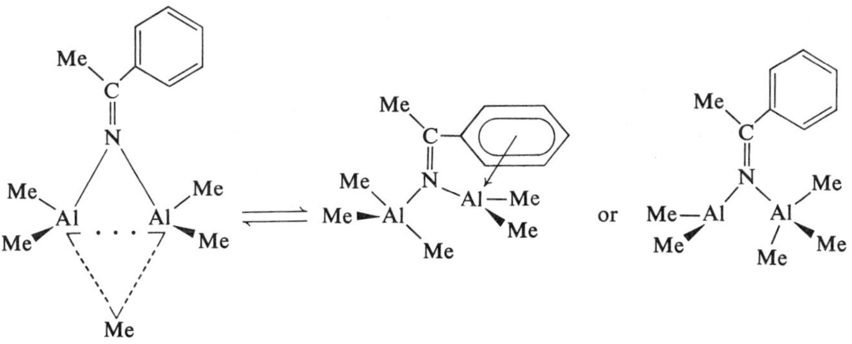

17 $\rightleftharpoons$ **18** or **19**

inert toward $GaMe_3$, and this result strengthens the already convincing evidence that the methyl exchange between $GaMe_3$ and Al_2Me_6 requires complete dissociation of the latter to 2 $AlMe_3$.

The inertness of Me_5Al_2O-*t*-Bu toward $GaMe_3$[31] or of $Me_5Al_2OCMePh_2$ toward $AlMe_3$[32] contrasts sharply with the reactivity of $Me_4Al_2(C{\equiv}C{-}Ph)_2$ toward $AlMe_3$ (Section III,D).[25] This gross difference in reactivity is easily understood if the mechanism involves rate-determining attack of the trialkylmetal on the bridging phenylethynyl group to form intermediates **11** and **12**, a process which would be sterically hindered with the *t*-butoxy group. It seems unlikely that there should be such a large difference if direct attack of the trialkylmetal at a terminal Al—Me bond (structure **10**) were the rate-determining process.

The high resistance of Me_5Al_2O-*t*-Bu (**15**) to complete dissociation demonstrates the ability of nucleophilic bridging groups to stabilize transition states for electrophilic displacements at carbon. Assuming that **15** is a good structural model for transition states involving alkyl transfers between boron, mercury, tin, and other less reactive organometallic compounds, it appears that the usual mechanism would begin with formation of a nucleophilic bridging link between the two metal atoms, followed by transfer of an alkyl group from one metal to the other.

In earlier work, Mole had shown that tetraalkylaluminum dialkoxides do not disproportionate[33] or dissociate.[34] The reaction of $Et_4Al_2(O$-*t*-$Bu)_2$ (**20**) with Al_2Me_6 requires minutes to hours at room temperature and gradually yields $Me_nEt_{(4-n)}Al_2(O$-*t*-$Bu)_2$. If the reaction proceeded by way of slow dissociation to Et_2AlO-*t*-Bu, those fragments should be completely converted to Me_2AlO-*t*-Bu by the excess of Al_2Me_6 before finding each other again to form another stable dimer, which would then be $Me_4Al_2(O$-*t*-$Bu)_2$. Since $Me_4Al_2(O$-*t*-$Bu)_2$ is not the initial product, this cannot be the mechanism. Mole proposed direct attack of $AlMe_3$ on the terminal Me—Al bond of the alkoxide,[34] though attack at the alkoxide bridge could yield the same result.

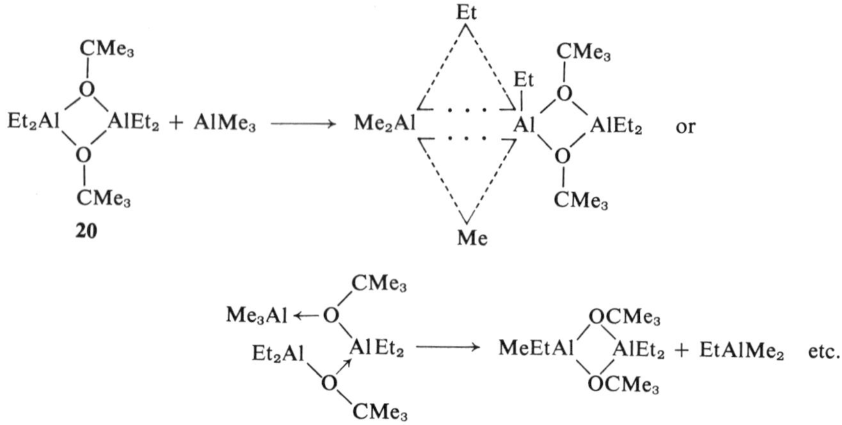

$$Et_2Al(\mu\text{-}OCMe_3)_2AlEt_2 \;(\mathbf{20}) + AlMe_3 \longrightarrow Me_2Al(\mu\text{-}Et)(\mu\text{-}Me)Al(\mu\text{-}OCMe_3)_2AlEt_2 \quad \text{or}$$

$$Me_3Al \leftarrow O(CMe_3)\text{—}[Et_2Al(\mu\text{-}OCMe_3)AlEt_2] \longrightarrow MeEtAl(\mu\text{-}OCMe_3)_2AlEt_2 + EtAlMe_2 \quad \text{etc.}$$

IV. Group II

A. Group IIb

Zinc, cadmium, and mercury alkyls often behave much like gallium, indium, and thallium alkyls except that the reactions of the former group are usually slower. Exchange reactions involving alkylmercury compounds are usually too slow to study by nmr line broadening and will be deferred until Chapter 3. For example, methyl exchange between dimethylmercury and either dimethylzinc or dimethylcadmium is too slow to measure by nmr at 25°C.[1] An exception is mercury migration around the ring in cyclopentadienylmercury compounds.[35] From the nmr spectra of cyclopentadienylmercuric chloride (**21**) at $-44°$ to $-113°C$, Rausch and co-workers found that the mercury is σ-bonded to the ring and equilibrates to all five positions. Relative rates of mercury migration are $(C_5H_5)_2Hg > C_5H_5HgI > C_5H_5HgBr > C_5H_5HgCl$. These migrations are, of course, intramolecular alkyl exchanges at mercury.

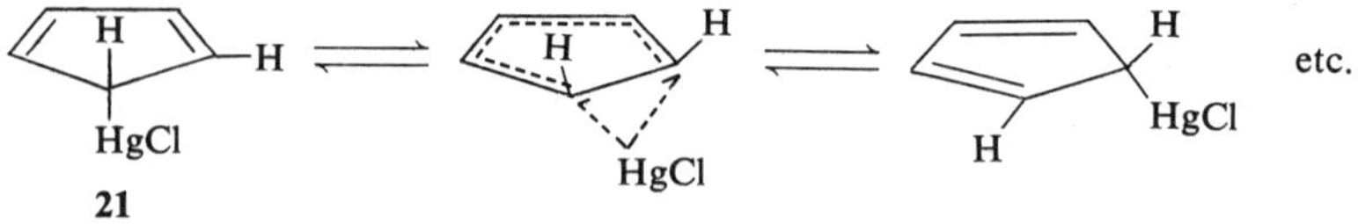

$$C_5H_5HgCl\;(\mathbf{21}) \rightleftharpoons [\text{transition state}] \rightleftharpoons C_5H_5HgCl \quad \text{etc.}$$

Methyl group exchange between two molecules of $CdMe_2$ proceeds at a convenient rate for nmr measurements somewhat above room temperature. The exchange is followed in this case by collapse of the ^{111}Cd and ^{113}Cd satellites of the methyl proton signals, which occurs as a result of the uncoupling of the H and Cd spins whenever a methyl group is transferred. Transfer of a methyl group between two cadmium atoms in the same spin state will not contribute to line broadening and would have to be corrected for,[5] except that the proportions of the isotopes yielding the splittings are small and collisions of two of a kind are rare. It is particularly important to exclude oxygen, which forms impurities that greatly accelerate the rate of methyl exchange.[36] When this precaution is taken, k for methyl exchange at 120°C in toluene is about 19 liters/mole second.[36] For neat $CdMe_2$, ΔH^* values found are 15.1 ± 0.2 kcal/mole[37] and 15.6 kcal/mole,[3] and both ΔS^* values agree at -9.8 eu.[37,38] These ΔS^* values are surprisingly small for a bimolecular reaction, though there is no doubt considerable freedom in the angle of approach to form intermediate or transition state **22** from two linear molecules of $CdMe_2$.

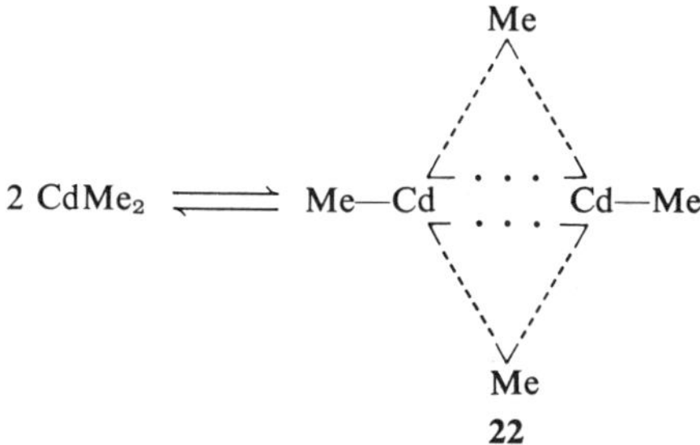

22

Exchange of methyl groups between $CdMe_2$ and $ZnMe_2$, $GaMe_3$, or $InMe_3$ has been studied by Oliver and co-workers.[38,39] These reactions are all bimolecular. For $CdMe_2$–$ZnMe_2$ exchange in methylcyclohexane, $\Delta H^* =$ 17 kcal/mole and $\Delta S^* = -3$ eu. It is surprising that ΔH^* should be higher than for self-exchange of $CdMe_2$ and that ΔS^* should be even less negative. It appears that the experiments were carefully done and reproducible[38,39] and the rate data were interpreted with the aid of a computer program which matched observed and calculated line shapes. Perhaps the most expedient way to ease the mild discomfort caused by contemplation of these anomalous results is to move on to the next topic. For the $CdMe_2$–$GaMe_3$ exchange in methylene chloride, $\Delta H^* = 6.7$ kcal/mole and ΔS^* is a very reasonable -19 eu.[39] For $CdMe_2$–$InMe_3$, $\Delta H^* = 7.8$ kcal/mole and $\Delta S^* = -14$ eu.

In order for methyl exchange to occur, the metal atoms must each have a vacant site available for methyl bridging. Accordingly, nucleophilic solvents greatly retard reactions involving $GaMe_3$, $InMe_3$,[39] or alkylaluminum compounds (Section III,F). However, a properly chosen nucleophile need not occupy the last vacant site on a Group II metal alkyl, and nucleophilic solvents greatly accelerate the self-exchange of $CdMe_2$ methyl groups.[37,39] For this exchange, diethyl ether lowers ΔH^* to 5.0 kcal/mole (from 15) and ΔS^* to -38 eu (from -10).[39] In tetrahydrofuran, ΔH^* is 6.3 kcal/mole and ΔS^* is -28.3 eu.[37] It may be imagined that the ether functions by bridging between two cadmium atoms as in structure **23** or perhaps a less symmetrical, more ionic structure.

$$2\,CdMe_2 + Et_2O \rightleftharpoons \text{Me—Cd(}\mu\text{-Me)}_2\text{(}\mu\text{-OEt}_2\text{)Cd—Me}$$

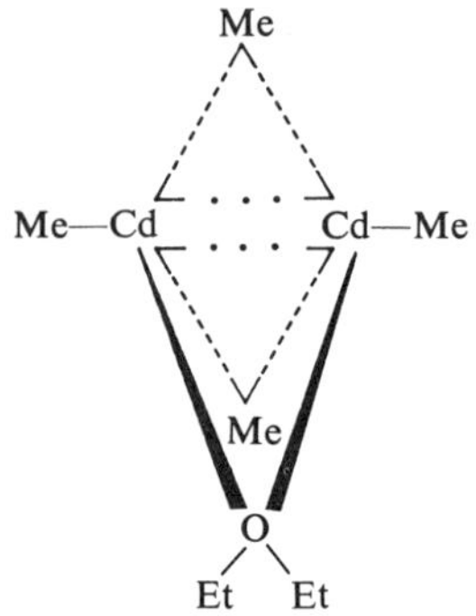

23

Jeffery and Mole have found that methylzinc and methylcadmium alkoxides are tetrameric in toluene solution.[40] The tetramers are relatively unreactive and cannot be the species responsible for the observed accelerating effect of oxygen on the self-exchange of $CdMe_2$.[36] Exchange between $(MeCdOMe)_4$ and $CdMe_2$ is too slow to observe by nmr, and exchange between $(MeZnOMe)_4$ and $ZnMe_2$ in toluene is rapid at 60°C, gives a single broadened line at 25°C, and is slow enough to give resolved methyl peaks at −5°C.[40] Alkyl exchange is very slow between $(MeZnO\text{-}t\text{-}Bu)_2$ and $ZnEt_2$, showing no reaction in several hours. Exchange between $ZnEt_2$ and the less-hindered $(MeZnO\text{-iso-}Pr)_4$ only requires about 2 minutes, which is faster than dissociation of the alkoxide tetramer.

Self-exchange of $ZnMe_2$ should be the fastest methyl group exchange in the Group IIb series, but it was the most elusive to prove. Roberts and coworkers have found that the ^{13}C nmr spectrum of $ZnMe_2$ shows no $^{13}C—Zn—C—^1H$ coupling, which confirms that methyl exchange is rapid.[41]

Prior to the nmr evidence that exchange is rapid between zinc alkyls there were several reports to the contrary,[42–44] but these are undoubtedly erroneous. "Propylbutylzinc" was reported as boiling at 38°–43°C (3 mm) and yielding the correct analyses for zinc and the hydrolysis products propane and butane.[42] A mixture of $ZnPr_2$ and $ZnBu_2$ was said to separate into two fractions boiling at 30° and 60°C (3–4 mm). No spectral evidence that the "compound" and the "mixture" were different was offered. The skeptical research director might point out that if the student expects PrZnBu and distils it quickly to prevent decomposition, that is what he will get, because a 1:2:1 mixture of equilibrating $ZnPr_2$, PrZnBu, and $ZnBu_2$ is not easily separated by distillation. However, when the student knows that both $ZnPr_2$ and $ZnBu_2$ are in there, he can distil the same mixture more slowly and get them apart. As for the reported isolation of crystalline dioxane complexes of phenylthienylzinc,[43] the data are insufficient to prove the structure, and in any event one component might be crystallized out of a rapidly equilibrating mixture. The reported failure to observe exchange between $ZnMe_2$ and $ZnEt_2$ or $Zn(t\text{-}Bu)_2$ by nmr[44] appears to rest on an anticipated chemical shift difference between the methyl groups of $ZnMe_2$ and those of MeZnEt or MeZn-*t*-Bu which did not materialize, but the difference could well be very small. The conclusion that there was no exchange[44] also seems to have been conditioned by faith in the unsymmetrical RZnR′ preparations.[42,43] For a scientist, faith is fine for motivation, but doubt is more likely to lead to the right answer.

B. Magnesium Alkyl Exchange with Retention of Configuration

Before discussing alkyl exchange reactions it is necessary to consider the structures of alkylmagnesium compounds. Dimethylmagnesium in the solid state is a polymer linked by Mg—CH_3—Mg bridges (Chapter 1, Section III,C) and this solvent-free polymer has been obtained by crystallization from ether.[45] Parris and Ashby have obtained 100 MHz proton nmr evidence that $MgMe_2$ dimerizes or polymerizes to some extent in ether at low temperatures.[46] The CH_3 singlet is at δ -1.45 (τ 11.45) at +30°C. It moves upfield to -1.70 on cooling to -100°C, which suggests that the terminal methyl groups of $(Et_2O)_2MgMe_2$ dominating the spectrum at +30°C enter into a rapid equilibrium in which they become bridge methyl groups of $(Et_2O)_2(MgMe_2)_n$ where $n = 2$ at -100°C. However, this is not the only possible reason for a change in an nmr spectrum with temperature and the evidence is therefore not conclusive. In tetrahydrofuran, $MgMe_2$ remains monomeric at -76°C.

Alkylmagnesium halides can form a bewildering variety of complexes, dimers, and disproportionation products. Chemists have assumed for a long time that the Schlenk equilibrium, Eq. (2-16), occurs in solutions of Grignard

$$2\ RMgX \rightleftharpoons R_2Mg + MgX_2 \tag{2-16}$$

reagents. However, the literature shortly after 1957 shows considerable confusion as a result of widespread acceptance of an erroneous report that R_2Mg and $^{28}MgX_2$ failed to exchange magnesium,[47] which together with other evidence led to the conclusion that the structure of the Grignard reagent was $R_2Mg \cdot MgX_2$. Careful rechecking of the work by Dessy and co-workers indicates that exchange does occur after all.[48] Of course, the fog lifted very quickly on this structure question when Stucky and Rundle reported the X-ray structure of phenylmagnesium bromide dietherate to be PhMgBr with a distorted tetrahedral magnesium atom having two ether molecules coordinated[49] (see Chapter 1, Section II,A, Fig. 1-1). The controversies concerning the fundamental structure of the Grignard reagent are now of only historical interest and have been summarized in a review by Ashby.[50]

Although the general structure RMgX is well established, there remain many unresolved details about its degree of aggregation in solution and the extent of the Schlenk equilibrium to R_2Mg and MgX_2. These details are strongly dependent on the alkyl group R, the halogen X, and the type of ether or amine solvating the magnesium.

At $-100°C$, mixtures of $MgMe_2$ and MeMgBr in ether show two separate methyl signals, $MgMe_2$ at $\delta -1.70$ and MeMgBr at $\delta -1.55$.[46] At lower temperatures $MgBr_2$ precipitates and $MgMe_2$ concentrates in the solution as a result of equilibration, which still proceeds at an appreciable rate even though it is slow on the nmr time scale. In tetrahydrofuran, $MgMe_2$ absorbs at $\delta -1.76$, MeMgBr at $\delta -1.70$, and MeMgBr with added $MgBr_2$ at $\delta -1.66$, and the spectrum of MeMgBr is temperature-dependent. Parris and Ashby interpreted these results as indicating a roughly statistical equilibrium between $MgMe_2$, MeMgBr, and $MgBr_2$ in tetrahydrofuran.[46]

With *t*-BuMgCl in diethyl ether, a single proton signal appears at $+30°C$ but two partially resolved peaks are seen at $-36°C$, the upfield peak being the smaller.[46] Evidently both *t*-BuMgCl and $Mg(t\text{-Bu})_2$ are present, with the former predominating. In tetrahydrofuran, separate peaks for *t*-BuMgCl and $Mg(t\text{-Bu})_2$ are seen even at $+65°C$. At 65°C the *t*-BuMgCl signal is stronger and at 33°C the $Mg(t\text{-Bu})_2$ is stronger.

Jensen and Nakamaye have measured the conformational preferences of cyclohexylmagnesium bromide and dicyclohexylmagnesium in dimethyl and diethyl ether at $-80°C$ and found that the magnesium has a stronger preference for the equatorial position in $C_6H_{11}MgBr$ than in $Mg(C_6H_{11})_2$.[51] This, of course, shows also that $C_6H_{11}MgBr$ and $Mg(C_6H_{11})_2$ are different in solution.

From ^{19}F nmr spectra Evans and Khan showed that *p*-fluorophenylmagnesium bromide exists in solution as FC_6H_4MgBr rather than $(FC_6H_4)_2Mg$ and $MgBr_2$.[52] The crystalline $Me_2NCH_2CH_2NMe_2$ complex of FC_6H_4MgBr has been isolated and found to exchange fluorophenyl groups relatively slowly.[53]

Tetracoordinate magnesium evidently predominates in diethyl ether,[49,54] but tetrahydrofuran is less bulky and allows coordination numbers up to 5 or 6. Crystalline $MeMgBr(C_4H_8O)_3$ has a trigonal bipyramidal magnesium atom with tetrahydrofuran occupying both axial and one equatorial site and the methyl group and bromine atom each in an equatorial site.[55] Hexacoordinate magnesium is found in $MgBr_2(C_4H_8O)_4$.[56]

A number of workers have concluded that Grignard reagents dimerize or polymerize to halide bridge structures in solution, but proof of this has been difficult. It should be kept in mind that saturated solutions contain appreciable proportions of $PhMgBr(Et_2O)_2$ and $EtMgBr(Et_2O)_2$, since these are the species that crystallize.[49,54] Toney and Stucky were able to crystallize a dimer of ethylmagnesium bromide (**24**) by using a bulky ligand, triethylamine, which can only coordinate easily with one site of the magnesium, and they proved the structure by X-ray crystallography.[57]

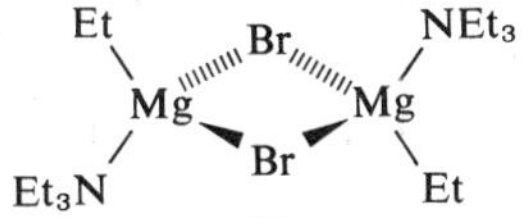

$$\begin{array}{ccccc} Et & & Br & & NEt_3 \\ & Mg & & Mg & \\ Et_3N & & Br & & Et \end{array}$$

24

Walker and Ashby have confirmed by ebullioscopic measurements that alkylmagnesium bromides are monomeric in dilute ether or tetrahydrofuran solutions, and have reviewed earlier literature in this field critically.[58] Alkylmagnesium chlorides form chloride-bridged dimers even in dilute solutions. Walker and Ashby concluded that the bromides also dimerize at higher concentrations. However, even though they were well aware of the non-ideality of the solutions and attempted to take that into account, the evidence is questionable. For example, a 2 molal solution of "$EtMgBr \cdot Et_2O$" (which was in the range examined) would contain 386 g of solute per 1000 g of solvent, unless they guessed wrong about the formula and it was really 534 g of $EtMgBr \cdot 2Et_2O$ per 852 g of "free" ether. The activity coefficient of the unbound ether in the very polar solution might be much greater than unity, and the boiling point elevations correspondingly lowered, the apparent molecular weight of the solute raised. Without relying too heavily on these results, it seems reasonable to expect some association of RMgBr at high concentrations in view of the results of Toney and Stucky,[57] but it should not be forgotten that what crystallizes from saturated ethereal solutions is $RMgBr \cdot 2Et_2O$.[49,54]

From all the observations of alkyl and halide bridging in organomagnesium compounds it is reasonable to postulate that the Schlenk equilibrium occurs by way of an intermediate or transition state having one halide bridge and one alkyl bridge between the two magnesium atoms involved, even though there are no data regarding the mechanistic details.

Alkyl exchange reactions of organomagnesium compounds have been studied in detail by House *et al.*[59] Methylmagnesium bromide in ether gives a single sharp methyl peak if it is prepared from specially purified magnesium. Ordinary Grignard grade magnesium contains traces of manganese, which under some conditions broaden the methylmagnesium signal as a result of formation of soluble paramagnetic methylmanganese compounds which exchange methyl groups rapidly with the methylmagnesium compound. Methylmagnesium alkoxides appear to consist of at least three different methylmagnesium species, two of which do not equilibrate readily, but the structures were not defined by the evidence. Dimethylmagnesium and phenylmethylmagnesium in ether exchange methyl groups very rapidly, only a slight broadening of the single average methyl peak being visible at −79°C.

The most interesting results were obtained with cyclopentadienylmagnesium alkyls.[59] Crystalline, ether-free methylcyclopentadienylmagnesium, $MeMgC_5H_5$ (**25**), was obtained from mixtures of dimethylmagnesium and dicyclopentadienylmagnesium. One virtue of **25** is that the magnetic anisotropy of the cyclopentadienide ring shifts the methyl absorption upfield by 0.4 δ from the position for dimethylmagnesium, which is enough separation for accurate rate studies. Another is that the rates of exchange between $MeMgC_5H_5$ and $MgMe_2$ fall in a convenient range for measurement.

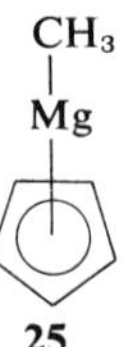

25

The rate of methyl exchange between $MeMgC_5H_5$ (**25**) and $MgMe_2$ depends strongly on the nucleophilicity and complexing ability of the solvent. The ΔH^* and ΔS^* relationships appear complex at first glance but are readily explained by variations of a single type of mechanism. Some of the data are summarized in Table 2-2.

TABLE 2-2

ACTIVATION PARAMETERS AND RATE CONSTANTS FOR METHYL EXCHANGE BETWEEN $MeMgC_5H_5$ AND $MgMe_2$ IN VARIOUS SOLVENT MIXTURES[a]

Solvents (mole fractions)		ΔH^* (kcal/mole)	ΔS^* (eu)	k_2 at −55°C (liters/mole-sec)
THF Ether	0.72 0.28	13.4	+12.9	125
THF Ether	0.64 0.36	11.8	+6.6	200
$MeOCH_2CH_2OMe$ Ether	0.72 0.28	7.0	−14.9	320[b]
Ether THF $(Me_2NCH_2)_2$	0.44 0.30 0.26	11.7	−5.3	0.63
THF Ether Et_2N	0.50 0.26 0.24	11.0	+4.8	500

[a] Data from H. O. House, R. A. Latham, and G. M. Whitesides, *J. Org. Chem.* **32**, 2481 (1967). [b] Revised figure.

House and co-workers suggested that the first step is rapid and reversible loss of one tetrahydrofuran molecule from the solvated dimethylmagnesium (**26**) in tetrahydrofuran–ether solutions, followed by rate-determining methyl exchange between the tricoordinate species **27** and cyclopentadienylmethylmagnesium.[59] In view of the isolation of the bis(tetrahydrofuran) complex of

$$\underset{\mathbf{26}}{Me_2Mg(OC_4H_8)_2} \xrightleftharpoons{\text{fast}} \underset{\mathbf{27}}{Me_2Mg(OC_4H_8)} + C_4H_8O$$

$$\underset{\mathbf{27}}{Me_2Mg(OC_4H_8)} + Me^*MgC_5H_5 \xrightleftharpoons{\text{slow}} [C_4H_8O\text{–}Mg(Me)(\mu\text{-}Me^*)(\mu\text{-}Me)Mg\text{–}C_5H_5] \rightleftharpoons Me^*MeMg(OC_4H_8) + MeMgC_5H_5$$

cyclopentadienylphenylmagnesium,[59] it seems possible that solvation of the cyclopentadienylmethylmagnesium may also be involved, even though $MeMgC_5H_5$ crystallizes without solvent.

From Table 2-2 it can be seen that increasing the nucleophilicity of the solvent increases the heat of activation and accordingly slows the methyl exchange. Chelating agents also slow the methyl exchange, but by making the entropy of activation negative without increasing the enthalpy beyond the normal range for the type of functional group involved. The 800-fold difference in rates between the triethylamine and 1-2-bis(dimethylamino)ethane complexes owes more to the 10 eu difference in ΔS^* than to the 0.7 kcal difference in ΔH^*. The low ΔH^* value for dimethoxyethane is probably comparable to that for diethyl ether, in which rates are too fast to measure for comparison. The dimethoxyethane complex reacts faster than the tetrahydrofuran complex at low temperatures, but above about -40°C the THF complex reacts faster.

With the chelate compounds such as **28**, dissociation of one nucleophilic group amounts to a ring opening. The total ΔS^* contains a small positive entropy term from the ring opening and a much larger negative entropy term for the bimolecular methyl exchange. With nonchelating solvents the

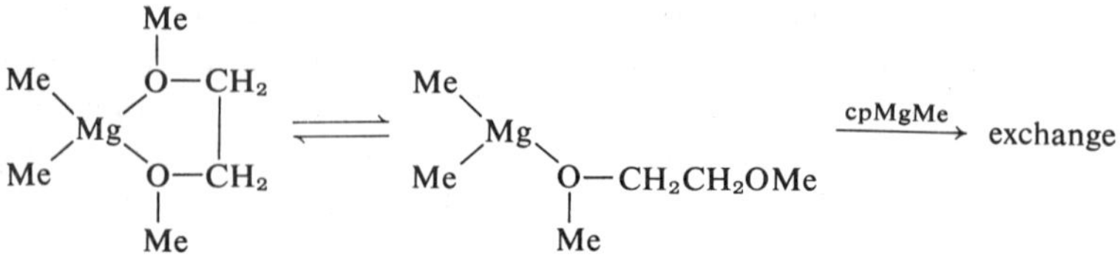

Me Me O—CH₂ Mg O—CH₂ Me Me ⇌ Me Me Mg O—CH₂CH₂OMe Me $\xrightarrow{\text{cpMgMe}}$ exchange

28

positive entropy of solvent dissociation from Me_2Mg (and perhaps $MeMgC_5H_5$ as well) dominates the total ΔS^*.

From studies of the exchange between $Mg[CH_2CH_2C(CH_3)_3]_2$ and $cpMgCH_2CH_2C(CH_3)_3$ it was found that the alkyl exchange with retention of configuration is 10^4 or 10^5 times faster than the inversion of the magnesium-bound carbon.[59] The details of the inversion process will be described in Section IV,C.

C. Magnesium Alkyl Exchange with Inversion

Although alkyl exchange usually proceeds by way of a cyclic three-center-bonded alkyl bridge between the two metal atoms, and therefore must proceed with retention of configuration at the exchange site, Roberts and co-workers have succeeded in measuring the much slower rates of exchange with inversion by the sophisticated choice of an appropriate alkyl group for proton nmr measurements.[60,61]

Inversion was detected with the aid of the neohexyl group, Me_3C—CH_2CH_2—. The *trans* conformation of a neohexylmagnesium compound (**29**) is favored sterically over the *gauche* conformation (**30**). Because the molecule spends more time in the *trans* conformation (**29**), the coupling constant between a given proton of the CH_2CH_2 group and the β proton *trans* to it differs from the coupling constant with the other β proton having a *gauche* relationship. As a result, the nmr spectrum shows an AA′XX′ pattern for the CH_2CH_2 group with 24 theoretically expected lines, 20 of which proved

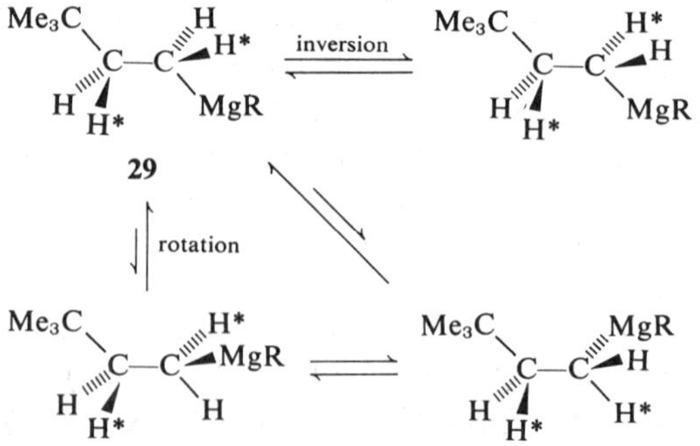

Me₃C, H, H*, C—C, H*, H, MgR (**29**) $\xrightarrow{\text{inversion}}$ Me₃C, H, H*, C—C, H*, H, MgR

rotation

Me₃C, H, H*, C—C, H*, MgR, H (**30**) ⇌ Me₃C, H, H*, C—C, MgR, H, H*

29

30

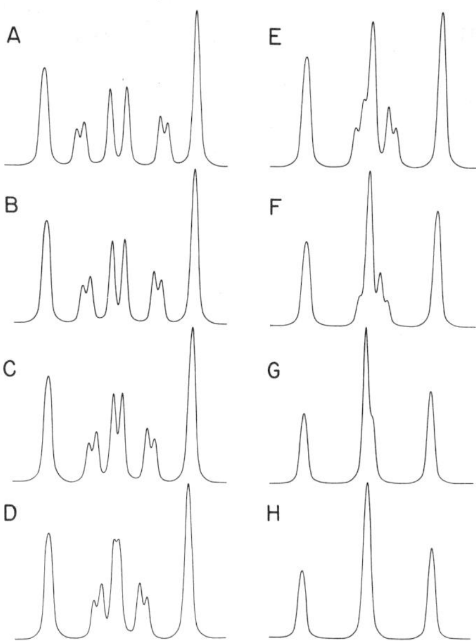

Fig. 2-5. Calculated spectra for the A part of an AA′XX′ spectrum for the internal rotation mechanisms. [From G. M. Whitesides, M. Witanowski, and J. D. Roberts, *J. Amer. Chem. Soc.* **87**, 2860 (1965).]

strong enough to observe.[60] As the solution of a neohexylmagnesium compound is warmed, the AA′XX′ spectrum gradually collapses to an A_2X_2 pattern with its familiar two sets of three lines each.

The nmr spectra distinguish clearly between inversion and rotation processes. Achievement of the high-temperature A_2X_2 spectrum by populating the *trans* (**29**) and two *gauche* (**30**) conformations equally should follow the set of calculated curves shown in Fig. 2-5. Instead, the curves calculated for the inversion process, Fig. 2-6, closely match the experimentally determined curves for dineohexylmagnesium, Fig. 2-7.

Several details regarding the distinction between the inversion process and the mere equalizing of rotamer populations may be noted. First, the inversion process interchanges two protons, resulting in an abrupt switch in their couplings to the other pair of protons. This exchange process results in typical line broadening (Figs. 2-6 and 2-7). In contrast, varying the rotamer populations would shift the coupling constants gradually to time-averaged values without ever uncoupling or exchanging anything, and no line broadening would result (Fig. 2-5). Second, note that changing the rotamer populations gradually shifts the positions of the peaks in the multiplet, and the two in the center coalesce first (Fig. 2-5). In contrast, the inversion process does

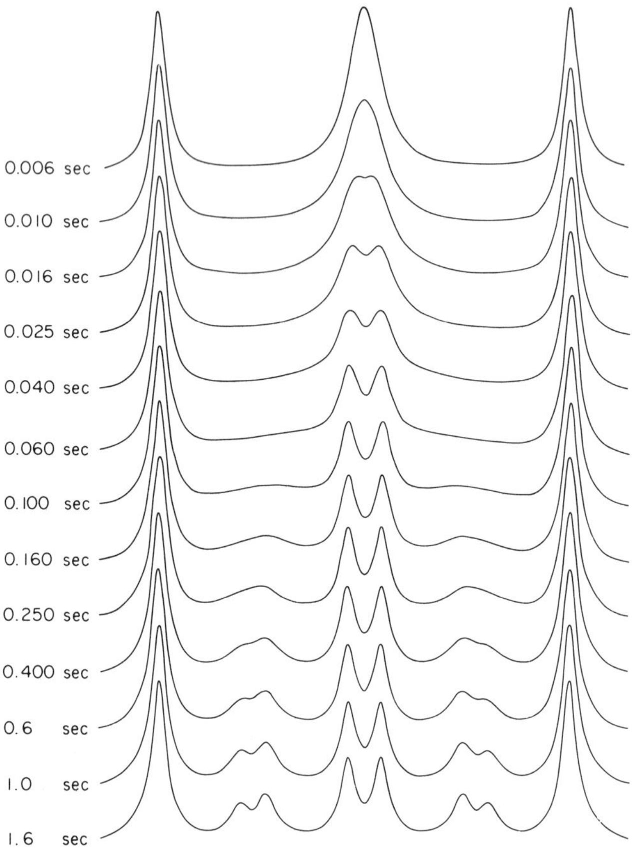

Fig. 2-6. Calculated spectra for the A part of an AA′XX′ spectrum for the inversion mechanism. [From G. M. Whitesides, M. Witanowski, and J. D. Roberts, *J. Amer. Chem. Soc.* **87**, 2859 (1965).]

not change the positions of any peaks except by broadening and time averaging (Figs. 2-6 and 2-7). The relevant quantities which can be observed directly from the spectra are $L = (J_{AX} - J_{AX'})$ and $N = (J_{AX} + J_{AX'})$. For the inversion process, the theory requires that N remain constant regardless of the rate, in accord with the observed constancy of the line separations. If varying populations of rotamers were involved, the negative quantity L would increase to zero at the limit of the A_2X_2 spectrum, the quantity $(\frac{3}{2}N + \frac{1}{2}L)$ would remain constant, and N would decrease. The sort of change which would occur would reduce N from 18.2 down to 14.7 Hz with increasing temperature in the case of dineohexylmagnesium. This magnitude of change is far outside the limit of experimental error.

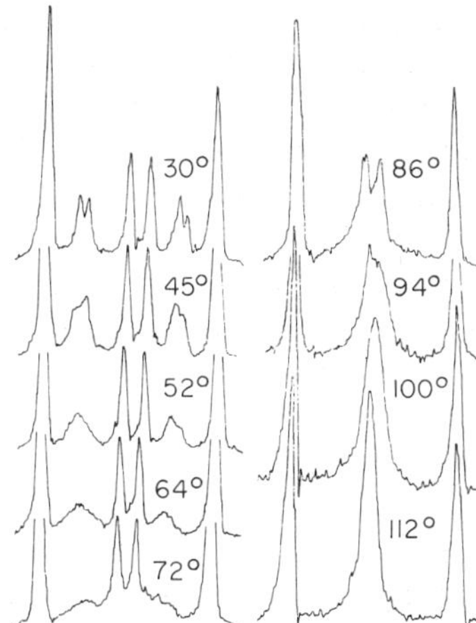

Fig. 2-7. Observed spectra of the α-CH_2 group of dineohexylmagnesium at 60 MHz (5 mole % in ether) in the temperature range 30°–112°C. [From M. Witanowski and J. D. Roberts, *J. Amer. Chem. Soc.* **88**, 738 (1966).]

Whitesides and co-workers have studied several neohexyl compounds which equilibrate conformers and do not invert. In the case of Me_3C—CH_2CH_2—$SiMe_3$ the energy separation between the *trans* and *gauche* rotamers was found to be 2.3 kcal/mole from the temperature dependence of the nmr spectra.[62]

The following activation parameters were obtained from the rate data for inversion in neohexylmetal compounds.[61] For neohexylmagnesium chloride, ΔH^* is about 11 kcal/mole and ΔS^* is -17 eu. Dineohexylmagnesium has $\Delta H^* = 20$ kcal and $\Delta S^* = -1$ eu. The values for neohexyllithium are $\Delta H^* = 15$ and $\Delta S^* = -1$, and for dineohexylzinc are $\Delta H^* = 26$ and $\Delta S^* = 13$. The errors in the ΔH^* values are all estimated at ± 2 kcal, those in the ΔS^* values are ± 7 eu. The general trend in the ΔH^* values is what would be expected from the relative electronegativities of the metals. It is no surprise that dineohexylmercury is configurationally stable in ether up to 163°C. Trineohexylaluminum also shows no sign of dissociation up to 150°C, contrary to expectation based on the behavior of the zinc compound. It may be noted that most of the ΔS^* values are zero or negative, which implies that the inversion is not the result of a simple dissociation to a carbanion and metal cation.

A limited set of experiments indicated that the inversion rate was first order in dineohexylmagnesium or neohexyllithium (presumably the tetramer), and ionization to a carbanion and metal cation was therefore postulated as the mechanism.[61] However, Fraenkel and co-workers have recently reported that the inversions of 2-methylbutylmagnesium compounds are second order in RMgBr, RMgI, or MgR_2 and first order in the dimeric $(RMgCl)_2$.[63] A possible mechanism for bimolecular inversion of 2-methylbutylmagnesium bromide (**31**) by way of bis(2-methylbutyl)magnesium (**32**) is illustrated. Once the MgR_2 is formed with inversion, it will most probably revert to RMgBr by the much faster retention mechanism for concerted alkyl and bromide exchange that establishes the Schlenk equilibrium (see Section IV,B). The principle of microscopic reversibility requires that an equal number of inversions take place by the reverse direction, formation of MgR_2 with retention and reversion to RMgBr with inversion. It is possible to write alternate pathways to that shown, for example, ionization of 2RMgBr to RMg^+ and $RMgBr_2^-$ followed by displacement to form MgR_2 and $MgBr_2$, and there is no information on which to base a choice. For bimolecular inversion of MgR_2 by itself, RMg^+ and MgR_3^- would be involved. There is no possibility of writing a symmetrical pathway for any inversion mechanism that does not release a free carbanion, since the geometry of inversion prohibits reciprocal exchange of ligands.

2 Et(H)(Me)C—C(H)(H*)MgBr (**31**) ⇌ [Br(CH₂CHMeEt)Mg···Et(H)(Me)C—C(H)(H*)···MgBr] ⇌

Et(H)(Me)C—C(H)(H*)—Br^-Mg—$CH_2CHMeEt$ + $MgBr^+$ ⇌ Et(H)(Me)C—C(H*)(H)Mg—$CH_2CHMeEt$ (**32**) + $MgBr_2$

It is also conceivable that the inversion mechanism does involve a carbanion, but that the dimeric magnesium compounds $(RMgCl)_2$, $(RMgBr)_2$, etc., ionize much faster than the corresponding monomers, perhaps because $R(MgCl)_2^+$ spreads the positive charge over two magnesium atoms. However, it should be kept in mind that the carbanion ought to be solvated, and the only species present which could solvate it are the magnesium compounds RMgCl, etc., which would lead to R_2MgCl^- rather than free R^-. The ether solvent should be capable of solvating RMg^+, $ClMg^+$, etc., effectively enough. Thus, consideration of acid–base relationships points toward bimolecular displacement rather than ionization to a carbanion.

The heats and entropies of activation for these inversion processes vary over a wide range, and about the only explanation that comes readily to mind is varying degrees of solvation of the transition state. Values found for

$(RMgCl)_2$ are $\Delta H^* = 6.2$ kcal/mole and $\Delta S^* = -28$ eu; for RMgBr, $\Delta H^* = 12.0$ and $\Delta S^* = -11$; for RMgI, $\Delta H^* = 20.4$ and $\Delta S^* = +11$; and for MgR_2, $\Delta H^* = 18.2$ and $\Delta S^* = +5$. It seems particularly strange that $(RMgCl)_2$ should show the most negative ΔS^*.

D. Beryllium

The position of the Schlenk equilibrium for MeBeCl in ether is not far from statistical.[64] Ashby and co-workers mixed 2 moles of $BeMe_2$ with 1 mole of $BeCl_2$ and observed two narrowly separated proton nmr peaks at −85°C, one for $BeMe_2$ and one for MeBeCl. One broadened peak was seen at −45°C, indicating rapid methyl exchange. Treatment of mixtures of $BeMe_2$, $BeEt_2$, or $BePh_2$ with $BeBr_2$ in ether with dioxane precipitated RBeBr, not $BeBr_2$.

Dimethylberyllium in dimethyl sulfide solution shows rapid equilibration between the solvated monomer $(Me_2S)_2BeMe_2$, the dimer $(Me_2S)_2(BeMe_2)_2$, and possibly higher polymers.[65] Kovar and Morgan observed separation of the methyl nmr line into a complicated set of several broadened peaks as the solution was cooled below −45°C. Evidently the solution contains both terminal and bridge methyl groups and at least three different molecular species to give rise to the number of absorptions found. One possibility that seems to have been overlooked is that the dimer $(Me_2S)_2(BeMe_2)_2$ might have two isomers, one with methyl bridges and the other with sulfur bridges. In any case, the system is too complicated for complete mechanistic analysis. Kovar and Morgan also found that mixing $BeMe_2$ with $BeCl_2$ in Me_2S yields MeBeCl.[65]

V. Lithium

A. Alkyllithiums

Alkyllithiums generally exist as tetramers or hexamers (see Chapter 1, Section III,F). T. L. Brown and co-workers have investigated the mechanism of alkyl exchange between these aggregates with the aid of ^7Li nmr and have shown that dissociation of the tetramers to dimers is generally involved.[66] It may be noted that lithium is even more extreme than aluminum in its tendency to favor bridge bonding. Reactions of aluminum compounds (this chapter, Section III) involve equilibration between bridged and unbridged species, but these organolithium reactions involve species in which the carbon bridges to three lithium atoms dissociating to species in which it only bridges two.

The reaction of $(t\text{-BuLi})_4$ with $(Me_3SiCH_2Li)_6$ in cyclopentane to form a mixture of aggregates $(t\text{-BuLi})_n(Me_3SiCH_2)_{4-n}$ requires several hours at room temperature and can be followed by the changes in the ^{7}Li nmr spectrum.[67,68] The initial reaction is first order in (t-BuLi)$_4$ and zero order in ($Me_3SiCH_2Li)_6$. The probable mechanism is shown in Eqs. (2-17) through (2-19).

$$(Me_3SiCH_2Li)_6 \xrightleftharpoons{\text{rapid}} (Me_3SiCH_2Li)_4 + (Me_3SiCH_2Li)_2 \quad (2\text{-}17)$$

$$(t\text{-BuLi})_4 \xrightleftharpoons{\text{rate-determining}} 2\ (t\text{-BuLi})_2 \quad (2\text{-}18)$$

$$(t\text{-BuLi})_2 + (Me_3SiCH_2Li)_2 \xrightleftharpoons{\text{rapid}} (t\text{-BuLi})_2(Me_3SiCH_2Li)_2 \quad (2\text{-}19)$$

The rate constant for the dissociation of (t-BuLi)$_4$, Eq. (2-18), is 1×10^{-5} sec^{-1} at 20°C in cyclopentane, ΔH^* is 24 ± 6 kcal/mole, and ΔS^* is near 0 eu. The reaction is about 20 times faster in toluene and is also greatly accelerated by small quantities of triethylamine. The rate of isotopic exchange between ^{6}Li and ^{7}Li-labeled samples of (t-BuLi)$_4$ found by mass spectral analysis was the same as that based on the nmr measurements.[68]

Once the t-butyllithium has been converted to mixed clusters with the trimethylsilylmethyllithium, equilibration becomes much faster. Exchange rates in the range accessible to nmr line broadening occur between −50° and +30°C, though the spectra are too complex for easy kinetic analysis.

These results require that both the dissociation of t-butyllithium tetramer and the recombination of t-butyllithium dimer be slow. If recombination of the dimer were rapid, the equilibrium would be shifted to favor (t-BuLi)$_4$ rather than the mixed species actually observed. Also, the kinetics being zero order in the other compound implies that the dissociation of the t-butyllithium tetramer is essentially irreversible under the reaction conditions.

Ether greatly accelerates the dissociation of alkyllithium tetramers, presumably by preferential solvation of the dimers.[66] Structure **33** is a reasonable guess for solvated methyllithium dimer. Ether must also solvate the tetramer, which accounts for tetramer being favored over hexamer.

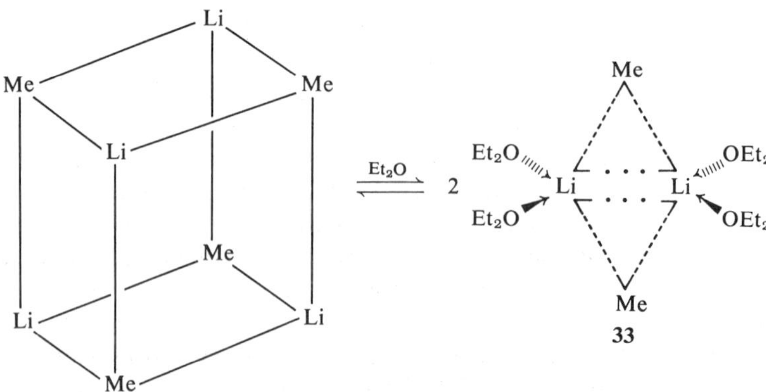

Mixtures of methyllithium and ethyllithium in ether equilibrate rapidly above $-50°C$, yielding a single line in the ^7Li nmr spectrum.[69] Four lines appear at lower temperatures. The peaks associated with environments high in ethyl groups broaden and coalesce at lower temperatures than those high in methyl groups, or in other words, the more ethyl substituents in the tetramer, the faster it dissociates. The mechanism is thought to involve dissociation of $(MeLi)_4$ and $(EtLi)_4$ to $(MeLi)_2$ and $(EtLi)_2$ followed by recombination of the dimers to form $(MeLi)_2(EtLi)_2$, etc. The activation energy for the dissociation of $(MeLi)_4$ could not be determined from these exchange experiments, but was found to be ~11 kcal/mole from the reactions described in Section V,B.

Phenyllithium is dimeric in ether.[70] This can be rationalized on the basis of the previously described orbital arrangement of the phenyl group, which favors bridging between two atoms rather than coordination with the triangular face of a polyhedron. In mixtures of phenyllithium and methyllithium, the species present at equilibrium are $(PhLi)_2$, $PhLi_2Me$, $PhLi_4Me_3$, and $(MeLi)_4$.

B. Bridged Lithium–Aluminum and Related Alkyls

Methyllithium and trimethylaluminum react in ether to form $LiAlMe_4$ (**34**), a compound with methyl bridge bonds between the lithium and aluminum,[3] and probably with ether filling the vacant coordination sites on

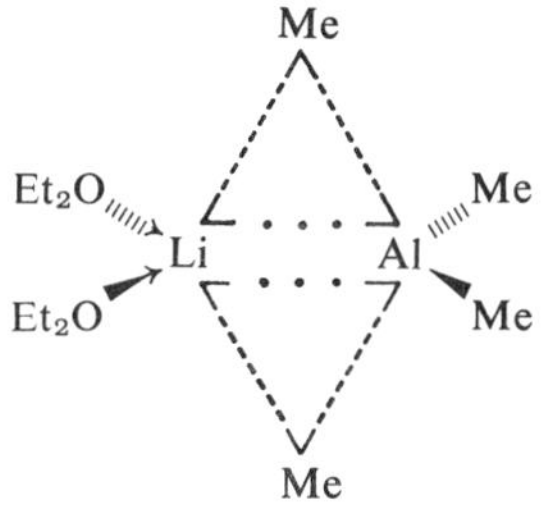

Me

Et_2O Li Al Me

Et_2O Me

Me

34

the lithium. Similar structures may be assigned to such compounds as $LiBMe_4$ and $LiGaEt_4$. Lithium ions but not methyl groups exchange rapidly between $LiAlMe_4$ and $(MeLi)_4$. The exchange is first order in $(MeLi)_4$ and zero order in $LiAlMe_4$, and the rate-determining step is the dissociation of $(MeLi)_4$ to 2 $(MeLi)_2$.[3] The probable mechanism is outlined in Eqs. (2-20) and (2-21).

$$(LiMe)_4 \rightleftharpoons 2\,(LiMe)_2 \tag{2-20}$$

$$(LiMe)_2 + LiAlMe_4 \rightleftharpoons \left[Me{-}Li_3{-}Me \right]^+ + AlMe_4^- \tag{2-21}$$

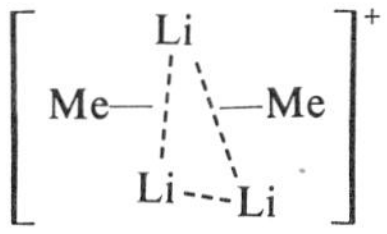

The lithium ions are scrambled on reversal of the second step. Exchange between $LiBMe_4$ and $(MeLi)_4$ occurs with approximately the same rate and activation energy, in accord with the dissociation of $(MeLi)_4$ being the rate-determining step. The value of ΔH^* is about 10.8 kcal/mole and ΔS^* is -4 eu, and the low ΔS^* value indicates participation of the ether solvent in the dissociation process. It may also be noted that ΔH^* for the dissociation of $(t\text{-BuLi})_4$ in cyclopentane is much higher, ~24 kcal, and that this difference results from solvation.[66] Increasing the size of the alkyl groups promotes dissociation, and $(t\text{-BuLi})_4$ would dissociate faster than $(MeLi)_4$ if both could be measured in the same solvent.

$LiAlMe_4$ exchanges methyl groups with $AlMe_3 \cdot Et_2O$. The reaction is first order in $LiAlMe_4$ and zero order in $AlMe_3 \cdot Et_2O$, with $\Delta H^* = 8.7$ kcal/mole and $\Delta S^* = -24$ eu. Williams and Brown formulated the rate-determining step as conversion of an intimate ion pair (of which structure **34** is an example, semantics and bond symbols aside) to a solvent-separated ion pair.[3] In view of the very negative ΔS^*, solvent molecules must become bound in this dissociation, which may alternatively be viewed as a nucleophilic displacement of $AlMe_4^-$ from Li^+ by Et_2O. The probable mechanism is outlined in Eqs. (2-22) and (2-23). The actual number of solvent molecules

$$(Et_2O)_2LiAlMe_4 + 2\,Et_2O \rightleftharpoons (Et_2O)_4Li^+ + AlMe_4^- \text{ (ion pair)} \tag{2-22}$$

$$(Et_2O)_4Li^+ + AlMe_4^- + AlMe_3^* \cdot Et_2O \rightleftharpoons (Et_2O)_4Li^+ + AlMe_4^{*-} + AlMe_3 \cdot Et_2O \tag{2-23}$$

involved is, of course, not known, and an arbitrary guess based on valence rules is presented for clarity. Several details of the process are missing, and transition state structures would have to be purely conjectural.

The bridge and terminal methyl groups of $LiAlMe_4$ or $LiBMe_4$ evidently exchange very rapidly, since a single line is seen in the nmr even at -60°C. The bridge bonds to the lithium would have considerable ionic character, and it may be easy to break one of them and form a new bond to another methyl group without involving enough solvent to dissociate the $LiAlMe_4$ molecule. At $+50$°C the proton nmr of $LiBMe_4$ shows a quartet resulting from ^{11}B coupling, which collapses at lower temperatures to a single broadened peak.[3] One possible explanation for this peculiar behavior would be that the boron atom is in an effectively more symmetrical environment (faster bridge–terminal equilibration) at the higher temperatures. Quadrupole relaxations of spin–spin couplings become more effective as the symmetry of the quadrupolar atom is decreased.

Methyllithium and dimethylmagnesium in ether form complexes Li_2MgMe_4 and Li_3MgMe_5,[71] and the analogous zinc compounds Li_2ZnMe_4 and Li_3ZnMe_5 have also been shown to be present in solution. With these complexes, methyllithium exchanges both ^7Li and methyl groups at the same rate and activation energy. The rate-determining process for the zinc system appears to be the dissociation of $(MeLi)_4$ to 2 $(MeLi)_2$, but the magnesium compound exchanges somewhat more slowly. Methyl exchange between $ZnMe_2$ and Li_2ZnMe_4 is too fast to measure even at $-107°C$. Methyl exchange between $MgMe_2$ and Li_2MgMe_4 is measurable below $-50°C$, with ΔH^* about 15 kcal. This result requires that ΔS^* also be positive in order to have a rapid reaction, and the rate-determining step must be a dissociation rather than a bimolecular collision. A likely candidate is the loss of ether from dimethylmagnesium dietherate. It is easy to write bridged structures which account for the methyl exchange,[71] though the probable occurrence of these after the rate-determining step means there is no kinetic evidence one way or the other about them.

Atwood and Stucky have shown by X-ray crystallography that the magnesium–aluminum alkyl $Me_2AlMe_2MgMe_2AlMe_2$ has a methyl-bridged structure (Chapter 1, Section III,C, Fig. 1-5). It shows reactivity similar to the lithium compounds.[72] Bridge–terminal methyl exchange is evidently very rapid, as there is only one line in the proton nmr spectrum at $-60°C$ in cyclopentane. Methyl exchange with Al_2Me_6 appears to involve reaction of $AlMe_3$ with the $Mg(AlMe_4)_2$. There is also an equilibrium between $Mg(AlMe_4)_2$ and $(MgAlMe_5)_2 + Al_2Me_6$. The exchange mechanisms obviously involve the making and breaking of three-center methyl-bridge bonds, but the available kinetic evidence provides no information about the details.

C. Allylic Carbanions

It has already been noted that alkyllithiums are ordinarily covalent compounds, but if the carbanion from dissociation is a relatively stable type, such as the benzyl anion, nucleophilic solvents can cause ionization to the carbanion and the solvated lithium ion (see Chapter 1, Section II,A, Fig. 1-2). Freedman *et al.* have shown by nmr that the same carbanion is formed from either *cis*- or *trans*-PhCH=CHCH$_2$Ph with butyllithium.[73] Evidently there is a low-energy path for interconversion of the *cis* and *trans* forms of the carbanion $PhCHCHCHPh^-$.

West, Purmort, and McKinley have observed that at $-87°C$ in ether or tetrahydrofuran the proton nmr spectrum of allyllithium shows an AA′BB′C pattern.[74] At higher temperatures, $-50°C$ up to $+37°C$, this collapses to an AB_4 pattern, in which the original A and B sets exchange rapidly. The species observed is evidently the allyl anion (**35**). The terminal protons are at δ1.7–2.5 in the low-temperature form, and show as a doublet at δ2.2 in the high-temperature form. The central proton is at δ6.4 in both.

It is apparent that rotation around the carbon–carbon bonds is restricted

35

at low temperatures but occurs at higher temperatures. The mechanism for the rotation is not known, but if allyllithium exists as an ion pair, one likely mechanism would be brief collapse of the ion pair to covalent allyllithium, $CH_2{=}CH{-}CH_2Li$, which would show free rotation about the carbon–carbon single bond.

VI. Group IV

Group IV elements lack any vacant site that could participate in bridge bonding and consequently their alkyl exchange reactions are generally very slow if they occur at all. Cyclopentadienyl compounds are, of course, an exception (compare mercury, Section IV,A). For example, Davison and Rakita have shown that rapid intramolecular migration of the trimethylsilyl group around the ring in $Me_3SiC_5H_4CH_3$ occurs above room temperature and that similar migration of the trimethylstannyl group occurs in $Me_3SnC_5H_4CH_3$ at lower temperatures.[75] The tin remains coupled with the ring protons under conditions of rapid migration. This verifies that the rearrangement is intramolecular, which will give an averaged coupling constant, and not intermolecular, which would have uncoupled the spins.

Exchange of methyl groups between $PbMe_4$ and Al_2Me_6 is too slow to cause nmr line broadening, but $PbMe_4$ and Al_2Et_6 do exchange methyl and ethyl groups as shown by the nmr spectrum of the product mixture.[76]

References

1. C. R. McCoy and A. L. Allred, *J. Amer. Chem. Soc.* **84**, 912 (1962).
2. J. E. Leffler and E. Grunwald, "Rates and Equilibria of Organic Reactions," p. 89. Wiley, New York, 1963.
3. K. C. Williams and T. L. Brown, *J. Amer. Chem. Soc.* **88**, 4134 (1966).
4. H. S. Gutowsky and C. H. Holm, *J. Chem. Phys.* **25**, 1228 (1956).
5. J. P. Maher and D. F. Evans, *J. Chem. Soc.* 5534 (1963).
6. K. C. Williams and T. L. Brown, *J. Amer. Chem. Soc.* **88**, 5460 (1966).
7. H. D. Visser and J. P. Oliver, *J. Amer. Chem. Soc.* **90**, 3579 (1968).
8. M. B. Smith, *J. Organometal. Chem.* **22**, 273 (1970).
9. A. W. Laubengayer and W. F. Gilliam, *J. Amer. Chem. Soc.* **63**, 477 (1941).
10. M. B. Smith, *J. Phys. Chem.* **71**, 364 (1967).
11. J. N. Hay, P. G. Hooper, and J. C. Robb, *J. Organometal Chem.* **28** 193, (1971).
12. K. C. Ramey, J. F. O'Brien, I. Hasegawa, and A. E. Borchert, *J. Phys. Chem*, **69**, 3418 (1965).

13. (a) E. A. Jeffery and T. Mole, *Aust. J. Chem.* **22**, 1129 (1969); (b) E. A. Jeffery and T. Mole, *Aust. J. Chem.* **26**, 739 (1973).
14. D. S. Matteson, *Inorg. Chem.* **10**, 1555 (1971).
15. T. L. Brown and L. L. Murrell, *J. Amer. Chem. Soc.* **94**, 378 (1972).
16. O. Yamamoto and K. Hayamizu, *J. Phys. Chem.* **72**, 822 (1968).
17. E. G. Hoffmann, *Trans. Faraday Soc.* **58**, 642 (1962).
18. D. A. Sanders and J. P. Oliver, *J. Amer. Chem. Soc.* **90**, 5910 (1968).
19. J. P. Oliver, *Advan. Organometal. Chem.* **8**, 167 (1970).
20. D. Moy, M. T. Emerson, and J. P. Oliver, *J. Amer. Chem. Soc.* **86**, 371 (1964).
21. T. Mole and J. R. Surtees, *Aust. J. Chem.* **17**, 1229 (1964).
22. E. A. Jeffery and T. Mole, *J. Organometal. Chem.* **11**, 393 (1968).
23. E. A. Jeffery, T. Mole, and J. K. Saunders, *Chem. Commun.* 696 (1967).
24. J. F. Malone and W. S. McDonald, *Chem. Commun.* 444 (1967).
25. N. S. Ham, E. A. Jeffery, and T. Mole, *Aust. J. Chem.* **21**, 2687 (1968).
26. C. P. Poole, Jr., H. E. Swift, and J. F. Itzel, Jr., *J. Chem. Phys.* **42**, 2576 (1965).
27. J. W. Moore, D. A. Sanders, P. A. Scherr, M. D. Glick, and J. P. Oliver, *J. Amer. Chem. Soc.* **93**, 1035 (1971).
28. T. Mole, *Aust. J. Chem.* **18**, 1183 (1965).
29. E. A. Jeffery and T. Mole, *Aust. J. Chem.* **21**, 1497 (1968).
30. V. R. Magnuson and G. D. Stucky, *J. Amer. Chem. Soc.* **91**, 2544 (1969).
31. E. A. Jeffery and T. Mole, *Aust. J. Chem.* **23**, 715 (1970).
32. E. C. Ashby, J. Laemmle, and G. E. Parris, *J. Organometal. Chem.* **19**, P24 (1969).
33. T. Mole, *Aust. J. Chem.* **19**, 373 (1966).
34. T. Mole, *Aust. J. Chem.* **19**, 381 (1966).
35. P. West, M. C. Woodville, and M. D. Rausch, *J. Amer. Chem. Soc.* **91**, 5649 (1969).
36. N. S. Ham, E. A. Jeffery, T. Mole, J. K. Saunders, and S. N. Stuart, *J. Organometal. Chem.* **8**, P7 (1967).
37. W. Bremser, M. Winokur, and J. D. Roberts, *J. Amer. Chem. Soc.* **92**, 1080 (1970).
38. J. Soulati, K. L. Henold, and J. P. Oliver, *J. Amer. Chem. Soc.* **93**, 5694 (1971).
39. K. Henold, J. Soulati, and J. P. Oliver, *J. Amer. Chem. Soc.* **91**, 3171 (1969).
40. E. A. Jeffery and T. Mole, *Aust. J. Chem.* **21**, 1187 (1968).
41. F. J. Weigert, M. Winokur, and J. D. Roberts, *J. Amer. Chem. Soc.* **90**, 1566 (1968).
42. M. H. Abraham and J. A. Hill, *J. Organometal. Chem.* **7**, 23 (1967).
43. N. I. Sheverdina, I. E. Paleeva, and K. A. Kocheshkov, *Izv. Akad. Nauk SSSR, Ser. Khim.* 587 (1967).
44. J. Boresma and J. G. Noltes, *J. Organometal. Chem.* **8**, 551 (1967).
45. G. E. Coates, R. Ridley, J. A. Heslop, and M. E. Redwood, *J. Chem. Soc. A* 1118 (1968).
46. G. E. Parris and E. C. Ashby, *J. Amer. Chem. Soc.* **93**, 1206 (1971).
47. R. E. Dessy, G. S. Handler, J. H. Wotiz, and C. A. Hollingsworth, *J. Amer. Chem. Soc.* **79**, 3476 (1957).
48. R. E. Dessy, S. Green, and R. M. Salinger, *Tetrahedron Lett.* 1369 (1964).
49. G. Stucky and R. E. Rundle, *J. Amer. Chem. Soc.* **85**, 1002 (1963); **86**, 4821 (1964).
50. E. C. Ashby, *Quart. Rev.* (*London*) **21**, 259 (1967).
51. F. R. Jensen and K. L. Nakamaye, *J. Amer. Chem. Soc.* **90**, 3248 (1968).
52. D. F. Evans and M. S. Khan, *J. Chem. Soc. A* 1643 (1967).
53. D. F. Evans and M. S. Khan, *J. Chem. Soc. A* 1648 (1967).
54. L. J. Guggenberger and R. E. Rundle, *J. Amer. Chem. Soc.* **86**, 5344 (1964); **90**, 5375 (1968).
55. M. Vallino, *J. Organometal. Chem.* **20**, 1 (1969).

56. F. Schröder and H. Spandau, *Naturwissenschaften* **53**, 360 (1966).
57. J. Toney and G. D. Stucky, *Chem. Commun.* 1168 (1967).
58. F. W. Walker and E. C. Ashby, *J. Amer. Chem. Soc.* **91**, 3845 (1969).
59. H. O. House, R. A. Latham, and G. M. Whitesides, *J. Org. Chem.* **32**, 2481 (1967).
60. G. M. Whitesides, M. Witanowski, and J. D. Roberts, *J. Amer. Chem. Soc.* **87**, 2854 (1965).
61. M. Witanowski and J. D. Roberts, *J. Amer. Chem. Soc.* **88**, 737 (1966).
62. G. M. Whitesides, J. P. Sevenair, and R. W. Goetz, *J. Amer. Chem. Soc.* **89**, 1135 (1967).
63. G. Fraenkel, C. E. Cottrell, and D. T. Dix, *J. Amer. Chem. Soc.* **93**, 1704 (1971).
64. J. R. Sanders, Jr., E. C. Ashby, and J. H. Carter, II, *J. Amer. Chem. Soc.* **91**, 6385 (1969).
65. R. A. Kovar and G. L. Morgan, *J. Amer. Chem. Soc.* **91**, 7269 (1969).
66. T. L. Brown, *Accounts Chem. Res.* **1**, 23 (1968).
67. G. E. Hartwell and T. L. Brown, *J. Amer. Chem. Soc.* **88**, 4625 (1966).
68. M. Y. Darensbourg, B. Y. Kimura, G. E. Hartwell, and T. L. Brown, *J. Amer. Chem. Soc.* **92**, 1236 (1970).
69. L. M. Seitz and T. L. Brown, *J. Amer. Chem. Soc.* **88**, 2174 (1966).
70. L. M. Seitz and T. L. Brown, *J. Amer. Chem. Soc.* **89**, 1607 (1967).
71. L. M. Seitz and T. L. Brown, *J. Amer. Chem. Soc.* **88**, 4140 (1966).
72. J. L. Atwood and G. D. Stucky, *J. Amer. Chem. Soc.* **91**, 2538 (1969).
73. H. H. Freedman, V. R. Sandel, and B. P. Thill, *J. Amer. Chem. Soc.* **89**, 1762 (1967).
74. P. West, J. I. Purmort, and S. V. McKinley, *J. Amer. Chem. Soc.* **90**, 797 (1968).
75. A. Davison and P. E. Rakita, *J. Amer. Chem. Soc.* **90**, 4479 (1968).
76. M. Boleslawski, S. Pasynkiewicz, and K. Jaworski, *J. Organometal. Chem.* **30**, 199 (1971).

CHAPTER 3

Electrophilic Displacement: Replacements of Metal Cations

I. Transition State Models

A. Topics Covered

This chapter covers the reactions that would be traditionally classified as electrophilic displacements and that have been studied by more or less classical kinetic and stereochemical methods. For the most part this excludes exchange reactions, which have been covered in Chapter 2, except that isotopic exchanges of mercury(II) are included here along with other displacements involving mercury(II).

Reactions which involve neighboring-group interactions are reserved for Chapter 4. However, three-center bonding or ligand bridging between the electrophile and the electrofuge is not considered to be a neighboring-group effect for classification purposes, and such interactions often occur in the reactions to be discussed in the present chapter. If the reaction involves a direct electron-pair bond between the electrophile and the electrofuge, a cyclopropane ring opening or closure, or an allylic rearrangement (S_E2' reaction) it is placed in Chapter 4. The latter two types have something in common with electrophilic additions to double bonds and eliminations, which are covered in Chapter 5.

Because of the large number of examples provided by the voluminous literature, the rather small number of different classes of mechanisms would be obscured by burying them in the main body of the long text of this chapter. Accordingly, selected examples of the various classes of mechanism are described first in this section on transition state models, with the supporting evidence and citing of numerous related examples being reserved for the main body of the chapter.

B. Displacement with Retention of Configuration

The most common mechanism for replacement of one metal cation by another at carbon is a variant of the alkyl exchange mechanism discussed in Chapter 2. Instead of exchanging two alkyl groups, the metal cations exchange one alkyl group and an anionic ligand. It should be noted at the outset that free metal cations are not involved, that there are always anionic or alkyl ligands firmly bound to both metal cations, and that as far as mechanistic function is concerned a nonionic compound such as boric acid behaves in the same manner as a complexed metal cation.

A good model for this type of transition state is provided by the stable compound μ-diphenylamino-μ-methyl(tetramethyldialuminum) (**1**). Details of the bonding in this compound have been discussed in Chapter 1, Section III,C and the X-ray structure is pictured in Fig. 1-6. The bridge–terminal methyl exchange of a related compound having a *t*-BuO$^-$ ligand bridge in place of the Ph_2N^- ligand has been discussed in Chapter 2, Section III,G.

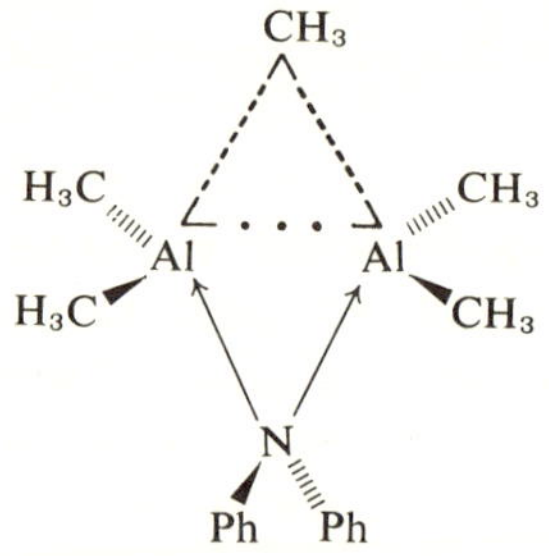

CH_3

H_3C Al Al CH_3

H_3C CH_3

N

Ph Ph

1

Winstein, Traylor, and Garner first proposed the bridged transition state structure **2** in 1955, with somewhat different symbolism but the same meaning, to account for retention of configuration in the displacement of a neophylmercuric ion from (β-methoxycyclohexyl)neophylmercury by a chloromercuric ion from mercuric chloride.[1] The structural analogy between the stable compound **1** and the transition state **2** should be readily apparent.

Several details of the bonding in transition state **2** are not defined by the available evidence. The postulated ligand bridge, Hg—Cl—Hg, might contribute more or less energy than the alkyl bridge, Hg—C—Hg, to the stability of the transition state. It is not known which bridge is formed first, but this would probably be the more stable one. It seems unlikely that there is no chloride bridge, Hg—Cl—Hg, in the transition state because the products RHgCl and R′HgCl are more stable than the ions $RHgCl_2^-$ and RHg^+ that would have to be formed if the chloride did not bridge. In considering the question of ligand bridging, it is well to remember that the oxygen bridge in Me_5Al_2O-*t*-Bu is much stronger than the methyl bridge

H
Hg—CH_2CMe_2Ph
+ $HgCl_2$ ⟶
H
OMe

Cl
Hg
H
Cl
Hg
H
CH_2CMe_2Ph
OMe
⟶
H
HgCl
H
OMe

2

+ Cl—Hg—CH_2CMe_2Ph

(Chapter 2, Section III,G), but with weakly complexing anions in highly polar solvents there may be mercury(II) displacements in which the ligand bridges are weak or absent. It is also possible that structure **2** is an unstable intermediate rather than a transition state, but if so its energy and therefore its structure (Hammond's postulate[2]) are not far from that of the true transition state, which would merely have one or more bonds stretched a bit. The strength of mercury–mercury bonding in the three-center bridge bond is inherently inseparable from the total bridge bond energy, except perhaps by a detailed computer treatment (see Chapter 1, Section III,B), but it provides a good topic for chemists to argue about over beer. All these details vary with the structure of the particular system, and all possible combinations and permutations of bond strengths may be expected to be found if one looks long enough.

Considerable distortion of the fundamental transition state model as represented by structures **1** and **2** must occur if the attacking electrophile is not a metal cation but an electronegative species such as a halogen or a proton. These reactions are often highly exothermic, and in such cases the transition state will occur early along the reaction coordinate before the electrophile has formed a strong bond to the carbon.[2] Ligand bridging is probably present in most cases, though Winstein and Traylor noted that the facile protodemercuration of the bridgehead mercury compound bis(4-camphyl)mercury in perchloric acid implies that ligand bridging is not necessary in order for displacement with retention to occur.[3] Jensen and co-workers provided one of the earlier well-studied examples of electrophilic displacements of this class, bromodemercuration, which was shown to proceed with retention.[4,5] The transition state they suggested for the bromodemercuration of 2-butylmercuric bromide is represented, with revised symbolism to be consistent with usage in this book, by structure **3**.

$$\text{H, Et, Me}{-}C{-}HgBr + Br_2 \longrightarrow \underset{\mathbf{3}}{[\text{H, Et, Me}{-}C \cdots Br{-}Br \cdots Hg{-}Br]} \longrightarrow \text{H, Et, Me}{-}C{-}Br + HgBr_2$$

It should be emphasized that the various bond orders represented by the solid, broken, dotted, and partially missing lines in structure **3** are speculative. They correctly represent the net orbital symmetries and the number of electron pairs involved in the bonding, assuming the choice of nuclear positions is qualitatively correct. They seem a reasonable qualitative representation of the likely bond orders, and at least serve as reminders of the considerable complexity required to represent a transition state with its nonstandard bond lengths and strengths. They fail to indicate electron density, and there is no direct information available anyway on how much electron density has been transferred to the bromine atom by the time this transition state is reached.

Further discussion of transition states for electrophilic displacement with retention will be postponed until they are encountered in the remainder of this chapter, of which they form the bulk of the examples to be covered.

C. Concerted Displacement with Inversion of Configuration

Inversion is known to occur in a few special circumstances, but in the known cases where one metal cation displaces another, the observable inversion processes are several orders of magnitude slower than displacement with retention (Chapter 2, Section IV,C). An orbital picture for the inversion process has already been drawn, and it was noted that the best available stable model compound, $LiB(CH_3)_4$, contains linear $B—CH_3—Li$ bridges which do not approach the required transition state geometry very closely, in contrast to the bent B—C—Li bridges which are good models for displacement with retention (Chapter 1, Section III,E).

With nonmetallic electrophiles, inversion is actually preferred over retention in some reactions. An early example was found by Matteson and Waldbillig in a reaction which closes a cyclopropane ring as a boronic acid group is displaced.[6] This reaction is rather complex and is discussed in detail in Chapter 4, Section VI,B.

Simpler examples are provided by certain halodemetallations. For example, Jensen and Davis have shown that the cleavage of 2-butyl(trineopentyl)tin by bromine in methanol yields predominantly inverted 2-bromobutane,[7] presumably by way of a transition state resembling structure **4**.

Several other bromodemetallations are known to proceed with preferential inversion. Some of these were reported before the example cited, but the

$$Br_2 + (H)(Me)(Et)C{-}Sn(CH_2CMe_3)_3 \longrightarrow \overset{\delta-}{Br}{\cdots}Br{\cdots}C(Me)(H)(Et){\cdots}\overset{\delta+}{Sn}(CH_2CMe_3)_3 \;(\mathbf{4}) \longrightarrow Br^- + Br{-}C(H)(Me)(Et) + \overset{+}{Sn}(CH_2CMe_3)_3$$

simplest example has been chosen for illustrative purposes and the others will be covered in detail in the main body of the chapter.

D. Carbanions

The extensive work of Cram and co-workers on electrophilic displacements involving considerable carbanion character in the transition state is not organometallic chemistry in the usual sense, but the fundamental significance of the work and its relevance to certain organometallic reactions requires its mention here. The solvolytic cleavage of certain *t*-alkoxide ions to carbanions and ketones leads to hydrocarbons by protonation of the carbanions with either retention, racemization, or inversion, depending on the solvent.[8] For example, in solvents of low polarity but substantial proton-donating ability,

$$(Me)(Et)(Ph)C{-}C(O^-K^+)(Me)(Et) \longrightarrow (Me)(Et)(Ph)C^- \;\; K^+ \;\; O{=}C(Et)(Me) \xrightarrow{t\text{-BuOH}}$$

$$(Me)(Et)(Ph)C^- \;\; K^+ \;\; H{-}O{-}t\text{-Bu} + Et\overset{O}{\overset{\|}{C}}Me \;(\mathbf{5}) \longrightarrow (Me)(Et)(Ph)C{-}H + K^+ \; {}^-O{-}t{-}Bu$$

$$(Me)(Et)(Ph)C{-}C(O^-K^+)(Me)(Et) + HOCH_2CH_2OH \longrightarrow (Me)(Et)(Ph)C{-}C(O^-)(Me)(Et) + \text{solvated } K^+ \longrightarrow$$

$$HOCH_2CH_2\overset{\delta-}{O}{\cdots}H{\cdots}\overset{\delta-}{C}(Me)(Et)(Ph){\cdots}C(\overset{\delta-}{O})(Me)(Et) \;(\mathbf{6}) \longrightarrow HOCH_2CH_2O^- + H{-}C(Me)(Et)(Ph) + Et\overset{O}{\overset{\|}{C}}Me$$

such as *t*-butyl alcohol, solvated ion pairs such as **5** are favored and ~90% net retention results. In ethylene glycol, which has high polarity, high proton-donating ability, and the ability to chelate potassium ions, it becomes much easier to break up the tight ion pairs, and the transition state **6** for inversion is moderately favored. Addition of a chelating cyclic polyether, dicyclohexyl-18-crown-6 ether, to tie up the potassium ion in *t*-butyl alcohol also resulted in inversion (15% net inversion, the rest racemization).[9] If the solvent is polar but a poor proton donor, such as dimethyl sulfoxide, the carbanion becomes separated from intimate association with any particular electrophile and complete racemization results.

It is not possible to do justice to the extensive and detailed work of Cram and co-workers in the space that can be allotted here. In the main text of the chapter, there is an example of a base-catalyzed protodeboronation that probably proceeds by a related mechanism, with boron in place of carbon as the departing electrofuge, and some net inversion was found in water. Other organometallic examples are likely to turn up, though the difficulty of generating carbanions in solvents which lack proton sources and the ease of direct reaction of most organometallic compounds with electrophiles is likely to limit such carbanion-forming reactions to protodemetallations.

A few examples of "carbanion" generation from organomercury and other relatively unreactive organometallic compounds have been reported and will be cited in the main body of the chapter. In general, the so-called carbanions generated by such processes are extremely stabilized by electron-withdrawing substituents, to the point where they are better described as olefins having electron-donating substituents. There have also been some reports of carbanion generation from simple alkylmercury compounds, but these have been proved wrong. Carbanion generation is not a characteristic reaction of ordinary organometallic compounds.

E. The S_E1–S_E2–S_Ei Classification Scheme (and Its Difficulties)

In view of the applicability of the S_N1—S_N2—S_Ni classification scheme to a large number of nucleophilic displacements, it is understandable that C. K. Ingold and co-workers should look for parallel behavior in the electrophilic displacement series. However, the corresponding S_E1—S_E2—S_Ei scheme does not really work. Ingold and co-workers studied organomercury reactions and attempted to distinguish between simple bimolecular S_E2 displacement and cyclic ligand-bridged S_Ei displacement.[10] The possibility of distinction is questionable, errors were made in the interpretation, and, of course, they did not know about the undiscovered class of electrophilic displacements which invert the carbon atom. The details will be discussed in Section II,A on mercury(II) displacements. In spite of the unfortunate interpretation problems, it should be noted that the experimental work is an important and useful contribution to the field.

The S_E1 category fared worse. Both Ingold's group[11,12] and Reutov and co-workers[13] found first-order kinetics in reactions of organomercury and organothallium compounds in dimethyl sulfoxide or dimethylformamide. Sometimes it almost seems that chemicals can be willfully mean to kineticists who forget to check the identity of their products. Jensen and Heyman showed that the alleged S_E1 reaction of Et_2TlBr with $(2\text{-}Bu)_2Hg$[12] is actually an air oxidation of the mercury compound.[14] Jensen and Rickborn have attacked the alleged ionization of ethyl α-bromomercuriphenylacetate[11,13] on grounds that it is not possible to reconcile the data with any reasonable ionization–exchange mechanism consistent with the principle of microscopic reversibility, and that the conditions used would again be likely to permit air oxidation.[15a] Other alleged examples of S_E1 reactions will be deflated in the main body of the chapter.

Beletskaya *et al.* have defended the S_E1 concept in a recent review.[16] Some of the work cited has been questioned by Jensen and Rickborn,[15] who question with good reasons. Other examples are probably correct, but the carbanions formed are generally either fluorocarbons or other relatively weak bases, such as the ethyl 2-cyanohexanoate anion from its complex with triphenylphosphinegold(I) reported by Gregory and Ingold.[17] However, the S_E1 label tends not to fit in most cases, since nucleophilic attack on the metal cation normally precedes dissociation, and it was proposed that the category be renamed $S_E1(N)$.[16]

Abraham and Hill have attempted to elaborate the S_E2 and S_Ei categories into a better description of the facts.[18] They proposed S_EC, where C stands for concerted, for cyclic transition states in which the nucleophilic properties of the ligands on the electrophile dominate the reactivity sequence, reserving S_Ei for reactions in which the electrophilic properties dominate, and noted that there would be intermediate categories such as S_EiC and other categories such as S_E-(alkyl bridge).

The problems with all these attempts at taxonomy of mechanisms are not in lack of examples to fit into the preconceived categories, but failure of the categories to cover all the known types of mechanisms and to divide them in a realistic way. One would not attempt to classify all dogs as spaniels (S), cocker spaniels (CS), and Mexican hairless (MH), then start adding categories such as S_{gs} because it turns out that some spaniels look more like German shepherds. Even the familiar S_N1 and S_N2 labels are much too limited and rigid to describe the details of nucleophilic displacements adequately, as Swain once pointed out in Bible-citing frustration in an argument with Ingold.[19] In addition the proliferation of arcane symbolism is not an aid to communication. A simple structural formula of the proposed transition state can reveal so much so clearly, including which points of structure the proposer is more or less ignorant about.

F. H_3^+ AND CH_5^+ AS TRANSITION STATE MODELS

For understanding the bonding in transition states it is useful to consider simplified models on which calculations can be carried out. The molecular orbital description of bridged aluminum compounds in Chapter 1, Section III,B, is actually an elaboration of the H_3^+ model. The CH_5^- model for nucleophilic displacement has been cited in another context in Chapter 1, Section IV,A.

The H_3^+ and H_3^- models for the transition states for electrophilic and nucleophilic displacement, respectively, provide qualitative insight into the stereochemical preferences. Such simplified models cannot be expected to predict anything, but should and do correlate with the gross trends that have been observed. Simple Hückel calculations on cyclic H_3^+ are the same as for the π electrons of the cyclopropenium cation, a stable species familiar to organic chemists.[20] For linear H_3^+ the calculations are the same as for the π electrons of the allyl cation. In this model, the only difference between electrophilic and nucleophilic displacement is in the filling of the molecular orbitals, one electron pair being involved in electrophilic displacement and two electron pairs in nucleophilic displacement. Hückel wave function coefficients and energies for linear and cyclic H_3^+ and H_3^- are summarized in Table 3-1.

As can be seen from Table 3-1 (and is probably well known to most readers), cyclic H_3^+ is a closed-shell, aromatic structure with considerable binding energy (4β), and cyclic H_3^- is calculated to be a biradical, anti-aromatic structure with less favorable binding energy (2β). For linear H_3^+ and H_3^- the binding energies are the same (2.8β) and in between the values for the cyclic structures. These calculations suggest that the cyclic structure, which requires retention of configuration, is favored for electrophilic displacement, and the linear structure, which requires inversion when steric factors are taken into account, is favored for nucleophilic displacement. This is not a bad guess for such a crude approximation.

Extrapolating this rough model to carbon and electrophilic or nucleophilic species introduces several complications. The electronegativities of carbon and the other atoms will differ, but it has already been noted in Chapter 1, Section III,B, that in a two-electron system the cyclic arrangement of orbitals is favored over the linear. Conversely, with four electrons the linear system is favored. Carbon adds a further complication in that the cyclic arrangement uses an sp^3 or similar hybrid orbital for the cyclic arrangement but a p orbital for the linear. The s character gives the hybrid a lower energy, favoring species in which the carbon bears some negative charge, while the p orbital

is nothing available in terms of simplistic bond representations that would work any better.

The same calculations on CH_5^- (nucleophilic displacement) favor D_{3h} symmetry (inversion) over C_s symmetry by 0.64 eV (14.75 kcal/mole).[21] Most nucleophilic displacements involve elements more electronegative than hydrogen, and localization of the antibonding electron pair toward these elements will tend to increase the repulsion between them.

When applying the CH_5^+ and CH_5^- models to other systems the relationship between hybridization and electronegativity of orbitals at carbon should be taken into account. In both CH_5^+ and CH_5^- the carbon is nearly neutral, with the more loosely bonded pair of hydrogens bearing the greatest charges.[21] In CH_5^+, all hydrogens bear some positive charge and the carbon is slightly negative. In organometallic reactions the amount of negative charge on the carbon would usually be greater than in CH_5^+, which would favor sp^3 over p character in the orbital used for bonding to the metal atoms and hence retention over inversion, as with the H_3^+ model. However, a highly electronegative electrophile such as a halogen cation would tend to withdraw electron density from the carbon and thus favor p character (inversion) to whatever extent this factor can contribute. Also, nucleophilic displacements generally involve electronegative heteroatoms, low electron density on carbon (carbonium ion character), and hence more p character and bias toward inversion than the CH_5^- model would predict. It seems reasonable to expect such effects to amount to several powers of 10 in favorable cases, inasmuch as a linear correlation of the pK_a's of hydrocarbons with percent s character in the C—H bond yields a difference of ~23 pK units (31 kcal/mole) between a pure p and an sp^3 orbital.[24]

Returning to the other calculations on CH_5^+, the CNDO/2 method indicates the C_s symmetry (**8**) (retention) to be favored over the D_{3h} symmetry (**7**) (inversion) by ~10 kcal/mole,[22] and semiempirical self-consistent field calculations yield a similar result.[23] An *ab initio* Hartree-Fock calculation has indicated that **7** and **8** represent energy minima and that **8** (retention) is favored very slightly.[25]

It would be unrealistic for the confirmed experimentalist to snort that the calculations disagree and therefore cannot predict anything. In fact, the calculations do all agree within their expected deviations that there is no large and fundamental energy difference separating the inversion and retention mechanisms for electrophilic displacement, that secondary influences might well govern the stereochemistry in particular cases, and that examples of both types of stereochemical preference are "expected," in accord with the experimental facts.

In view of all this discussion of CH_5^+, it may be noted that this species is not a figment of some computer's imagination but has been detected by mass spectrometry[26] as well as solution chemistry.[23] The experimental evidence does not indicate which geometry is preferred, but suggests flexibility. Olah and co-workers found that methane-d_4 undergoes deuteron–proton exchange with FSO_3H–SbF_5 and eliminates hydrogen (D_2 or HD) in the process to form CD_3^+, which ultimately undergoes exchange and condensation to the *t*-butyl cation, $(CH_3)_3C^+$.[23] Evidence that displacement with retention is possible is provided by adamantane, which undergoes proton–deuteron exchange at the bridgeheads in DF—SbF_5 without suffering the usual C—C bond cleavage (transition state **9**).[27] Similarly, nitronium hexafluorophosphate nitrates adamantane at the bridgehead.[28] These very strong

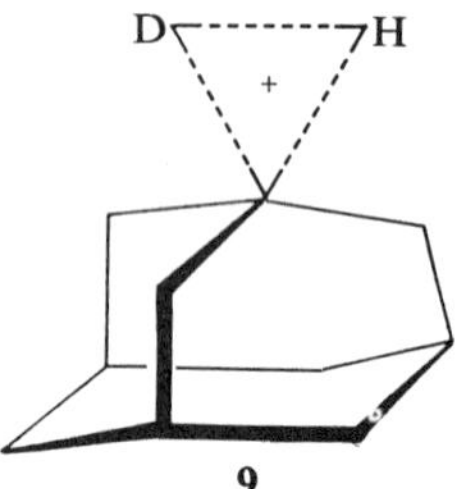

9

electrophiles cleave the carbon–carbon bond in ethane, nitration yielding 25% nitroethane and 75% nitromethane,[28] and this sort of electrophilic displacement thus differs sharply from organometallic examples.

Because BH_5 is isoelectronic with CH_5^+ and would be less strongly bonded because of the lower central nuclear charge, it is of interest that BH_5 and related molecules appear to exist as finite-lived intermediates in the hydrolysis of borohydrides. For example, the hydrolysis of BH_3CN^- is first order in H^+ and first order in BH_3CN^-, incorporation of D^+ is 15 times faster than hydrolysis, and the hydrolysis is faster in D_2O than in water.[29] These results imply that a covalent BH_4CN is an intermediate, not a transition state, since it may lose either H^+ or H_2. There is no way of deciding the preferred geometry, except that two of the hydrogens can obviously move close together with little difficulty. The hydrolysis of BH_4^- likewise appears to involve BH_5 as the intermediate, inasmuch as BD_4^- reacts faster than BH_4^- (incompatible with rate-determining attack at a proton)[30] and cleavage by Me_3NH^+ yields only a few percent of trimethylamine borane (incompatible with nucleophilic attack of nitrogen at boron in the transition state).[31] In the hydrolysis of pyridine arylboranes, k_H/k_D values for a series of four compounds correlate with B—H stretching frequencies, not B—H bending frequencies, but these results do not contradict the idea of a pentacovalent boron intermediate provided its energy is fairly insensitive to the BH_2 bond angle.[32]

The evidence that CH_5^+ and BH_5 can have finite lifetimes suggests that many of the "transition states" written for electrophilic displacements are actually short-lived intermediates, with the actual transition states being slightly bent and stretched versions of these structures. This would hardly be surprising in view of the existence of stable bridge-bonded compounds such as trimethylaluminum dimer and others (Chapter 1, Section III). However, investigators working with the less reactive organometallics such as mercury, tin, or boron compounds have not generally looked for any evidence of metastable intermediates containing pentacoordinate carbon. Such intermediates would not be expected in exothermic halogenations, but the possibility of finding them in metal exchange reactions seems a suitable question for future research.

II. Mercury(II) Electrophiles

A. Introduction

Mercury(II) and organomercury(II) salts are active electrophiles capable of displacing metal cations from most other organometallic compounds. The organomercury products are generally fairly stable, can be handled in air and water, and (with luck) easily crystallized. Organomercury compounds are toxic, but except for particularly volatile compounds such as dimethylmercury, the hazard is easily taken care of by a suitable hood and other normal precautions. It is easy to see why organomercury compounds have long been popular with chemists who wanted to study organometallic reaction mechanisms.

There is much organomercury literature which is fairly old, directed toward synthetic ends, superficially mechanistic, aimed toward poorly chosen mechanistic questions, or done in ignorance of the complexity of misbehavior these compounds can exhibit. Other reviews of mercury chemistry are available[15,33,34] and need not be duplicated here. Jensen and Rickborn's book[15] is a critical, comprehensive coverage of mechanistic organomercury chemistry. Reutov and Beletskaya have provided an extensive survey of the literature on organometallic reaction mechanisms, including much information on mercury and on the Russian literature.[34]

B. Mercury(II) as Electrofuge

Except for being much slower and thus accessible to study by classical kinetic methods, these are analogous to the metal exchange reactions discussed in Chapter 2. Wright first showed that the reaction of R_2Hg with $HgCl_2$ yields 2 RHgCl with retention of configuration where R is *cis*-2-methoxycyclohexyl,[35] and the significant more comprehensive work on *cis*-2-methoxycyclohexylmercury compounds and theoretical interpretation by Winstein *et al.*[1] followed shortly (see Section I,A).

Whenever there are two asymmetric centers, and especially when there is a potentially reactive neighboring group such as β-methoxy, it is possible that the stereochemical outcome depends largely on asymmetric induction or other secondary effects. To remove this doubt, the optical resolution of 2-butylmercuric salts was carried out by Jensen and Gale[5] and by Charman, Hughes, and Ingold.[36] Cleavage of optically active di-2-butylmercury (**10**) by mercuric bromide was then shown to yield 2-butylmercuric bromide (**11**) with complete retention of configuration.[37,38] It was only practical to intro-

$$\underset{\mathbf{10}}{(CH_3)(H)(C_2H_5)C{-}Hg{-}CH(CH_3)(C_2H_5)} + HgBr_2 \longrightarrow \underset{\mathbf{11}}{(CH_3)(H)(C_2H_5)C{-}HgBr} + BrHg{-}CH(CH_3)(C_2H_5)$$

duce one resolved center into the di-2-butylmercury (**10**) which was prepared from optically active 2-butylmercuric bromide (**11**) and (necessarily) racemic 2-butylmagnesium bromide. Thus, the material examined was a 50–50 mixture of *R*,*R* and *R*,*S* isomers, which on cleavage with mercuric bromide yielded $\frac{3}{4}$ *R*- and $\frac{1}{4}$ *S*-2-butylmercuric bromide, that is, 50% *R* and 50% racemic. Consequently, the rotation of the cleavage product was just half that of the original *R*-2-butylmercuric bromide (which need not have been 100% optically pure for the foregoing description to be valid). The constitution of the di-2-butylmercury was further checked by acid cleavage to butane and 2-butylmercuric bromide, which again had half the optical activity of the starting material.[37] Thus, it is unlikely that the rotations observed result from any fortuitous cancellation of opposing asymmetric inductions, and retention of configuration in the electrophilic displacement is 100%. Similar work has been carried out by Reutov and Uglova on 5-methyl-2-hexylmercury compounds.[39]

Jensen and Landgrebe found that reduction of *R*-2-butylmercuric bromide with magnesium led to *R*,*R*-di-2-butylmercury, which could be cleaved by mercuric bromide to *R*-2-butylmercuric bromide with 90% of the original optical activity.[15,40]

Charman, Hughes, and Ingold noted that mercuric nitrate and acetate react much faster than mercuric bromide with di-2-butylmercury in ethanol and concluded that ligand bridging (Hg—Br—Hg) is relatively unimportant in the transition state.[37] They accordingly classified this reaction as S_E2, which is a gross oversimplification at best. The data merely indicate that the greater electrophilicity of $Hg(NO_3)_2$ influences the rate more than does the greater ligand bridging ability of $HgBr_2$, which in no way proves ligand bridging to be absent (see also Jensen's book[15]). There is nothing in the data to contradict transition state structure **12** for this reaction, which is just one more example of the usual sort of mechanism for metal exchanges developed in Chapter 2.

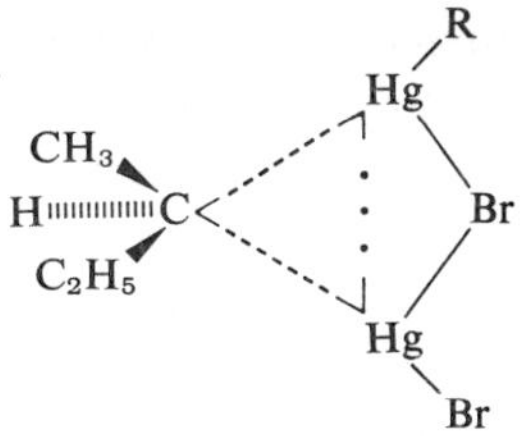

12

Exchange reactions involving identical reactants and products in the organomercury series are generally too slow to be studied by nmr, but were laboriously done by Ingold and co-workers using radioactive labeling before nmr became available for the study of other organometallics. For example, the exchange between 2-butylmercuric bromide and di-2-butylmercury in ethanol was followed in this manner and retention of configuration was proved.[41] The added salts lithium perchlorate, nitrate, and bromide accelerated the reaction, with the bromide being somewhat less effective than the other two. Relative exchange rates of $(2\text{-Bu})_2Hg$ were with 2-BuHgBr, 1; 2-BuHgOAc, ~10; 2-$BuHgNO_3$, ~1000. This is in accord with relative electrophilicities or abilities to ionize to alkylmercuric cation.

Exchanges of the form

$$RHgX + Hg^*X_2 = RHg^*X + HgX_2$$

were also studied.[10,42] Methylmercuric nitrate and mercuric nitrate exchange mercury much faster than the corresponding halides do, and the iodides exchange slightly faster than the bromides. Added salts increase the rates. Steric effects are slight, MeHgOAc reacting only 17 times faster than 2-BuHgOAc with $Hg(OAc)_2$. Only 10% of the optical activity of 2-BuHgOAc was lost in 8.7 exchange half-lives, presumably by way of a slight decomposition reaction having nothing to do with the electrophilic displacement, and this reaction thus proves the stereospecificity of displacement with retention to a higher degree than do most of the other experiments. Hughes, Ingold, Thorpe, and Volger concluded that ". . . the transition state is an open one, apart from its solvation"[42] (structure **13**), on the basis of the relative reactivities of the nitrates, acetates, and halides, but this conclusion rests on faulty reasoning, as will be shown in the following paragraphs.

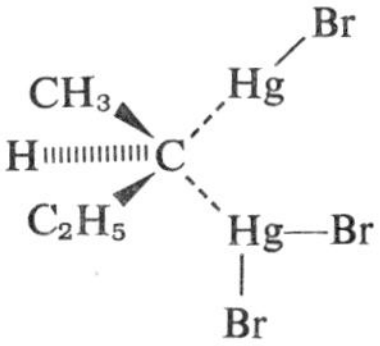

13

Bromide ion catalyzes mercury exchange between 2-BuHgBr and Hg^*Br_2.[10] This could hardly be attributed to the electrophilicity of $HgBr_3^-$ or $HgBr_4^{2-}$ being greater than that of $HgBr_2$, but suggests that bromide ion may either be bridging between both mercuries or bonded to the leaving mercury atom in the transition state. Noting that bromide ion complexes strongly with $HgBr_2$, weakly with RHgBr, and "not at all" (?) with R_2Hg, Charman, Hughes, Ingold, and Volger concluded that the mechanism involves attack of $HgBr_3^-$ or RHgBr, not $HgBr_2$ on $RHgBr_2^-$. "The bromide ion not only can, but must bridge, because it enters, as $HgBr_3^-$, with the lower Hg of formula [**14**], and leaves, as $HgBr_2$, with the upper one; and hence it must be on its way from one mercury to the other in the transition state."[10]

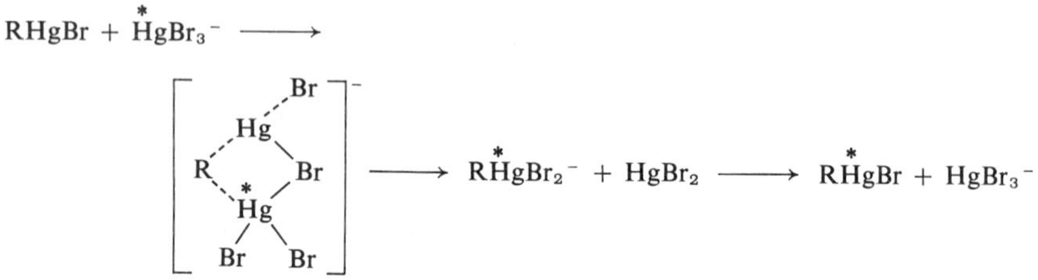

$$RHgBr + \overset{*}{H}gBr_3^- \longrightarrow$$

$$\left[\begin{array}{c} \text{R}\cdots\text{Hg}(\cdots\text{Br})\text{—Br—}\overset{*}{\text{Hg}}(\text{Br})(\text{Br})\cdots\text{R} \end{array}\right]^- \longrightarrow R\overset{*}{H}gBr_2^- + HgBr_2 \longrightarrow R\overset{*}{H}gBr + HgBr_3^-$$

14

(Reaction path violates microscopic reversibility)

The reasoning just quoted violates the principle of microscopic reversibility, as has been pointed out previously by Jensen and Rickborn[15a] and by the author.[43] The starting materials and products are identical, and therefore the potential energy surface must be fully symmetrical, with identical forward and reverse reaction paths. (The effect of the isotopic label is totally negligible in this case, and even if it were not, the electronic potential energy surface is unaffected by nuclear mass, as discussed in Chapter 1, Section IV,A.)

One way to patch up transition state **14** is to write reverse arrows at all stages of the reaction, and then the forward and reverse paths could proceed with equal rates. This assumption does not require some weird and improbable coincidence of k values, since the starting materials and products are at one energy level, the unsymmetrical transition state at another, and only the elevation gain counts in determining the rate. The energy surface is analogous to a branched mountain valley, where one might climb up the right-hand stream bed, cross over a pass behind the mountain to the left, and descend by way of the left-hand crookeder stream bed back down to the junction, or vice versa. A high wall at the head of the valley would separate it from other drainages.

Although such landscapes exist in nature, when mapping potential energy surfaces for reactions it should be kept in mind that every symmetry element is at a maximum or minimum on the multidimensional surface (see Chapter 1,

Section IV,E). There may be other maxima and minima, but the symmetry elements can be counted on. The foregoing version of transition state **14** would require that all the symmetrical structures lie along a high-energy wall separating two equivalent mirror-image sets of valleys interrelated by other scramblings of R, Hg, and Br. However, there is no obvious reason to attribute high energy to all symmetrical structures in this system. Structure **15** or **16** might be the transition state or an intermediate between two unsymmetrical transition states for the exchange reaction.

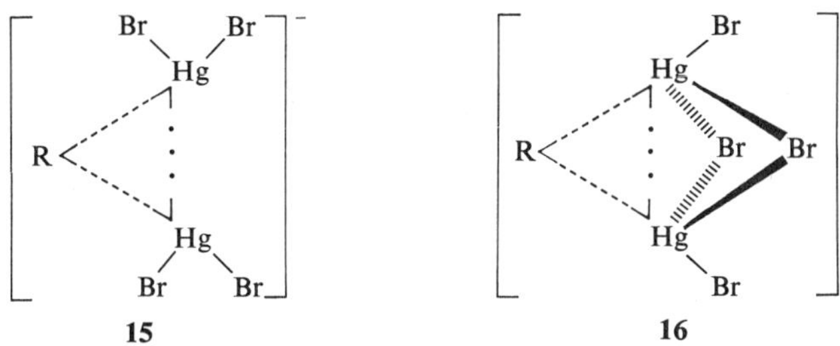

The symmetrical structure to be considered in the same way for the previously mentioned exchange between RHgBr and Hg^*Br_2 without added bromide ion is **17**. The fact that nitrates and acetates react faster than bromides does not rule out **17** as a transition state or intermediate, since electrophilicity (or ability to form an alkyl bridge, Hg—C—Hg) could outweight anionic ligand bridging (Hg—Br—Hg) in determining relative reactivities even though the anionic ligand bridging involves a considerable amount of energy.

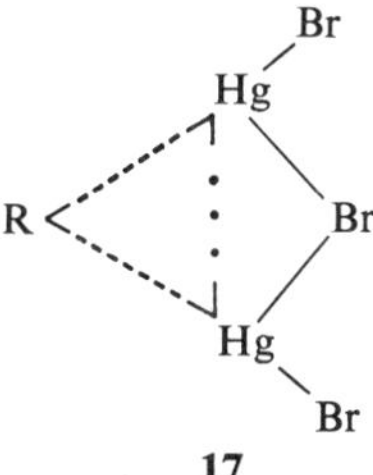

17

It is, of course, conceivable that symmetrical structures such as **17** are maxima with respect to all possible distortions, that is, represent a sort of mountain peak with reaction pathways going around both sides. However, such behavior is characteristic of poorly bonded structures such as antiaromatic systems (see Chapter 1, Section IV,E), and in view of the stability of ligand-bridged compounds such as $Me_5Al_2NPh_2$ (Chapter 1, Section III,C, Fig. 1-6 and Chapter 2, Section III,G) it appears that **17** and related structures should have considerable stability, at least to the point of being transition states.

For a detailed discussion of these mercury(II) isotopic exchange reactions and pathways consistent with the principle of microscopic reversibility the reader may consult Jensen and Rickborn's book.[15a]

Abraham and co-workers have published a polemical defense of Ingold's views, complete with contour maps of hypothetical potential energy surfaces.[44] The criticisms by Jensen and Rickborn[15a] and Matteson[43] are labeled "erroneous,"[44] but in fact much of the paper by Abraham and co-workers is in fundamental agreement with those criticisms, and what is not is itself erroneous. For the record, the author long ago recognized that the principle of microscopic reversibility allows a symmetrically related set of unsymmetrical pathways and pointed out this minor revelation in a footnote[45] three years before the commonly quoted paper by Burwell and Pearson.[46] The criticism of Ingold's conclusions is not that the gross mechanisms are wrong, but that some of the details are wrong, and that the mechanisms as originally written and discussed violate the principle of microscopic reversibility, as pointed out earlier in this section.

Abraham and co-workers ran into unsuspected logical contradictions when they constructed a potential energy surface for an alleged S_E1 reaction (their Figure 7).[44] Following the periphery of their proposed potential surface amounts to postulating the sequence of Eqs. (3-1) through (3-3) and the reverse. The catch here is that if the reaction is to be S_E1, it must be first

$$RHgCl \underset{k_{-1}}{\overset{k_1}{\rightleftharpoons}} R^- + HgCl^+ \tag{3-1}$$

$$R^- + Hg^*Cl_2 \underset{k_{-2}}{\overset{k_2}{\rightleftharpoons}} RHg^*Cl_2^- \tag{3-2}$$

$$RHg^*Cl_2^- + HgCl^+ \underset{k_{-3}}{\overset{k_3}{\rightleftharpoons}} RHg^*Cl + HgCl_2 \tag{3-3}$$

order in RHgCl and zero order in Hg^*Cl_2. This condition requires that k_1 be rate-controlling. However, Abraham and co-workers failed to notice the chemical equivalence of Hg^*Cl_2 and $HgCl_2$ in Eqs. (3-2) and (3-3). If these reactions are faster than that of Eq. (3-1), they accomplish isotopic exchange without waiting for that S_E1 rate-determining step, and the observed kinetics will show a dependence on the Hg^*Cl_2 concentration. That is what the criticism by Jensen and Rickborn is all about.[15a] The potential surface drawn[44] is wrong in that it separates things that cannot be separated and, one hates to say it, but it violates the principle of microscopic reversibility. Faith can lead one in the footsteps of the Master, but if he gets lost in the jungle, only doubt can break a new path out.

Numerous other studies of mercury(II) exchange have been reported. Dessy and Lee examined reactions of mercuric halides with dialkyl- and diarylmercuries.[47] Little difference in rates was found between ethyl, isopropyl, and *n*-propyl. Electron-donating groups greatly accelerated the cleavage of diarylmercuries, with Hammett's $\rho = 5.87$. Dessy and Lee concluded that the transition state generally has a four-center, ligand-bridged structure with a cyclic three-center CHg_2 bond, analogous to **17**. However, it should be pointed out here that electrophilic aromatic substitution contains a quantum mechanical symmetry element not present in the aliphatic series, namely, the π-bond system of the aromatic ring, which overlaps with the three-center molecular orbital that is antibonding with respect to the two mercury atoms (see Chapter 2, Section III,C). This overlap may partially or totally negate mercury–mercury bonding while strengthening carbon–mercury bonding and allowing considerable electron donation from the ring π orbital to the carbon–mercury bonds. In the limit, which is probably approached by mercury, the three-center bond becomes inappropriate and two σ bonds should be written in its place (**18**).

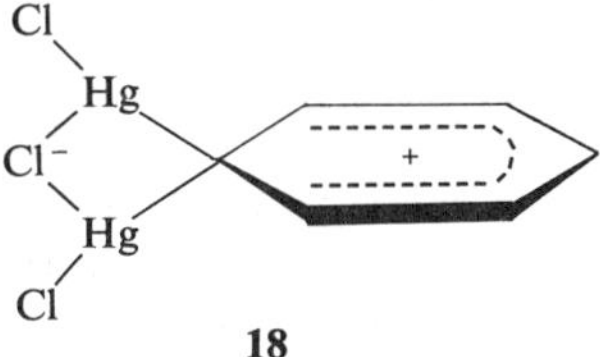

18

Other mercury exchange studies do not add anything new to what has already been discussed, and a thorough critical review is already available in Jensen and Rickborn's book.[15] Mercury exchanges involving the free metal are not electrophilic displacements and will be covered in Chapter 8, Section II,A.

C. Geometry of Bonding at Mercury

In the foregoing discussion it has been implied that mercury is just one more metal and that all the conclusions of Chapter 2 based on aluminum, zinc, magnesium, etc., apply here, too. It is well to look at a few possible bond angles and distances for verification. The mercury covalent radius is ~ 1.30 Å in alkylmercuric halides, and the Hg—Hg bond distance ranges from 2.43 Å in Hg_2F_2 to 2.69 Å in Hg_2I_2, which represents considerable flexibility.[48] In mercury metal, the nearest neighbors are 3.0 Å and the second nearest 3.46 Å, with the van der Waals radius probably falling between these figures. For a carbon–mercury distance of 2.07 Å and a mercury–mercury distance of 2.60 Å, the Hg—C—Hg bond angle becomes 78°, which is typical of three-center bonds, in a transition state or intermediate such as **17**. The actual bond distances would all be increased about 10% to allow for the looser three-center bond, as in $Al_2Me_5NPh_2$ (Chapter 1,

Section III,C). To avoid three-center bonding, one has to move the mercury atoms more than 3.3–3.5 Å apart, which requires an Hg—C—Hg angle greater than 90°–100° for a reasonable C—Hg distance of 2.30 Å. Such a wide angle is reasonable for aromatic substitution, but if it applied to aliphatic substitution it should lead to a high degree of steric compression and large steric effects, not the supreme indifference to the bulk of the alkyl group that was actually observed.[42,47] It is difficult to see any reason why three-center bonding would be avoided in transition states or intermediates for mercury exchange. The only difference between mercury and, for example, magnesium, is that in mercury compounds covalent two-center electron-pair carbon–metal bonds are energetically favored over three-center bridge bonds, while in magnesium dialkyls the reverse energy relationship prevails.

Another aspect of mercury bonding which is consistent with the reported kinetic data is that mercury tends to form two strong covalent bonds 180° apart, and any additional ligands are weakly and more ionically bonded at 90° to the main bond axis.[48] Thus, formation of a transition state or intermediate such as **17** requires moving one bromide ion out of the strong, covalent bonding position to the weak, ionic bonding site, where it can form a convenient ~90° Hg—Br—Hg bent bridge. This amounts to a partial ionization of the ligands, and is in accord with the observed salt effects[10] as well as the greater electrophilicity of the nitrate and acetate compared with the halides.

D. Boron(III) as Electrofuge

Mercury(II) salts can displace boron from a wide variety of organoboron compounds. The reaction of benzylboronic acid, $PhCH_2B(OH)_2$, with mercuric chloride to form benzylmercuric chloride, $PhCH_2HgCl$, was reported in 1909.[49] All three alkyl groups of primary trialkylboranes have been replaced by treatment with mercuric oxide freshly precipitated from mercuric chloride and sodium hydroxide,[50] but secondary alkaneboronic acids, $RB(OH)_2$, are relatively inert.[51,52] This type of reaction has a number of convenient properties for mechanistic studies. Much of the work in this field has been done by the author and co-workers, and naturally a fairly detailed account follows.

Retention of configuration has been shown to predominate in the reaction of 1-phenylethaneboronic acid with mercuric chloride.[53] Side reactions, especially the facile racemization of 1-phenylethylmercuric chloride, made it impossible to prove that the reaction is stereospecific, the observed net retention being only 14%, but it is improbable that there is a close balance in energy between the retention and inversion transition states, and the lack of sensitivity to steric effects subsequently found in other mercuri-deboronation reactions is additional evidence for the retention mechanism.

1-Phenylethaneboronic acid was partially resolved into optical isomers through the use of bis(demethyl)brucine, an optically active substituted catechol which forms a fairly stable chelate ester with the boronic acid but allows recovery of the resolved boronic acid as its dibutyl ester (**19**) by extraction with butanol from an acidified aqueous solution.[53] It was shown that (−)-dibutyl 1-phenylethaneboronate (**19**) has the *R* absolute configuration by oxidizing it with alkaline hydrogen peroxide to *R*-(+)-1-phenylethanol, a reaction which is known to proceed with retention.[54,55] The product (+)-1-phenylethylmercuric chloride (**20**) was also shown to have the *R* configuration by reaction with bromine in pyridine, which goes with retention[4,5] and yields *R*-(+)-1-phenylethyl bromide.

$$\underset{\mathbf{19}}{H\!-\!\overset{C_6H_5}{\underset{CH_3}{C}}\!-\!B(OC_4H_9)_2} \xrightarrow[\text{(to be proved)}]{HgCl_2} \underset{\mathbf{20}}{H\!-\!\overset{C_6H_5}{\underset{CH_3}{C}}\!-\!HgCl}$$

$$\downarrow H_2O_2 \text{ (retention known)} \qquad\qquad \downarrow Br_2 \text{ (retention known)}$$

$$H\!-\!\overset{C_6H_5}{\underset{CH_3}{C}}\!-\!OH \quad \text{(absolute configurations known)} \quad H\!-\!\overset{C_6H_5}{\underset{CH_3}{C}}\!-\!Br$$

The racemization, perhaps a radical process involving RHg^+, is slowed by chloride ion, which unfortunately also slows the desired electrophilic displacement. In order to obtain optically active 1-phenylethylmercuric chloride it was necessary to add glycerol to chelate the boron and make it a better leaving group. The rate acceleration is only about a factor of 20,[56] but it makes the difference between success and failure. A basic catalyst, sodium acetate, is also essential.

Kinetic measurements on this system proved difficult because of the instability of the organomercury product, and only rough data could be obtained.[57] It was found that a *p*-chloro substituent on the phenyl group retarded the rate by about a factor of 3, a *m*-trifluoromethyl by a factor of 8.

Much better kinetic data were obtained with benzylboronic esters, $ArCH_2B(OR)_2$.[56] The previous analytical method[52] was superseded by a much better one based on the complexing of mercuric chloride but not benzylmercuric chloride by EDTA.[56] The reaction is first order in $ArCH_2B(OR)_2$, $HgCl_2$, and OH^- (not acetate ion, which was the buffer used) in aqueous ethanol and glycerol. An analogous rate law has been reported for the reaction of $PhHg^+$ with $PhB(OH)_2$ in the presence of buffers. For a series of four synthetically accessible compounds, $ArCH_2B(OBu)_2$, relative rates at 30°C were *p*-CH_3, 0.7; H, 1.0; *p*-Cl, 2.1; *m*-CF_3, 2.5.[56] This order is opposite that found in the 1-arylethyl series,[57] but the most significant fact is that the substituent effects are small. Electron-withdrawing substituents increase the Lewis acidity of the boronic ester more than the rate of reaction with mercuric chloride. A Hammett correlation of the rates of reaction of $ArCH_2B(OR)_2$ with $HgCl_2$ and OR^- yields $\rho = 0.93$, but for $ArCH_2B(OR)_3^-$ with $HgCl_2$, $\rho = -0.55$ (Fig. 3-1). Thus, the electron density at the benzyl carbon in the transition state is between that in the boronic ester $ArCH_2B(OR)_2$ and its anion $ArCH_2B(OR)_3^-$.

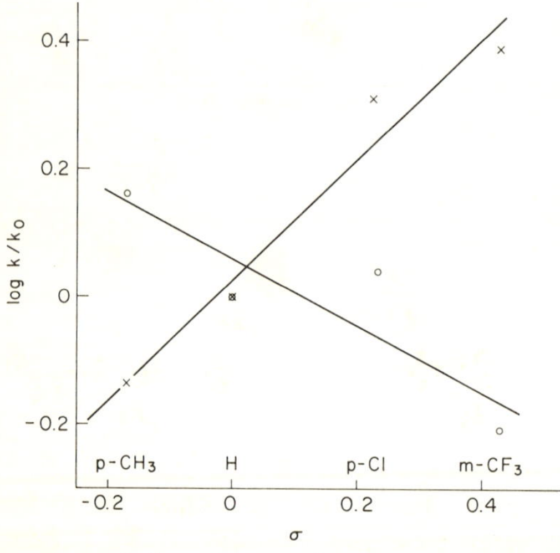

Fig. 3-1. Hammett correlation of rates of reaction of benzylboronic esters with mercuric chloride: ×, relative *k*'s for $ArCH_2B(OR)_2$; ○, *k*'s for $ArCH_2B(OR)_3^-$ derived by taking pK_a's into account. [From D. S. Matteson and E. Krämer, *J. Amer. Chem. Soc.*, **90**, 7261 (1968).]

Chloromercuric acetate attacked the benzylboronic esters about two orders of magnitude faster than did mercuric chloride,[56] as expected of a more electrophilic mercury(II) source. Second-order plots were curved and a satisfactory rate law was not discovered, but it was clear that in this case electron-withdrawing substituents retard the reaction, the unsubstituted compound reacting ~3 times as fast as the *p*-chloro- and *m*-trifluoromethyl-substituted compounds.

These results are all consistent with the postulated three-center bonding in the transition state **21**. It is likely but unproved that the hydroxide ion forms a ligand bridge between boron and mercury, and it is possible that **21** is a short-lived intermediate rather than a transition state. It cannot be proved whether **21** is formed from attack of $HgCl_2$ on $ArCH_2B(OR)_3^-$ (R = H or alkyl) or from Cl_2HgOH^- and $ArCH_2B(OR)_2$, since $HgCl_2$ may be as strong an acid or stronger than the boronic ester. In view of the behavior of aluminum compounds which contain both alkoxide and alkyl bridges (Chapter 2, Section III,G), it seems likely that the Hg—OH—B bridge is formed before the Hg—C—B bridge. [See also the discussion of $PhB(OH)_2$ mercuri-deboronation later in this section.] Because chloride does not complex appreciably with boron in aqueous solution, an Hg—Cl—B bridge is unlikely.

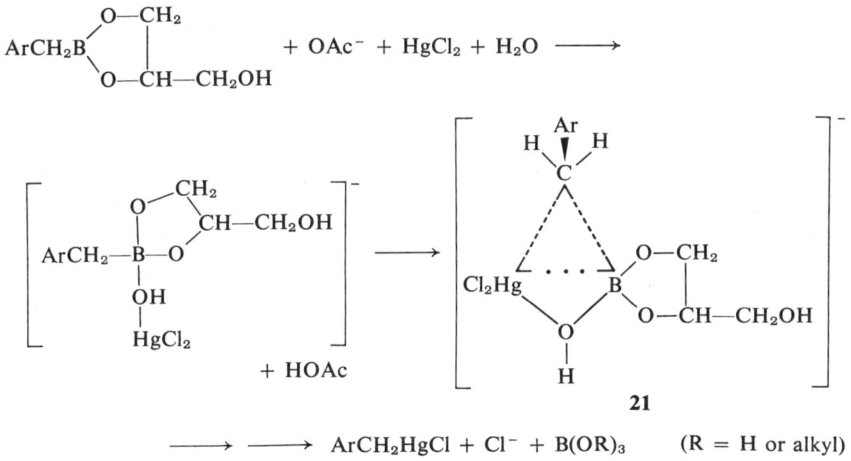

$ArCH_2B(O—CH_2—CH(CH_2OH)—O)$ (cyclic) $+ OAc^- + HgCl_2 + H_2O \longrightarrow$

$[ArCH_2—B(O—CH_2—CH(CH_2OH)—O)(OH—HgCl_2)]^- + HOAc \longrightarrow$

$[Cl_2Hg(\mu\text{-}CH_2Ar)(\mu\text{-}OH)B(O—CH_2—CH(CH_2OH)—O)]^-$

21

$\longrightarrow \longrightarrow ArCH_2HgCl + Cl^- + B(OR)_3$ (R = H or alkyl)

The small substituent effects are particularly inconsistent with any appreciable carbanion character in the transition state, but indicate a strictly concerted process. This is consistent with the properties of compounds containing three-center bridge bonds (Chapters 1 and 2), though it does not provide any definitive information about the boron–mercury bond order.

A further probe of the nature of the transition states for electrophilic displacements is made possible by neighboring group effects. It was noted that the reaction of ethane-1,1-diboronic acid, $CH_3CH[B(OH)_2]_2$, with mercuric chloride and alkali to produce ethane-1,1-dimercuric chloride, $CH_3CH(HgCl)_2$, was anomalously fast.[59] Later a wide range of multiple-boron-substituted carbon compounds became available through the reaction of di, tri-, or tetrachloromethanes with lithium and dimethoxyboron chloride,[60] which made possible a study of the effect of neighboring boron or mercury on the electrophilic displacement of boron.[61]

As usual, any time a new series of compounds is used in kinetic experiments, a whole new analytical approach has to be developed. The reaction of bis(dimethoxyboryl)methane, $CH_2[B(OMe)_2]_2$, with mercury(II) proved to be so rapid that complexing the mercury with EDTA failed to stop the reaction. Instead, a spectrophotometric method based on the complexing of mercury(II) by dithizone was devised. It was also found that the reactions proceeded satisfactorily with methyl esters in pure methanol, a simpler and theoretically more satisfying system than the water–ethanol–glycerol mixture used previously.[56] The stoichiometry of the reaction is given by Eq. (3-4).

$$ZCH_2B(OMe)_2 + HgCl_2 + OAc^- + MeOH = ZCH_2HgCl + B(OMe)_3 + HOAc + Cl^- \quad (3\text{-}4)$$

The rate law was shown to follow Eq. (3-5). An eightfold range of concentra-

$$-d[HgCl_2]/dt = \frac{k[ZCH_2B(OMe)_2][HgCl_2][NaOAc]}{[HOAc]} \quad (3\text{-}5)$$

tion of each reagent was tested with bis(dimethoxyboryl)methane [Z = $B(OMe)_2$] and lesser ranges were checked with other compounds used in the study. It was found that $CH_2[B(OMe)_2]_2$ reacts 200 times faster than $CH_3B(OMe)$, or 100 times when statistically corrected for the number of boron atoms. The rate constants are summarized in Table 3-2.

The order of reactivities $(MeO)_2B > ClHg > H > Ph$ for Z in the $ZCH_2B(OMe)_2$ series suggests that the vacant *p* orbitals of boron and mercury are directly involved in the neighboring group effect. The order of relative acidities is $Ph > H > (MeO)_2B$ (with ClHg not measurable because

TABLE 3-2

RATES OF REACTION WITH $HgCl_2$, NaOAc, AND HOAc AT 30°C in METHANOL[a]

Compound	k liters/mole second	k/k_0 per boron
$CH_3B(OMe)_2$	6.7×10^{-4}	1.0
$CH_2[B(OMe)_2]_2$	0.131	98
$ClHgCH_2B(OMe)_2$	2.0×10^{-3}	3.0
$HC[B(OMe)_2]_3$	0.27–0.60	135–300
$C[B(OMe)_2]_4$	0.016	6
$C_6H_5CH_2B(OMe)_2$	2.4×10^{-4}	0.36
n-$C_4H_9B(OMe)_2$	$\sim 3 \times 10^{-4}$	$\sim$0.5

[a] From D. S. Matteson and P. G. Allies, *J. Amer. Chem. Soc.* **92**, 1801 (1970).

the compound is destroyed by base). Thus, relative acidities cannot account for the rates, nor can any closely related phenomenon such as the ability of the neighboring group to stabilize carbanion character.

The most attractive explanation is that the vacant *p* orbital of the neighboring, stationary boron overlaps with the bonding orbitals of the carbon and the attacking mercury, as portrayed in the extended three-center bond structure **22** or the stylized orbital picture **23**.[61]

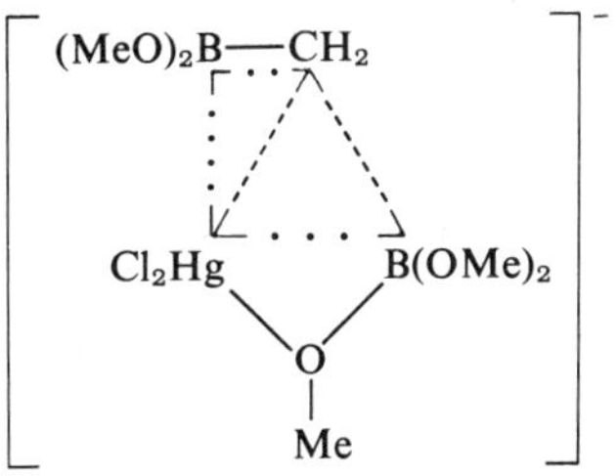

22

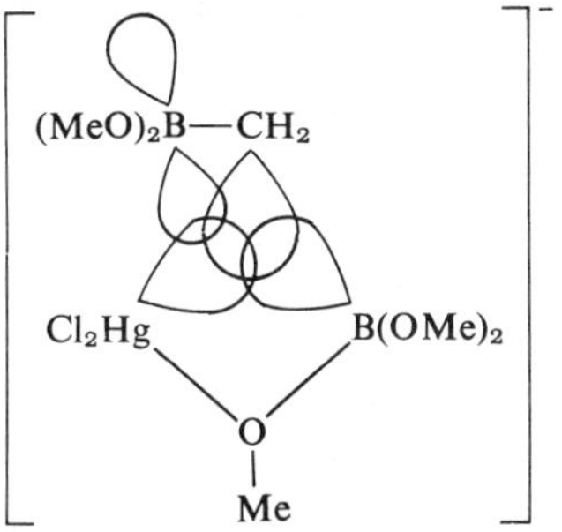

23

The activation parameters are consistent with transition state **22**. For $CH_2[B(OMe)_2]_2$, ΔH^* is about 8.8 kcal/mole and ΔS^* is -33 cal/deg mole. For $CH_3B(OMe)_2$, ΔH^* is 15.4 kcal/mole and ΔS^* is -22 cal/deg mole, and for $PhCH_2B(OMe)_2$, ΔH^* is 16.2 kcal/mole and ΔS^* is -21 cal/deg mole. The more negative entropy and lower heat of activation with $CH_2[B(OMe)_2]_2$ are what would be expected from tying down the second $(MeO)_2B$ group in a three-membered ring and increasing the bonding energy.

Why not involve the vacant orbital of the neighboring boron as a ligand bridging site, as in **24**? Mercury ought to form a better link than boron to the

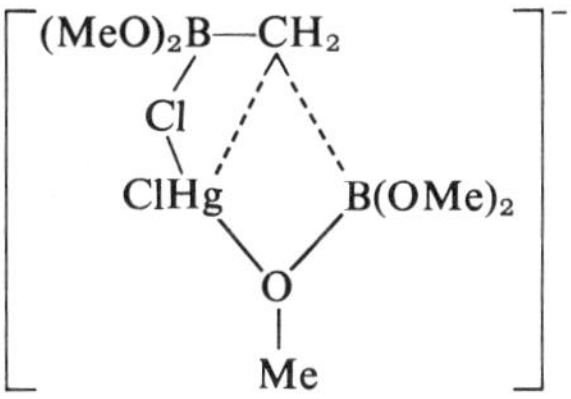

24
(inadequate explanation)

chloride ligand, but the ClHg group is less effective than $(MeO)_2B$ by a factor of 30 in promoting the reaction. Perhaps the mercury would be a more effective neighboring atom except for steric factors, but the reaction is notably insensitive to steric hindrance. In contrast to the large rate differences found in nucleophilic displacements, $C_4H_9B(OMe)_2$ reacts almost as fast as $CH_3B(OMe)_2$ with mercuric chloride, and an α-methyl substituent only slows the reaction of $PhCH_2B(OMe)_2$ by roughly a factor of 2. As usual, ligand bridging is not definitely excluded, but it does not appear to be the major contributor to the energy difference involved in the neighboring group effect.

The cyclic three-center bond symbol has been omitted from structure **24** to make a point. If ligand bridging alone could account for the stability of the transition state (**24**), then there would be little ground for indicating a boron–mercury bond order greater than zero to either boron atom, and there would be little reason other than quantum mechanical symmetry rules (which are a potent reason in themselves, Chapter 1, Section III,B) for writing the cyclic three-center bond symbol. However, the experimental results are in accord with expectation based on the three-center bond model.

It should further be noted that the neighboring boron atom is in a good position to interact with the incoming mercury with very little distortion of the bond angles. In fact, the three-center bonding would probably merely decrease the amount by which the neighboring group has to move out of the way.

Could the neighboring boron merely interact by π-bonding to the carbon, without involving the mercury? In the first place, some orbital overlap is inevitable, and it all leads to stabilization of the system. In the second place, such interaction would merely amount to stabilization of carbanion character, which is inconsistent with the observations in more ways than have been mentioned yet. Phenyl ought to be as good or better than boron or mercury if ordinary π-bonding to carbon is involved. If, even though the aryl compounds show no carbanion character in the Hammett plot, the system were on the edge of having such a tendency, two boron atoms ought to be far better than one at stabilizing the carbanion site. However, $HC[B(OMe)_2]_3$ is only a little more reactive than $CH_2[B(OMe)_2]_2$, and $C[B(OMe)_2]_4$ is somewhat less so, in contrast to the behavior observed where carbanions do appear to be involved.[60,62,63] Only the third or fourth boron atom will depart readily to leave a carbanion in the presence of base, and bis(dimethoxyboryl)-methanes, $R_2C[B(OMe)_2]_2$, resist further loss of boron. The foregoing may be taken as evidence against the sort of delocalization represented by the

$$\begin{array}{ccc} (MeO)_2B{-}CH_2 & & \\ & \diagdown & \\ Cl_2Hg & & \bar{B}(OMe)_3 \end{array} \longleftrightarrow \begin{array}{ccc} (MeO)_2B{-}CH_2 & & \\ \diagup & & \\ Cl_2\bar{H}g & & B(OMe)_3 \end{array} \longleftrightarrow$$

$$\begin{array}{ccc} (MeO)_2B{-}\bar{C}H_2 & & \\ Cl_2Hg & & B(OMe)_3 \end{array} \longleftrightarrow \begin{array}{ccc} (MeO)_2\bar{B}{=}CH_2 & & \\ Cl_2Hg & & B(OMe)_3 \end{array}$$

25

resonance structures **25**. It would be possible to add ligand bridging to **25**, but that is not the essential feature under consideration at the moment. It would also be possible to add two more resonance structures, **26** and **27**.

Structure **26** is what is missing from **25** to make the three-center Hg—C—B bond, and structure **27** introduces the neighboring boron–mercury interaction. With the methoxide bridge added, these would make **25** equivalent to

$$(MeO)_2B—\overset{+}{C}H_2 \qquad Cl_2\bar{H}g—\bar{B}(OMe)_3$$

26

$$(MeO)_2B—\overset{+}{C}H_2 \quad | \quad Cl_2\bar{H}g \qquad B(OMe)_3$$

27

22 expressed in the language of resonance. The fact that **26** and **27** look like lousy resonance structures to any experienced organic chemist may explain some of the reluctance to accept three-center bonding. However, the inability of simple resonance theory to cope with the difference between benzene and cyclobutadiene is too familiar to review here, and the special stability of a three-center or four-center two-electron system is likewise beyond the scope of resonance theory.

One other alternative explanation can be disposed of. Hyperconjugation from carbon–metal bonds has been shown to have very large effects on molecular properties[64] (see Chapter 5, Section II). However, in the present case this effect would be in the wrong direction. The carbon–metal bond can only donate electrons, and electron donors do not assist this reaction. If the hyperconjugation were a special case, the ClHg group ought to be as effective as the $(MeO)_2B$, but it is not. Therefore, structures **28** do not account for the rate acceleration.

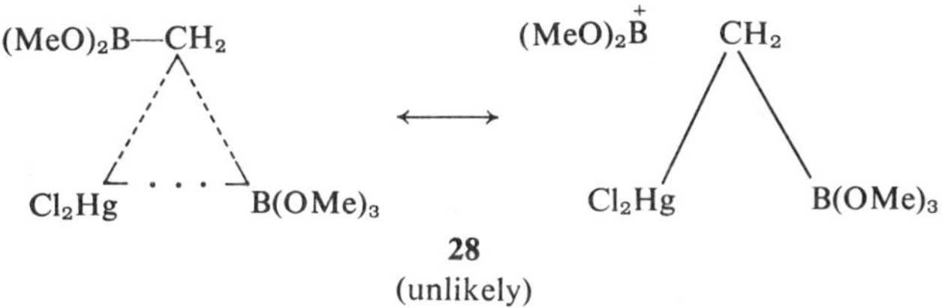

28
(unlikely)

A referee once dismissed all the foregoing arguments and proposed that obviously the transition states have similar structures but that the starting material $CH_2[B(OMe)_2]_2$ is destabilized relative to $CH_3B(OMe)_2$. This might have been a good point, except that in the closest available model system, $CH_3CH_2CH_2CH(BR_2)_2$ is thermodynamically a bit more stable than $R_2BCH_2CH_2CH_2CH_2BR_2$.[64a]

There is another reaction which fits in with the mechanistic picture that has been developed in the foregoing discussion. Mercuric acetate in ethanol efficiently converts $C[B(OMe)_2]_4$ to $C(HgOAc)_4$.[65] This reaction cannot involve carbanions, which would capture protons from the solvent rather than mercuric ion, and $C[B(OMe)_2]_4$ is stable in slightly acidic refluxing methanol. The four borons of $C[B(OMe)_2]_4$ are sterically crowded, and as a result this compound reacts more slowly than $HC[B(OMe)_2]_3$ with methanolic base in protodeboronation[60] or with mercuric chloride in the first stage of mercuri-deboronation[61] (Table 3-2). The four mercury atoms of $C(HgOAc)_4$ fit easily around the central carbon, the probable C–Hg distance being 2.07 Å,[48] the calculated Hg–Hg distance 3.4 Å for tetrahedral angles, and the Hg–Hg van der Waals distance about 3.1–3.3 Å.[48] This does not leave much room to squeeze out the last boron as the last mercury attacks in a concerted process. Perhaps some other mechanism such as a dissociation to radicals takes over that stage, but $C(HgOAc)_4$ and its derivatives show no signs of instability that might be expected if the $(XHg)_3C\cdot$ radical were especially easily formed. For example, $C(HgCl)_4$ is not reduced by aqueous stannous chloride, and the insoluble nitrate salt is not visibly attacked by hot nitric–hydrochloric acid mixture, though $C(HgBr)_4$ does react readily with bromine to form CBr_4 by an unknown mechanism that might be either concerted or two-step replacement. The available evidence indicates that at least some stages of the reaction of $C[B(OMe)_2]_4$ with mercuric acetate do involve direct electrophilic displacement, that these show some retardation by steric hindrance but the effects are remarkably small, and that transition states analogous to **22** with low steric requirements and direct assistance by the vacant orbital of a neighboring metal atom are consistent with the results.

In summary, all the data on mercuri-deboronation are consistent with the presence of a cyclic Hg—C—B three-center two-electron bond in the transition state, as well as an Hg—OR—B ligand bridge. It has not been proved experimentally that both the direct Hg—B interaction and the Hg—OR—B ligand bridge contribute important stabilization to the transition state, but the neighboring group effects strongly implicate one or the other or preferably both. Consideration of stable bridged systems such as $Al_2Me_5NPh_2$ (Chapter 1, Section III,C, and Chapter 2, Section III,G) together with quantum mechanical symmetry arguments lends support to both types of interaction.

Mercuri-deboronation of areneboronic acids[66] was discovered before that of benzylboronic acids, and was also the first to be studied mechanistically.[58] Direct boron–mercury interaction is unlikely to be significant here because of the interaction with the π orbitals of the benzene ring as mentioned earlier, which introduces an antisymmetric component into the wave function between the two metal atoms. On the other hand, ligand bridging of the Hg—OR—B type is likely, even though the probably greater Hg—C—B bond angle makes such bridging less favorable. Kuivila and Muller measured the kinetics, for practical reasons, on basic phenylmercuric perchlorate and benzeneboronic acid in buffered aqueous ethanol.[58] In the near-neutral pH range, where the predominant species are PhHgOH and $PhB(OH)_2$ (or ethyl ester), the reaction is insensitive to pH and the rate law is $-dx/dt = k[PhHgOH][PhB(OH)_2]$. It was also shown that PhHgX, where X is phosphate or acetate, is not measurably active as an electrophile. The two possible transition states are **29** and **30**.

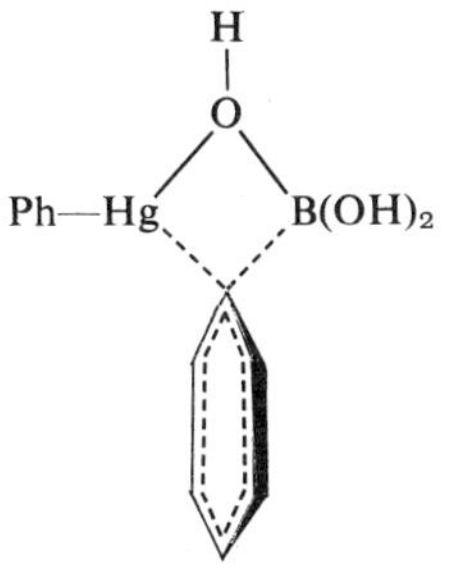

H
|
O
Ph—Hg B(OH)$_2$

29

$\delta+$ $\delta-$
Ph—Hg B(OH)$_3$

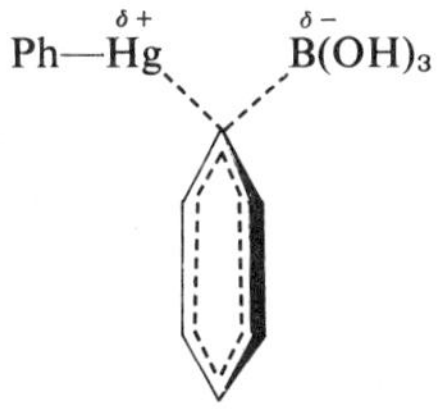

30

Kuivila and Muller pointed out that the kinetics allow either reaction of PhHgOH with PhB(OH)$_2$ or of PhHg$^+$ with PhB(OH)$_3^-$, the latter pair being minor species in equilibrium with the former. They did not speculate on which would be most likely. It is useful to consider the qualitative aspects of possible potential energy surfaces here. Even though a unique mechanism cannot be defined with certainty from the available data, the possibilities can be narrowed considerably.

The reaction path must accomplish two parallel things, the transfer of a phenyl group from boron to mercury and the transfer of a hydroxyl group from mercury to boron. At the start the stable state is PhHgOH + PhB(OH)$_2$, at the end PhHgPh + B(OH)$_3$. What species lie along the energy trough which climbs over the barrier between? Total transfer of a hydroxyl group first to form PhHg$^+$ and PhB(OH)$_3^-$ is energetically uphill from the starting materials. Total transfer of the phenyl group first to form Ph$_2$HgOH$^-$ and (HO)$_2$B$^+$ is obviously farther uphill, being grossly contrary to the ionization tendency of boric acid, and is clearly out of the energy trough. The most straightforward assumption is that the energy trough passes between these extreme ionizations by way of a concerted mechanism, transition state **29**. Partial transfer of the phenyl group is aided by partial transfer of the hydroxyl group, just as total transfer of the phenyl requires total transfer of hydroxyl. A hypothetical contour map of this situation is illustrated in Fig. 3-2.

The potential surface is drawn with the assumption that phenylmercuric hydroxide can form a metastable complex with benzeneboronic acid through a hydroxide bridge. It should also be noted that "PhHg$^+$" on this drawing is not the same species formed by ionization of PhHgOH in aqueous solution, which would be better represented as PhHgOH$_2^+$. There may be another energy trough lying off this hypothetical contour map for the reaction of PhHgOH$_2^+$ with PhB(OH)$_3^-$ by way of a hydrated version of transition

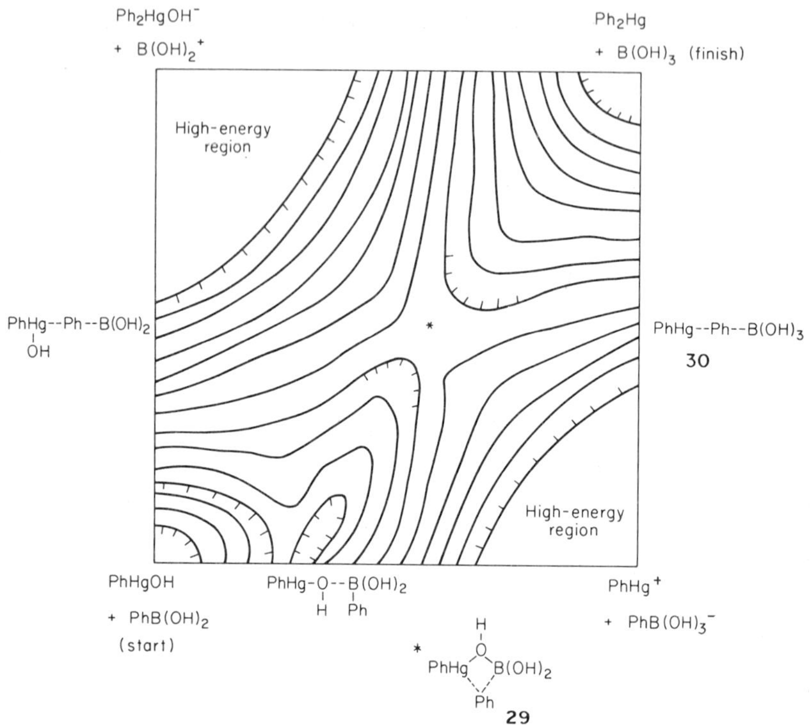

Fig. 3-2. Hypothetical potential energy surface for the reaction of PhHgOH with $PhB(OH)_2$. On the ordinate is plotted a function of the C—B/C—Hg distance ratio, on the abscissa a similar function of the O—Hg/C—B distance ratio. It is assumed that bond distances and angles are otherwise optimized and that the interatomic distance ratio function is in principle logarithmic, with the ratio small at the origin, 1 at the midpoint, and large at the end. This plot is strictly qualitative and is presented only as an aid to visualization, since its details far exceed any data one could hope to get.

state structure **30**. However, the reaction is relatively insensitive to changes in solvent polarity, showing only a fourfold rate increase on going from 60% ethanol to 30% ethanol, and it appears contrary to the experimental evidence to write a mechanism involving separated ions. It seems more likely *a priori* that the extra water molecule would get in the way of the reaction rather than catalyze it.

It cannot be overstressed that potential surface plots such as this one are strictly qualitative aids to visualization. There is at present no way to obtain enough data to make accurate plots. Also, a multidimensional surface is needed, and all that can be visualized is a cross section based on the assumption that two parameters could be separated for plotting while the multitude of the rest are optimized for each plotted point. Even with all these unknowns and limitations, the concept of potential energy surfaces is the fundamentally correct one to use in reasoning about reaction mechanisms, and mechanism chemists might avoid various difficulties in interpretation more often if they would habitually relate their proposed mechanisms to possible potential energy surfaces, though these can have traps of their own (see Section II,B).

As would be expected for any concerted displacement, the reaction of basic phenylmercuric nitrate with benzeneboronic acid shows an isotope effect, $k_{10_B}/k_{11_B} = 1.02$.[67] This result suggests that the carbon–boron bond is appreciably weakened in the transition state.

E. MAGNESIUM(II) AS ELECTROFUGE

Although mercury(II) will displace a wide variety of other metal ions from carbon, mechanistic studies have been severely limited by the general inconvenience of working with reactive organometallic compounds. Retention of configuration has been proved in an example of a reaction with a Grignard reagent. Norbornyl bromide and magnesium form a mixture of diastereoisomeric norbornylmagnesium bromides, from which the *exo* isomer can be removed by its more rapid reaction with benzophenone to leave the pure *endo*-2-norbornylmagnesium bromide (**31**) in the solution.[68] Reaction with mercuric bromide yields norbornylmercuric bromide that is 95% *endo*.

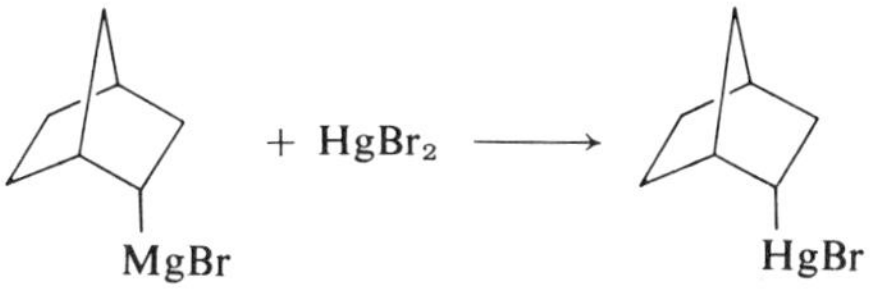

MgBr + $HgBr_2$ ⟶ HgBr

31

III. Halogen Electrophiles

A. INTRODUCTION

The reaction R—M + X_2 = R—X + M^+X^-, where R is alkyl or aryl, M is any metal, and X is a halogen, is a very general one. A common mechanism is straightforward electrophilic displacement, but inversion as well as retention processes have been observed, and some organometallics are prone to react by radical-chain mechanisms (Chapter 7). Three-center bonding is still a probable feature of the transition states for retention, but as noted in the first part of this chapter the effects of such bonding are likely to be small because the halogen cation differs grossly in electronegativity from the metal cation and the reactions are exothermic with transition states early along the reaction coordinate while exchange integrals are still small.

B. Mercury(II) as Electrofuge

Jensen and co-workers have studied the stereochemistry of halo-demercurations in considerable detail, taking special precautions to overcome the radical processes which compete.[4,5] The radical processes predominate in nonpolar solvents and are only suppressed completely in good donor solvents such as pyridine and at low temperatures. Two types of asymmetric center were utilized, the diastereoisomeric *cis*- and *trans*-4-methylcyclohexylmercuric bromides or iodides, and the optically active 2-butylmercuric bromide. With the former, complete retention of configuration was observed when the reaction was carried out with the pyridine–bromine complex in pyridine. Transition states **32** and **33** seem reasonable, though the existence of a ligand bridge has not been proved and the choice of bridging atom is arbitrary. Considerable retention was also observed with bromine in acetic acid. The

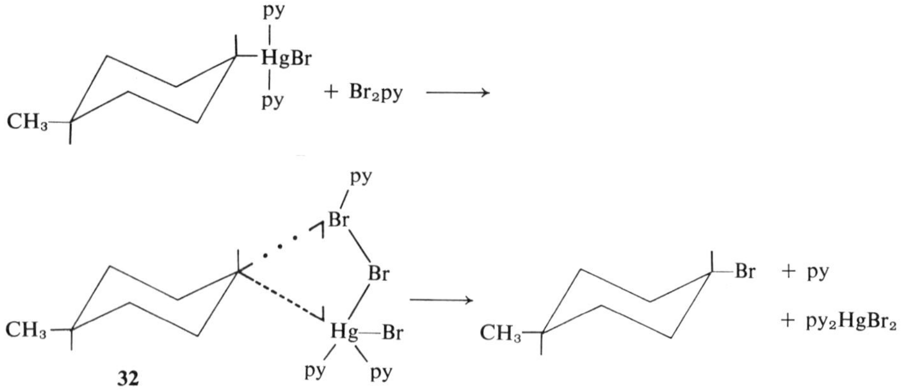

CH₃–C₆H₁₀–HgBr(py)₂ + Br₂py ⟶ **32** ⟶ CH₃–C₆H₁₀–Br + py + py_2HgBr_2

CH₃–C₆H₁₀–Hg(py)₂Br ⟶ **33** ⟶ CH₃–C₆H₁₀–Br + py + py_2HgBr_2

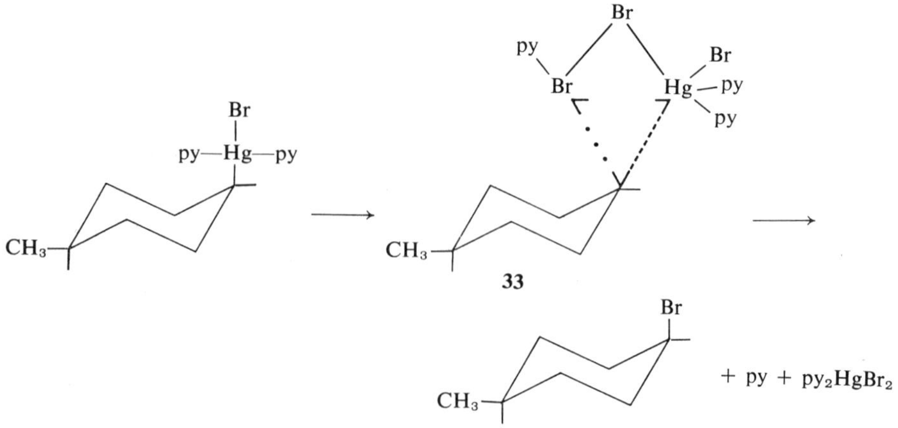

(py = pyridine, linked by the lone pair on nitrogen)

use of both diastereoisomers ensures that the reaction is stereospecific, not merely the result of some one-sided thermodynamic equilibrium along the way, and it further ensures that the spectroscopic methods used for deciding which isomer is which have been interpreted correctly.

With 2-butylmercuric bromide retention was also observed but the degree of stereospecificity was generally less, reaching 100% only in a pyridine–collidine mixture at $-65°C$.[4] The transition state structure **3** proposed in Section I,B becomes equivalent to **32** and **33** if complexing with pyridine is added, or is itself a reasonable representation of the reaction in CH_3OH–CH_2Cl_2, in which some net retention was observed but the methanol solvent is only weakly bound to the mercury. Such solvation is, by convention and because of the difficulty of proof, normally omitted from written structures, but that does not necessarily mean its effects are negligible.

Jensen and co-workers have examined other halogen cleavages of organomercury compounds and found retention to be favored where the electrophile is ICl, I_2, or Cl_2.[15c] Possible four-center transition states are discussed. No new concepts are involved, and a detailed review here would be repetitious.

C. Group IV Electrofuges

Halogens generally cleave one or two alkyl groups from tetraalkyltins or tetraalkylleads. No further reaction occurs under mild conditions, even with the more reactive aryl derivatives. Three classes of mechanisms are known, electrophilic displacement with retention of configuration at carbon, electrophilic displacement with inversion, and free-radical displacement by a chain process. The first two will be discussed in detail here, the third in Chapter 7, Section III,C.

Gielen, Nasielski, and co-workers carried out an extensive investigation of this field during the 1960s.[69–79] Their data are very useful and many of their conclusions are valid, but nature has a way of being more complex than the chemist expects it to be, and there is a high probability that some of their reactions proceed with inversion of the carbon atom,[7] which they failed to discover. Their proof of stereochemistry rests on studies with substituted cyclopropyltin compounds[79] and cyclopropyl groups are relatively resistant to inversion processes. On the other hand, the sterically hindered 2-butyltrineopentyltin used by Jensen and Davis[7] to demonstrate inversion in bromination in methanol might be relatively resistant to retention stereochemistry, and currently available information does not establish the stereochemistry of reactions of simple organotin compounds such as tetramethyltin with bromine in polar solvents. A further limitation on the theoretical interpretations by Gielen and Nasielski is their reliance on the S_E2 and S_Ei concepts of Hughes, Ingold, and co-workers.[10,36,37,42] If the reactions in polar solvents classified as S_E2 by Gielen and Nasielski actually proceed with inversion, then the S_E2 mechanism could, in principle, be distinguished clearly from the cyclic S_Ei transition state. However, if configuration at carbon is retained, all that can be determined is whether ligand bridging is relatively strong or relatively weak, with weakness to the point of insignificance or nonexistence being unprovable, as pointed out in Section II,B on mercury(II) exchanges.

The rate of bromination of tetraalkyltins in the nonpolar solvent chlorobenzene is not very sensitive to variation of the alkyl groups.[74] Relative rates are Me_4Sn, 1.0; Et_4Sn, 12.0; *n*-Pr_4Sn, 4.5; *iso*-Pr_4Sn, 13.0. This insensitivity to steric hindrance is consistent with the usual sort of cyclic transition state (**34**). Gielen and Nasielski were not concerned with quantum mechanical fine points and did not write three-center bonds, but inclusion of this feature here is a refinement of and not a disagreement with their conclusions. A term second order in bromine was also found in the rate law at higher

$Me_4Sn + Br_2 \longrightarrow$ [Me_3Sn(Br)(Br)(Me) cyclic transition state] $\longrightarrow Me_3SnBr + MeBr$

34

concentrations,[76] but it is not obvious where the second molecule of bromine might be placed in the transition state.

In polar solvents the brominations are generally much faster, with the solvent itself appearing to serve as a nucleophile, and the rates become more sensitive to steric hindrance. Relative rates of bromination in acetic acid are Me_4Sn, 80; Et_4Sn, 67; *n*-Pr_4Sn, 9.6; iso-Pr_4Sn, 2.0.[72,74,75] (It may be noted that iso-Pr_4Sn reacts more slowly in acetic acid than in chlorobenzene.) Iodine in acetic acid is more selective, the Me_4Sn/isoPr_4Sn rate ratio being 2500.[74] Iodinations are catalyzed by iodide ion in nonpolar solvents, in spite of the conversion of I_2 to the less electrophilic I_3^-. Pilloni and Tagliavini noted that iodination of tetramethyllead in carbon tetrachloride is accelerated as the square of the added methanol or acetonitrile concentration.[80] Unlike the tetraalkyltin cleavage, the tetramethyllead reaction is not catalyzed by iodide ion in acetone or acetonitrile.

Gielen and Nasielski correlated rates of displacement of tin and lead by halogens with the linear free-energy relationship of Eq. (3-6). The reference

$$\log k/k_0 = pX \qquad (3\text{-}6)$$

point k_0 is the rate constant in acetic acid, k is the rate constant in a different solvent, p is a constant characteristic of the electrophile and alkylmetal compound and defined as 1.00 for the Me_4Sn reaction with Br_2, and X is the "polarity" of the solvent as defined by this equation.[72,75] (Units of X are powers of 10 in the rate.) For acetic acid, $X = 0$; carbon tetrachloride, $X = -4.8$; chlorobenzene, $X = -1.9$; acetonitrile, $X = 0.04$; methanol, $X = 0.9$; dimethyl sulfoxide, $X = 1.6$. Such relationships as $k_{Me_4Sn} < k_{Et_4Sn}$ in nonpolar solvents change gradually with increasing polarity until $k_{Me_4Sn} > k_{Et_4Sn}$.

Gielen and Nasielski concluded that the reactions in polar solvents proceed by way of acyclic S_E2 transition states (**35**). As noted previously in the section on mercury(II) exchanges, it is not possible to rule out cyclic ligand bridging in such situations, even though the energy derived from improving the ligand bridge may be less than that obtainable by improving the electrophile. Accordingly, the cyclic (S_Ei) transition state **36** would always be in competition with the acyclic version **35**, and in view of the available models discussed in Chapter 2, Section III,G, **36** is probably generally lower in energy. However in view of the inversion observed by Jensen and Davis,[7] to be discussed next, transition state **37** probably is favored over **35** under some reaction conditions, and the two are probably in substantial competition under other conditions. In view of the smooth solvent correlations obtained, the stoichiometries and polarities of **36** and **37** are probably similar.

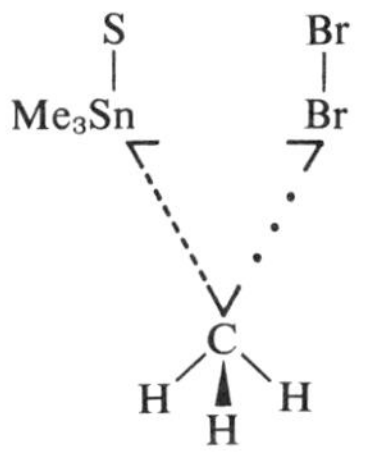

35 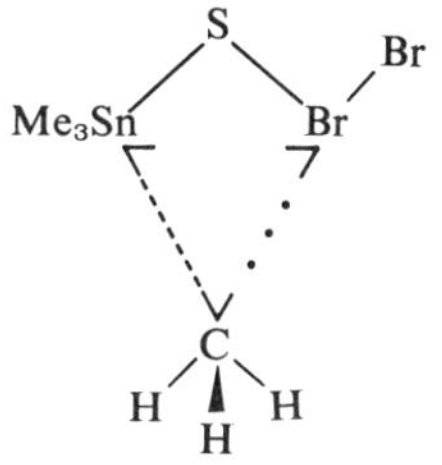 **36** 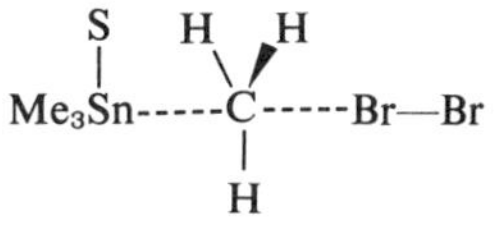 **37**

(S = nucleophilic solvent)

Jensen and Davis found that the reaction of 2-butyltrineopentyltin (**38**) with bromine in methanol yields 2-bromobutane with 35% net inversion of configuration under the conditions of the kinetic runs or 80% at higher concentrations.[7] These figures are lower limits because there are possible racemization mechanisms unrelated to the electrophilic displacement. A transition state model (**4**) for inversion has been drawn in Section 1,C. Cleavage of 2-butyltrineopentyltin (**38**) with bromine in carbon tetrachloride proceeded by a free-radical mechanism resulting only in racemization, but retention was observed with 2-butyldicyclohexyltin bromide (**39**) and bromine in carbon tetrachloride.

$$\text{H}\blacktriangleright\underset{\text{Et}}{\overset{\text{Me}}{\text{C}}}\blacktriangleleft\text{Br} + \text{Ph}_3\text{SnNa} \xrightarrow{\text{inversion}} \text{Ph}_3\text{Sn}\blacktriangleright\underset{\text{Et}}{\overset{\text{Me}}{\text{C}}}\blacktriangleleft\text{H} + \text{NaBr}$$

S-(+)-2-Bromobutane → *R*(−)-2-Butyltriphenyltin

$$(\text{Me}_3\text{CCH}_2)_3\text{Sn}\blacktriangleright\underset{\text{Et}}{\overset{\text{Me}}{\text{C}}}\blacktriangleleft\text{H}$$

38

R-(−)-2-Butyltrineopentyltin

$$\xrightarrow[\text{inversion}]{\text{Br}_2,\ \text{methanol}}\ \text{H}\blacktriangleright\underset{\text{Et}}{\overset{\text{Me}}{\text{C}}}\blacktriangleleft\text{Br}$$

S-(+)-2-Bromobutane

$$(\text{C}_6\text{H}_{11})_2\underset{\text{Br}}{\text{Sn}}\blacktriangleright\underset{\text{Et}}{\overset{\text{Me}}{\text{C}}}\blacktriangleleft\text{H}$$

39

R-(−)-2-Butyldicyclohexyltin bromide

$$\xrightarrow[\text{retention}]{\text{Br}_2,\ \text{CCl}_4}\ \text{Br}\blacktriangleright\underset{\text{Et}}{\overset{\text{Me}}{\text{C}}}\blacktriangleleft\text{H}$$

R-(−)-2-Bromobutane

The fact that both inversion and retention processes were observed in reaction sequences starting from 2-butyltriphenyltin provides extra assurance that the stereochemical assignments are correct. The nucleophilic displacement of Br^- by Ph_3Sn^- proceeds with inversion (Chapter 8), and the subsequent removal of the phenyl groups from the tin and replacement by alkyl groups would not affect the carbon–tin bond. The sequence used for these syntheses of **38** and **39** involved bromination to organotin dibromide and reaction with an alkyl Grignard reagent.[81]

$$\text{Ph}_3\text{Sn—2-Bu} + \text{Br}_2 \longrightarrow \text{Br}_2\underset{\text{Ph}}{\text{Sn}}\text{—2-Bu} \xrightarrow{\text{RMgBr}} \text{R}_2\underset{\text{Ph}}{\text{Sn}}\text{—2-Bu} \xrightarrow{\text{Br}_2}$$

$$\text{R}_2\underset{\text{Br}}{\text{Sn}}\text{—2-Bu} \xrightarrow{\text{RMgBr}} \text{R}_3\text{Sn—2-Bu}$$

39 (R = cyclohexyl)   **38** (R = neopentyl)

TABLE 3-3

RELATIVE RATES OF DISPLACEMENTS COMPARED TO THE ETHYL COMPOUND[a]

R	$(Me_3CCH_2)_3SnR$ Br_2, MeOH, 45°C	RX × Y^- nucleophilic	R_4Sn, I_2 MeOH	R_4Sn, Br_2 HOAc
Me	6.9	30	8.5	1.2
Et	1.00	1.00	1.00	1.00
Pr	0.28	0.4	0.13	0.14
iso-Pr	0.052	0.025	—	0.03
2-Bu	0.018	—	—	—
Me_3CCH_2	0.0024	0.00001	—	—

[a] Data compiled from references 7, 72–75, and 83.

The choice of alkyl groups on the tin was dictated by practical considerations. Phenyl is attacked much faster than alkyl by halogens. The rates of cleavage of cyclohexyl and 2-butyl groups are about the same, favoring the former by a factor of 1.25. Neopentyl is cleaved about 1/4 as fast as 2-butyl, and the ratio of 2-bromobutane to neopentyl bromide from the bromination of 2-butyltrineophenyltin in methanol is only 44:56. Radical cleavages were a severe problem with these highly hindered tin compounds in nonpolar solvents. The degree of steric hindrance in 2-butyltricyclohexyltin and 2-butyltrineopentyltin is sufficient to alter the conformation of the 2-butyl group and cause partial failure of the usual correlation between bond refraction and optical rotation.[82]

Jensen and Davis also measured the relative rates of cleavage of various alkyltrineopentyltins by bromine in methanol.[7] In general these parallel the relative rates of cleavage of tetraalkyltins and also the average relative rates of nucleophilic displacements on the alkyl groups.[83] These correlations are summarized in Table 3-3. A plot of log k for the R_4Sn–Br_2 reaction versus log k for reaction of RCl with Br^- (Fig. 3-3) yields a straight line with a slope of 0.55, where R is varied in steric bulk from methyl to neopentyl.

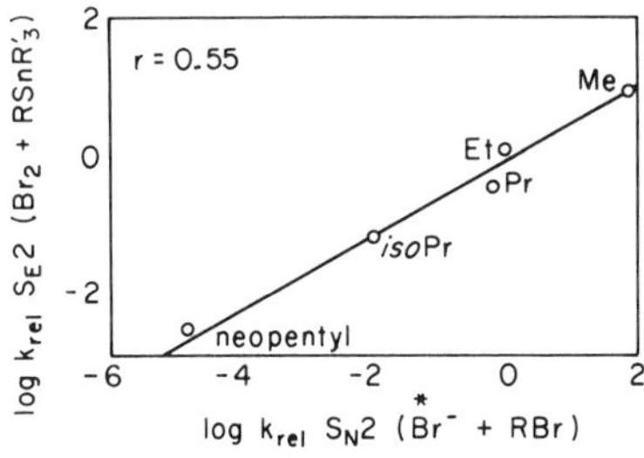

Fig. 3-3. Plot of log k for $R_3SnR' + Br_2 \rightarrow R_3SnBr + R'Br$ vs. log k for $R'Cl + Br^- \rightarrow R'Br + Cl^-$. [From F. R. Jensen and D. D. Davis, *J. Amer. Chem. Soc.* **93**, 4049 (1971).]

Jensen and Davis have proposed that electrophilic displacements which proceed with inversion can be distinguished from those with retention by their characteristic patterns of reactivity on varying the alkyl group.[7] For example, for cleavage of R_2Hg by HCl with retention in dioxane the relative rates are Me, 1; Et, 5.95; *n*-Pr, 3.2; and iso-Pr, 3.5, grossly different from the sterically controlled pattern found for displacements with inversion. Similar rate patterns are found for the cleavage of R_4Sn with Br_2 in PhCl,[74] which probably also proceeds with retention. However, due caution must be used in applying this principle. The relative bridging abilities of alkyl groups in Al_2R_6 are Me, 6; Et, 1; and iso-Bu, 0.35 (Chapter 2, Section III,C), which will give a reasonable log K versus log k plot with the average reactivities in nucleophilic displacement,[83] Me, 30; Et, 1; and iso-Bu, 0.03. The slope is between 0.3 and 0.55, appropriate for inversion, but these alkyl bridges are the model for displacement with retention.

Much detail about the transition state for inversion is unknown, but most of the unknown features are in the realm of solvation and not easily accessible to experimental testing. Perhaps some day it will be possible to define whether the tin atom retains its configuration, is inverted, or is racemized. Before that can be done, a much cleaner system will be needed and many difficult experimental problems will have to be solved.

Inversion at silicon has been found in the cleavage of silicon–carbon bonds by bromine.[84,85] The silicon atom of a *p*-methoxyphenylsilicon compound (**40**) is inverted even though the solvent is nonpolar (benzene or carbon tetrachloride) and retention at carbon is dictated by the aromatic ring.

$$p\text{-MeOC}_6\text{H}_4\text{—Si(Ph)(Me)}\blacktriangleleft\text{Np} + \text{Br}_2 \longrightarrow p\text{-MeOC}_6\text{H}_4\text{Br} + \text{Np}\blacktriangleright\text{Si(Ph)(Me)—Br}$$

40

D. Transition Metal Electrofuges

Although this book is not supposed to cover transition metals, the mechanism of electrophilic cleavage of certain alkyl transition metal compounds having strong back-bonding ligands on the metal is probably essentially the same as with alkyl nontransition metal compounds. Whitesides and Boschetto have shown that *threo*-1,2-dideutero-3,3-dimethylbutyl(cyclopentadienyl)iron dicarbonyl (**41**) reacts with mercuric chloride with retention of configuration but undergoes inversion with bromine or sulfur dioxide.[86] The stereochemistry is readily established by nmr in each case.

Jensen and co-workers have shown that the cleavage of carbon–cobalt bonds by halogens proceeds with inversion.[87] Both cyclohexyl and *sec*-butyl systems were used. The cobalt atom is complexed in the cobaloxime com-

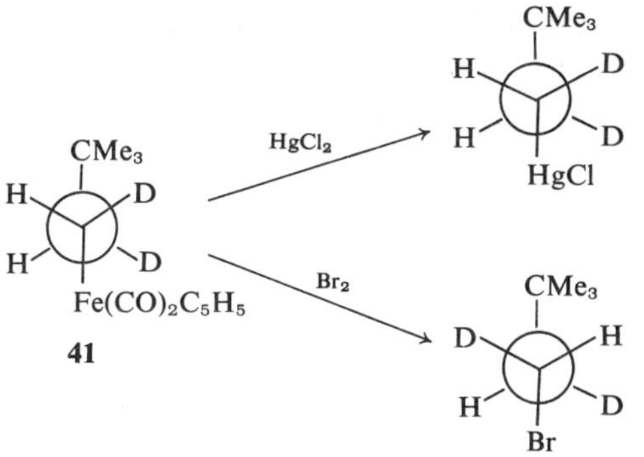

CMe_3 H D H D $Fe(CO)_2C_5H_5$ **41** $\xrightarrow{HgCl_2}$ CMe_3 H D H D HgCl ; $\xrightarrow{Br_2}$ CMe_3 D H H D Br

pounds used so that there are no vacant *d* orbitals and it behaves much like a sterically hindered nontransition metal.

E. BORON(III) AS ELECTROFUGE

Brown and co-workers have found that iodine readily cleaves two alkyl groups from trialkylboranes in the presence of sodium hydroxide to yield alkyl iodides.[88] A similar preparation of alkyl bromides has been achieved by the use of sodium methoxide as the base, and all three alkyl groups of the trialkylborane are converted.[89]

Brown and Lane have found that bromine in tetrahydrofuran with sodium methoxide as catalyst cleaves tris(*exo*-norbornyl)borane (**42**) with predominant inversion of configuration, yielding norbornyl bromide that is 75% *endo*.[90]

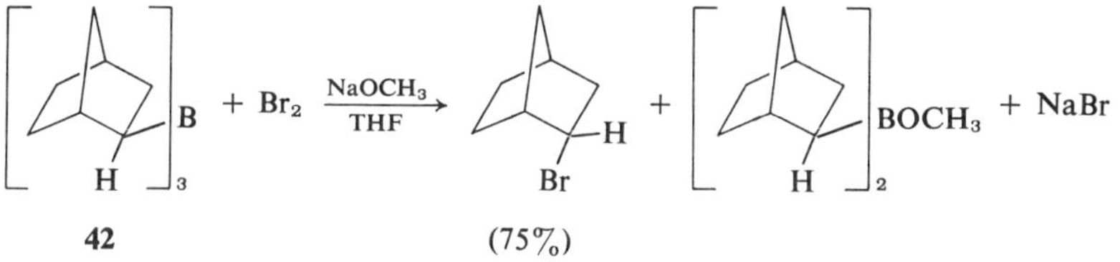

$$[\text{H}]_3\text{B} + Br_2 \xrightarrow[\text{THF}]{NaOCH_3} \text{(Br, H)} + [\text{H}]_2\text{BOCH}_3 + NaBr$$

42 (75%)

Arylboron compounds are considerably more reactive than the alkylboranes toward halogens. The mechanism of halogenation of areneboronic acids has been studied by Kuivila and co-workers.[91–96] Detailed kinetic studies on bromination in aqueous acetic acid indicate that the rate is $k[Br_2][ArB(OH)_3{}^-]$ in the absence of buffers, the rate being suppressed by added H^+. In the presence of moderately basic anions such as sulfate, hydrogen phosphate, acetate, or fluoride (which is particularly basic toward boron) the rate becomes $k[Br_2][ArB(OH)_2][\text{base}]$ and deviates widely from the Brønsted catalysis law, indicating that the base coordinates with boron rather than removing a proton from a hydroxyl group. Chelating agents which increase the acidity of the boron greatly accelerate the reaction, citric acid, glycolic acid, and *cis*-indane-1,2-diol being the most effective tested. Like any typical aromatic substitution, the reaction is promoted by electron-donating substituents. Relative rates for substituted boronic acids include *p*-MeO, 1.45×10^6; *p*-Me, 79; H, 1.0; *p*-CO_2Et, 0.010; and a dozen other consistent with this general trend. The mechanism, illustrated with fluoride ion as the base, evidently involves **43** or analogous transition states.

$$C_6H_5B(OH)_2 + F^- \longrightarrow C_6H_5\bar{B}(OH)_2F \xrightarrow{Br_2}$$

$$[\text{43}]^- \longrightarrow C_6H_5Br + Br^- + FB(OH)_2 \xrightarrow{F^-} BF_4^-$$

43

F. Lithium(I) as Electrofuge

The bromination of alkyllithiums is electrophilic displacement by definition, but there is considerable uncertainty about certain mechanistic details because of the strong tendency of alkyllithium compounds to generate free radicals in reactions with electrophiles (Chapter 7, Section IV). Reactions with halogens do not proceed mainly by a chain reaction involving free alkyl radicals, because such a mechanism would completely racemize the alkyl radicals, contrary to observation. However, radical pairs may be involved, or the reaction may be so exothermic that the transition state is an electronically excited state of the three-center bond. Referring back to Table 3-1, for retention one electron in the lowest energy level and one in the next would yield the energy $2\alpha + \beta$, and for the inversion the figure would be a more favorable $2\alpha + 1.414\beta$. The exothermic process would not be very discriminating between retention and inversion. The picture is further complicated by the existence of alkyllithiums as tetramers in equilibrium with more reactive dimers.

Applequist and Chmurny used the lithium reagents prepared from *exo*- and *endo*-2-norbornyl chloride (**44** and **45**) to demonstrate predominant inversion in the reaction with Br_2 and predominant retention with the other electrophiles tested, $ClCO_2CH_3$, CO_2, and $BrCH_2CH_2Br$.[97] Mixtures of the diastereomeric norbornyllithiums result even at the lowest workable temperature of preparation, 20°C in pentane, and yields are low, but enough stereospecificity was obtained to establish the stereochemical preference beyond reasonable doubt. The equilibrium mixture of norbornyllithiums obtained from either chloride in refluxing pentane appears to be about 83% *exo*, assuming the reaction with CO_2 is stereospecific. The *endo*/*exo* ratios observed are summarized in Table 3-4.

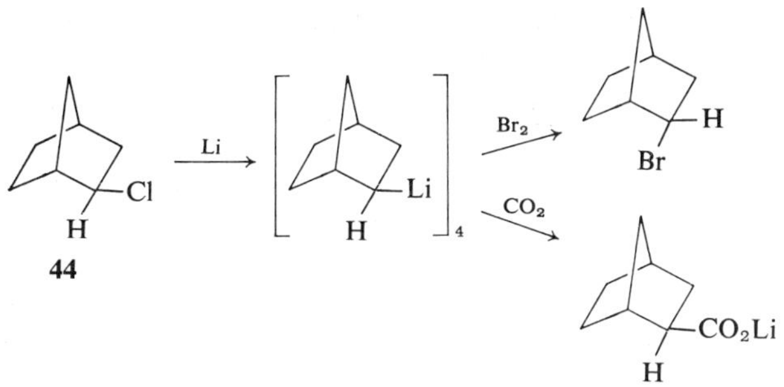

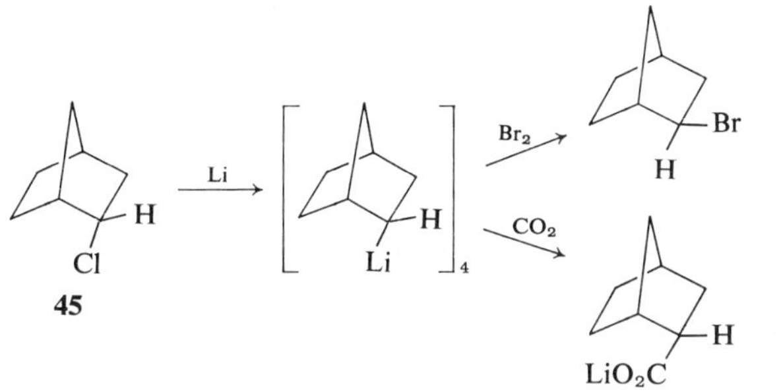

The evidence for inversion with Br_2 is considerably strengthened by the use of both diastereomers of norbornyllithium. The product ratios do not vary with the extent of partial reaction with electrophiles, indicating that the *exo* and *endo* RLi isomers are consumed at similar rates under the reaction conditions. The *exo* and *endo* RLi isomers both show about the same degree

TABLE 3-4

endo/*exo* RATIOS IN PRODUCTS FROM NORBORNYLLITHIUMS[a]

Reagent	Product (R = norbornyl)	*endo*/*exo* Ratio from: Equilibrium RLi	*exo*-RLi	*endo*-RLi
$ClCO_2CH_3$	RCO_2CH_3	0.40	0.28	—
CO_2	RCO_2Li	0.20	0.09	—
Br_2	RBr	1.6	2.8	0.64
$BrCH_2CH_2Br$	RBr	0.23	—	—

[a] From D. E. Applequist and G. N. Chmurny, *J. Amer. Chem. Soc.* **89**, 875 (1967).

of preference for retention with $ClCO_2CH_3$, implying that there is no large steric obstacle to *endo* attack at the 2-norbornyl site.

Glaze and co-workers have demonstrated predominant inversion in the reactions of menthyllithium and 4-*t*-butylcyclohexyllithium with bromine and iodine.[98] At Dry Ice temperature, inversion/retention ratios ranged from 64:36 to 79:21. Chlorine shows a similar preference for inversion with 4-*t*-butylcyclohexyllithium but favors retention slightly with menthyllithium. Other electrophiles favor retention. With bromine–pyridine complex the percentage of retention product is 75–80, with $BrCH_2CH_2Br$ it is 97, and with $ClSiMe_3$ it is 90–95.

A useful feature of this work is that the alkyllithium reagents could be crystallized from pentane and showed indefinite configurational stability in hydrocarbon solvents. Solutions of the lithium reagents in pentane were added to excess halogen to avoid coupling with the alkyl halide products.

Because it is a sterically hindered cyclohexyl system, menthyllithium is dimeric rather than tetrameric in cyclohexane or benzene. Menthyllithium reacts very rapidly with bromobenzene to produce menthyl bromide and phenyllithium under conditions where the half-life of 1-butyllithium is 19 hours and that of 2-butyllithium is 19 minutes.[99] Evidently the exchange with bromobenzene requires dissociation of $(RLi)_4$ to $2(RLi)_2$. However, the very rapid reactions with halogens do not allow time for such dissociation (see Chapter 2, Section V for kinetics of dissociation).

Cyclopropyllithiums tend to retain configuration on reaction with halogens.[100,101] In view of the general resistance of the cyclopropyl group to inversion processes (for example, nucleophilic displacement[83]) it is not surprising that retention should be preferred in these systems. Retention in the halogenation of cyclopropyltin compounds[79] under conditions where other alkyltin compounds yield inverted products[7] has been noted in Section III,C.

IV. The Proton as Electrophile or Electrofuge

A. Introduction

The proton is usually the first electrophile learned (and the first forgotten) by the beginning organic chemistry student for the general reaction $RMgBr + X^+ \rightarrow RX + MgBr^+$. The net reaction normally converts hard-won organometallic compounds back to commonplace hydrocarbons. However, thermodynamics usually greatly favors protodemetallation reactions, and mechanism chemists are prone to join the thermodynamic trend and the lure of possibly simple mechanisms, so there are many studies of this sort of reaction in the literature. As usual, they range in quality from highly sophisticated to totally confused. Only a brief and selective review will be included here. Readers who want 132 references to this field should consult Reutov and Beletskaya.[34]

The most electrophilic metal cations, mercury(II) and thallium(III), are capable of displacing protons from aryl groups. These reactions are included in this section as the reverse of the more common protodemetallations.

B. Mercury(II) and Thallium(III) Electrophiles and Electrofuges

The cleavage of diarylmercury compounds by acids is a typical electrophilic aromatic substitution reaction. Dessy and co-workers found the reaction to be first order in Ar_2Hg and first order in HCl in 90% dimethyl sulfoxide and 10% dioxane, and the substituent effects on the rate could be plotted as $(\sigma + \sigma^+)/2$ versus $\log k$ to give a straight line with $\rho = -2.8$.[102,103] They concluded that molecular HCl was involved in a cyclic four-center transition state, but Jensen and Rickborn have pointed out the deficiencies in the evidence and the reasoning leading to this conclusion.[15d] There is earlier evidence that the anion of the acid does participate, since HBr reacts twice as fast as HCl with Ph_2Hg in aqueous methanol, and other strong acids react slowly if at all.[104] Some form of simultaneous attack of the proton on the benzene ring and the halide on mercury is probable, and the four-center cyclic transition state model is reasonable, especially in less polar solvents, even though the evidence is not at all definitive.[15d]

R. D. Brown and co-workers have reported that the rate law for the cleavage of arylmercuric chlorides by hydrochloric acid in aqueous 90% ethanol is $k[\mathrm{ArHgCl}][\mathrm{H}^+]$, with no dependence on the Cl^- concentration even though the reaction fails if the hydrochloric acid is replaced by sulfuric or perchloric acid.[105] Sufficiently high concentrations of chloride ion do begin to increase the rate, but perchlorate has a similar effect, suggesting a salt effect rather than complexing by chloride.[15d] Substituted phenylmercuric chlorides yield a Hammett correlation with σ^+ leading to $\rho = -2.44$, typical for electrophilic aromatic substitutions.[105] These results are inconsistent with any four-center cyclic transition state in ethanol, but suggest that the proton attacks the benzene ring first in the rate-determining step and chloride attacks the mercury in a second step which is very rapid, provided the chloride concentration exceeds some undefined low level. A short-lived intermediate **46** is required in this mechanism. The evidence summarized by

$$C_6H_5\text{—HgCl} + H^+ \xrightarrow[\text{determining}]{\text{rate-}} [C_6H_5(H)(HgCl)]^+ \;(\mathbf{46}) \xrightarrow[\text{rapid}]{Cl^-} C_6H_5\text{—H} + HgCl_2$$

46

Jensen and Rickborn on the cleavage of Ar_2Hg appears to be consistent with a similar mechanism.[15d]

Mercuration of aromatic compounds, the reverse of the cleavage reactions just discussed, is a general and very useful synthetic reaction. The equilibrium is shifted in favor of mercuration by using mercuric acetate, perchlorate, or other salts which yield relatively high activities of Hg^{2+}, not the halides. Kitching has reviewed this subject recently.[106]

Perrin and Westheimer have studied the kinetics of mercuration of benzene by aqueous mercuric perchlorate.[107] The aqueous solution was kept saturated by a separate benzene phase, and the concentration of benzene in the aqueous phase was measured by ultraviolet spectroscopy. The extent of conversion of Hg^{2+} to $ArHg^+$ was determined by titration with thiocyanate. The reaction is complicated by further mercuration of $PhHg^+$ to $C_6H_4Hg_2{}^{2+}$, etc., but the rate of the first stage was sorted out with the aid of an analog computer. The data are consistent with the rate law $k[PhH][Hg^{2+}]$. Hydrogen ion is not a catalyst. Sodium perchlorate and perchloric acid both accelerate the reaction, both equally effectively at moderate concentrations but the acid somewhat more effectively at high concentrations. The acceleration resulting from 7.5 M $NaClO_4$ is a factor of 300. Inverse correlation of the rate with the activity of the water was observed. The explanation is that the transition state (**47**) contains fewer moles of bound water than the aqueous mercuric ion. An arbitrary but reasonable number of water molecules is illustrated. The reaction is written in two stages because there is a strong primary deuterium

$$C_6H_6 + Hg(H_2O)_4{}^{2+} \longrightarrow [C_6H_6\text{-}Hg^+(OH_2)]^+ + 3\,H_2O \longrightarrow$$

$$[H_2O\cdots H\text{-}C_6H_5\text{-}Hg^+(OH_2)]^+ \ (\mathbf{47}) \longrightarrow C_6H_5Hg^+(OH_2) + H_3O^+$$

47

isotope effect, indicating proton transfer in the rate-determining step. (Note that in the analogous reaction in the reverse direction, the acid cleavage of PhHgCl just discussed, proton transfer also appears to have the highest energy barrier. The system of reaction coordinate curves is self-consistent if proton transfer is largely unaffected by the complexing of the mercury, the transition state for breaking the carbon–mercury bond is not greatly affected, and the thermodynamic balance is significantly shifted only as free aquated Hg^{2+} or $HgCl_2$ is approached.)

Assuming that the Hammett ρ value for this mercuration is about -5, typical of other mercurations, the σ^+ constant of the Hg^+ substituent is in the range 0.09 to 0.14, estimated from the amount of dimercuration which occurs.[107] This is surprisingly low for a positively charged group, but is evidently accounted for by the large size of the mercury atom and consequent lack of concentration of the charge.

H. C. Brown and McGary have studied the kinetics of mercuration of benzene and methylated benzenes with mercuric acetate in acetic acid.[108] In this system the reaction is acid-catalyzed, since perchloric acid removes acetate ion from covalent $Hg(OAc)_2$ to generate a more active mercury(II) cation. Mercuration was found to have a fairly large steric requirement, as shown by only 31% *ortho* substitution in toluene. Data from these mercuration reactions were used as part of the evidence in the origination of the concept of the σ^+ constant.[109]

Thallium(III), like mercury(II), is a sufficiently strong oxidizing agent to metalate aromatic compounds under proper conditions.[106] Taylor and co-workers have found thallium trifluoroacetate to be particularly useful for this purpose.[110,111] Methyl benzoate is unexpectedly attacked at the *ortho* position, suggesting intervention of some sort of direct complex between the thallium and the carboxylate ester function as in **48**. Benzyl alcohol and benzyl methyl ether are also attacked at the *ortho* position, phenylacetic acid and its

CO_2CH_3 (benzene) + $Tl(O_2CCF_3)_3$ ⟶ $CH_3OC(=O \rightarrow Tl(O_2CCF_3)_3)$ (benzene) ⟶

48

CO_2CH_3, $Tl(O_2CCF_3)_2$ (benzene, *ortho*) + HO_2CCF_3

derivatives are attacked mostly at the *ortho* and partly at the *para* position, but phenylpropionic acid is attacked mainly at the *para* position. Definitive mechanistic data are not yet available on this system, but in view of the potential utility of thallation reactions it appears to be a significant field for future research.

Cleavage of aliphatic organomercury compounds by proton donors has been studied extensively. Where the data are good, this appears to be another typical example of concerted electrophilic displacement at saturated carbon. Reports of S_E1 mechanisms for protodemercuration are erroneous and will be disposed of at the end of this section.

Winstein and Traylor found that the rates of cleavage of organomercury compounds by acetic acid fall in the order $Ph_2Hg > 2\text{-}Bu_2Hg > 1\text{-}Bu_2Hg > (PhCMe_2CH_2)_2Hg$.[3,112] The rate of reaction of a bridgehead mercury compound, bis(4-camphyl)mercury (**49**), which must cleave with retention, fell between those of the 1-butyl and neophyl compounds. Thus, protodemercuration where retention of configuration is required was concluded to be comparable in rate to cases where the stereochemistry is not known, implying that retention is possible in all cases even though it cannot be proved. It may be noted that the reactivity pattern implies retention by the criteria of Jensen and Davis.[7]

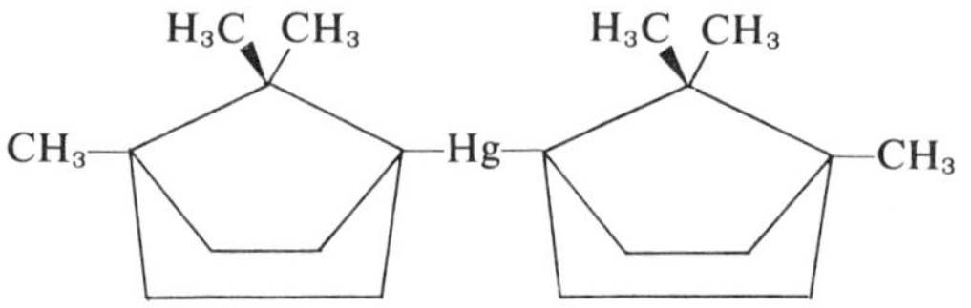

H3C CH3 H3C CH3

CH3— —Hg— —CH3

49

Perchloric acid cleaves all these dialkylmercuries readily, including **49**. Cyclic four-center bonding would be weak or nonexistent in transition states involving aqueous perchloric acid, and it may be concluded that cyclic interactions are incidental rather than a fundamental requirement for retention in electrophilic displacements.

Retention has been observed in the cleavage of alkylmercury compounds with HCl in dioxane.[113] Other studies of cleavages of alkylmercury compounds have been reported, but the details would not change the general description of the mechanism presented here, and Jensen and Rickborn have already provided a thorough critical review.[15d]

Mercuration of an alkyl group, the reverse of the reactions just discussed, is possible if there is a suitable activating substituent. Kitching and Wells have studied the mechanism of formation of $AcOHgCH_2CO_2H$ from mercuric acetate and acetic acid.[114] The reaction is first order in $Hg(OAc)_2$ and independent of added acetate. The order in the solvent acetic acid is of course unknown. Deuterium does not exchange into the methyl from the hydroxyl group of acetic acid under the reaction conditions, ruling out any mechanism involving the enol of acetic acid, $CH_2{=}C(OH)_2$. It is not clear from the published report whether the corresponding enol of mercuric acetate (**50**) can be positively ruled out, but there is no apparent reason to expect mercuric acetate to enolize more readily than the free acid. The isotope effect k_H/k_D is 3.3, indicating that the C—H bond is being broken in the rate-determining step. The most likely transition state appears to be **51**, in which the neighboring carboxyl group assists the electrophilic displacement much in the same way that a neighboring boron atom promotes mercuri-deboronation (Section II,D). Unassisted displacement of H^+ from an alkyl group by Hg^{2+} seems unlikely.

$$Hg(O_2CCH_3)_2 \xrightarrow[\text{determining}]{\text{rate-}} CH_3CO_2Hg{-}O{-}C(OH){=}CH_2 \xrightarrow{\text{rapid}} CH_3CO_2Hg{-}CH_2CO_2H$$

50

$$CH_3CO_2H + Hg(O_2CCH_3)_2 \longrightarrow [\text{transition state } \mathbf{51}] \longrightarrow H_2C(HgO_2CCH_3){-}C(=O){-}OH + HO{-}C(=O){-}CH_3$$

51

Variations of transition state **51** can also be written. Coordination of a carboxyl oxygen with mercury is likely, or it is possible to rearrange $Hg(OAc)_2$ without intervention of an additional molecule of HOAc. The interaction between the carbonyl carbon and mercury is based only on analogy to the boronic ester analog and is not necessarily justified. Kitching and Wells did not go beyond writing the usual sort of four-center transition state for electrophilic displacement.

Protodemercuration of benzylmercury compounds has two special points of interest. One is real, the incorporation of deuterium into the benzene ring as the reaction with HCl proceeds, but discussion will be deferred until Chapter 4, Section VII on S_E2' mechanisms. The other is yet another example of the ability of nature to deceive incautious chemists. First-order dependence on $(PhCH_2)_2Hg$ and zero-order dependence on HCl were observed in the presumed protodemercuration,[115] but the real reaction being followed has been shown to be an acid-catalyzed air oxidation.[116] Second-order kinetics for the HCl–$(PhCH_2)_2Hg$ reaction under oxygen-free conditions had been reported previously,[117] and the air oxidation of iso-PrHgI and *t*-BuHgI in the presence of perchloric acid had also been observed.[118] Another supposed example of an S_E1 reaction is thus nothing but a mistake, one more ghost to haunt the chemist who is in a hurry.

Coad and Johnson have provided a well-documented reaction which can be classified S_E1, the chloride-catalyzed dissociation of *N*-protonated γ-pyridylmethylmercuric chloride.[119] However, the intermediate **52** is not a carbanion but an amino-substituted olefin, the process leading to its formation is not merely first order in substrate but requires H^+ and Cl^-, and the reaction might just as well be classified as an S_E2' attack by the proton at the nitrogen or as a sort of addition–elimination process. β-Pyridylmethylmercuric chloride cannot yield an olefin analogous to **52**, only a zwitterion, and it is inert.[119]

$$\mathrm{NC_5H_4{-}CH_2HgCl} \xrightarrow{H^+,\ Cl^-} \mathrm{H{-}\overset{+}{N}C_5H_4{-}CH_2HgCl_2^-} \xrightarrow{-HgCl_2}$$

$$\mathrm{H{-}NC_5H_4{=}CH_2}\ (\mathbf{52}) \xrightarrow{H^+} \mathrm{H{-}\overset{+}{N}C_5H_4{-}CH_3}$$

Further studies have shown that the aminoolefin intermediate **52** reacts with mercury(II) in competition with H^+ and, from the rate of incorporation of labeled mercury, that second-order mercury exchange competes with the first-order dissociative mechanism.[120] What is essentially the reverse reaction, mercuration of γ-pyridylacetic acid, is first order in protonated γ-pyridylacetic acid and zero order in mercuric nitrate.[121] In accord with the aminoolefin intermediate, α-pyridylacetic acid reacts similarly but the β isomer is inert.

It may be concluded that real examples of S_E1 reactions can be found, provided one rigs the system appropriately and stretches the definition a bit.

C. Boron and Group IV Electrofuges

It is convenient to group boron with silicon and the Group IV metals, which show considerable similarity in protodemetallation reactions. There is abundant literature on the cleavages of arylmetal compounds, which proceed by typical electrophilic aromatic substitution mechanisms, and only a few selected papers are reviewed here. The synthetic chemist would normally regard these cleavages as nuisance reactions, though there might be some potential utility of the $-B(OH)_2$ or $-SiMe_3$ functions as blocking groups in exotic syntheses.

Kuivila and Nahabedian have studied protodeboronation of areneboronic acids in considerable detail.[122–124] With *p*-methoxybenzeneboronic acid in aqueous perchloric, sulfuric, or phosphoric acid the rates correlate with the Hammett acidity function H_0, except that phosphoric acid provides some extra catalysis, making the reaction several times faster at a given H_0 value.[122] In formic acid the rate does not correlate with acidity, being unaffected by the addition of sodium formate, though added sulfuric acid increases the rate.

The H_0-insensitive process must involve general acid catalysis by a molecule of formic acid rather than attack of hydrogen ion (H_3O^+, $HCO_2H_2^+$, etc.) on the substrate, with the latter mechanism entering at higher acidities. Proton transfer is in progress during the transition state, as shown by the isotope effect $k_H/k_D = 3.7$ for *p*-methoxybenzeneboronic acid in 6 *M* sulfuric acid at 25°C and $k_H/k_D = 1.65$ for 2,6-dimethoxybenzeneboronic acid in 0.1 *M* aqueous perchloric acid at 60°C.[123] The boron atom activates the adjacent carbon toward proton attack by an order of magnitude or more, comparing $ArB(OH)_2$ with ArD.[124] This implies substantial boron-to-carbon electron donation, consistent with some degree of boron–carbon bond breaking or weakening in the transition state.

From a plot of log *k* vs. σ^+ for nine substituted benzeneboronic acids in 74.5% H_2SO_4 ($H_0 = -6.12$) at 60°C, $\sigma = -5.2$, and for five compounds in 55.4% H_2SO_4 ($H_0 = -3.66$), $\sigma = -5.0$,[124] values typical of electrophilic aromatic substitutions. Two distinct mechanisms seem to be indicated by the plots of log *k* vs. H_0, which generally show slopes around unity or above in the high acidity region ($H_0 = -6$ or more negative) and somewhat lower slopes, decreased sharply by ~0.3 unit, at lower acidities. The more acidic (less reactive) boronic acids tend to have lower slopes in these plots. It may be concluded that nucleophilic participation by the solvent or acid anion is more important in the transition state at lower acidities of the medium, and also more important with more strongly acidic boronic acids,[124] which is a logical and satisfying result.

Kuivila and Nahabedian proposed mechanisms such as those proceeding through states **53**, **54**, and **55**.[124] For reactions at high acidities, it seems reasonable to consider the possibility that the intermediate **56** might decompose without nucleophilic assistance to the $B(OH)_2^+$ cation. There is some

$$C_6H_5B(OH)_2 + H_3O^+ \longrightarrow [H_2O\cdots H^+\cdots C_6H_5B(OH)_2] \longrightarrow$$

53

$$\mathbf{56}\ (H,\ B(OH)_2) \xrightarrow[\text{rapid}]{H_2O} C_6H_5H + (HO)_2B{-}OH_2^+ \xrightarrow{H_2O} B(OH)_3 + H_3O^+$$

56

natural reluctance among experienced boron chemists to write free $B(OH)_2^+$, for which there is no direct evidence, and for which there is contrary evidence in dilute acid, the hydrated species $(HO)_2BOH_2^+$ being strongly preferred. However, 75% sulfuric acid might change the thermodynamic balance, and the $B(OH)_2^+$ leaving group is consistent with the greater than unit slopes of the log k vs. H_0 plots. The nature of the leaving group might be foreshadowed by two modes of solvation of the transition state (**53** or **55**), or the more solvated boron might arise from an analog of the cyclic transition state **54**. If hydration had to occur before the boron atom could separate from carbon, at high acidities the critical step should become nucleophilic attack on boron rather than proton attack on carbon, and the slope of the log k vs. H_0 plot

$C_6H_5B(OH)_2$ + HCO_2H ⇌ [HC(=O→B(OH)₂)OH complex] ⟶ **54** ⟶

C_6H_6 + $HC(=O)$—O—$B(OH)_2$ $\xrightarrow{H_2O}$ $HCO_2H + B(OH)_3$

and analogous mechanisms with H_2SO_4 and H_3PO_4

$C_6H_5B(OH)_2$ + H_2SO_4 ⟶ [$HOSO_3^-$ ··· H^+ $B(OH)_2$] **55** ⟶ [H $B(OH)_2$, +] **56**

$\xrightarrow[\text{rapid}]{H_2O \text{ or } HSO_4^-}$ C_6H_6 + $(HO)_2B$—OSO_3H, etc.

$\xrightarrow{\text{uncatalyzed}}$ C_6H_6 + HO—$\overset{+}{B}$—OH $\xrightarrow{H_2O}$ $B(OH)_3 + H_3O^+$

should decrease. That the slope increases around $H_0 = -5$ or -6 implies that when the proton activity becomes high enough, a mechanism not involving nucleophilic attack on boron begins to predominate over one which does.

Activation parameters do not assist in choosing between transition state models in this system. The ΔS^* values generally fall in the range -15 to -25 eu. Perchloric acid, which would not form a cyclic transition state (analog of **54**), was only tested on *p*-methoxybenzeneboronic acid, where ΔS^* is an unusual -5 eu (30% $HClO_4$), which may be compared with -12 eu for the same boron compound in 30% H_2SO_4, -22 eu for other boronic acids in 30% H_2SO_4. There are not enough data to yield a definite conclusion, and the ΔS^* values do not change much in the high acidity range, ΔH^* being the source of the faster rates. Further complicating any interpretation is the possibility that the species $ArB(OH)_2$ may be in equilibrium with mixed acid anhydrides such as $ArB(OH)OSO_3H$ in concentrated acid.

Kuivila *et al.* have also studied the base-catalyzed and "uncatalyzed" protodeboronation of areneboronic acids.[125] The "uncatalyzed" reaction appears to involve attack of H_3O^+ on $ArB(OH)_3^-$ and the base-catalyzed reaction to involve proton transfer from water to $ArB(OH)_3^-$. General base catalysis was not found, suggesting that $ArB(OH)_3^-$ present in equilibrium with $ArB(OH)_2$ and OH^- is the reactive species, and other $ArB(OH)_2X^-$ species do not compete effectively. A Hammett plot for the base-catalyzed reaction works with ordinary σ, not σ^+, with $\rho = -2.32$. Although still an electrophilic aromatic substitution, this reaction must be promoted by a high degree of carbon–boron bond breaking in the transition state (**57**).

$C_6H_5B(OH)_2 + OH^- \rightleftharpoons C_6H_5B(OH)_3^- \xrightarrow{H_2O}$ [H–O···H···C_6H_5···$B(OH)_3$]$^-$ (**57**) $\longrightarrow C_6H_6 + B(OH)_3 + OH^-$

57

Acid cleavage of arylsilicon and other aryl–Group IV compounds has also been studied extensively. The ranges of mechanisms and substituent effects observed closely parallel those of the boron compounds in several respects. In the boron series, the carbon–boron strength and effective electronegativity of the boron both decrease with increasing nucleophilic coordination. With the Group IV metals the carbon–metal bond strength and the electronegativity both decrease as the atomic number increases, yielding similar effects on the characteristics of the reaction. Nucleophilic coordination also must affect the group IV arylmetal compounds in a secondary way, but the strength of the coordination is probably less than with boron to begin with, changes in coordination on moving down the periodic table will merely parallel other trends, and there is no positive information regarding the solvation of the R_3M^+ leaving groups. Where *M* is silicon, one tightly bound water molecule to form $R_3SiOH_2{}^+$ seems likely, and where M is tin or lead there may be two or three water molecules but weaker coordination.

Eaborn and co-workers have carried out the most extensive studies of protodemetallation of Group IV aryls. These include kinetics and modified Hammett correlations for a series of $ArSiMe_3$ in methanol with aqueous perchloric acid at 51.7°C,[126] $ArGeEt_3$ in methanol (5 volumes) and concentrated aqueous perchloric acid (2 volumes) at 50°C,[127] $ArSnR_3$ (R = Me and R = cyclohexyl) in ethanol (50 volumes) and perchloric acid (2 volumes) at 50°C,[128] and $ArPbR_3$ (R = cyclohexyl) in ethanol (10 volumes) and perchloric acid (1 volume) at 25°C.[129] Relative rates were determined for a series of $ArMEt_3$, where M is Si, Ge, Sn, or Pb, in 5:1 ethanol–perchloric acid at 50°C.[130] Isotope effects k_H/k_D were measured in dioxane containing HCl and 25% H_2O or D_2O with the substrates *p*-$MeOC_6H_4SiMe_3$, *p*-$MeOC_6H_4GeEt_3$, $PhSnEt_3$, and $PhPbEt_3$.[131] All these minor variations in the conditions are presumed to have merely minor effects on the numbers obtained, and were obviously done for experimental convenience or necessity. The results are summarized in Table 3-5.

Neither σ nor σ^+ yielded good Hammett correlations with the $ArMR_3$ compounds, but the compromise between the two provided by the Yukawa-Tsuno equation[132] did. For this correlation substituent constants are taken

TABLE 3-5

COMPARISONS OF PROTODEMETALLATIONS OF $ArMR_3$ COMPOUNDS[a]

M	Yukawa-Tsuno ρ	Yukawa-Tsuno r	Relative rate	k_H/k_D
Si	−5.0	0.7	1	1.55
Ge	−4.4	0.6	36	1.73
Sn	−3.8	0.4	3.5×10^5	2.55
Pb	−2.5	0.4	2×10^8	3.05

[a] From R. W. Bott, C. Eaborn, and P. M. Greasley, *J. Chem. Soc.*, 4804 (1964).

as $\sigma + r(\sigma^+ - \sigma)$, where r is an adjustable parameter which may range from 0 (ionization of benzoic acids) to 1 (ionization of $ArCMe_2Cl$) and even somewhat higher as the degree of concentration of the positive charge in the π system of the benzene ring increases. As may be seen from Table 3-5, the value of r is relatively large for $ArSiMe_3$, where formation of the C—H bond largely precedes breaking of the C—Si bond, and r decreases with increasing atomic number, since breaking the C—Pb bond is concerted with formation of the C—H bond and little positive charge goes to the aromatic ring. Though the boron compounds were not explicitly correlated on this basis, the change in r between $ArB(OH)_2$ and $ArB(OH)_3^-$ would parallel that between $ArSiR_3$ and $ArPbR_3$, the change in Hammett ρ values is similar, and the changes in rates of reaction with H^+ are within a few orders of magnitude.

The change in the isotope effects in Table 3-5 is thought to reflect the position of the transition state along the reaction coordinate with respect to proton transfer. The relatively low k_H/k_D values for silicon and germanium suggest nearly completed proton transfer from water to the aromatic ring by the time the transition state is reached. With tin and lead the higher isotope effects imply that the transition state occurs earlier along the proton transfer coordinate. However, from the other evidence it is later with respect to the carbon–metal bond breaking coordinate, that is, the reaction appears to be a one-step process with tin or lead (transition state **58**) but a two-step process with silicon (transition state **59**, intermediate **60**). The similarity between

$PhPbR_3 + H_3O^+ \longrightarrow$ [$H_2O \cdots H^+ \cdots PbR_3$ on ring] **58** $\longrightarrow$ PhH $+ PbR_3^+ + H_2O$

$PhSiR_3 + H_3O^+ \longrightarrow$ [$H_2O \cdots H \cdots SiR_3$ on ring, +] **59** $\longrightarrow$ [H, SiR_3 on ring, +] **60** $\longrightarrow PhH + SiR_3^+ + H_2O$

the PbR_3^+ and $B(OH)_3$ leaving groups may be noted in proposed transition states **58** and **57**, between SiR_3^+ and $B(OH)_2^+$ in **59** and **53**. The sort of reaction coordinate curves consistent with the mechanistic evidence are qualitatively illustrated in Fig. 3-4.

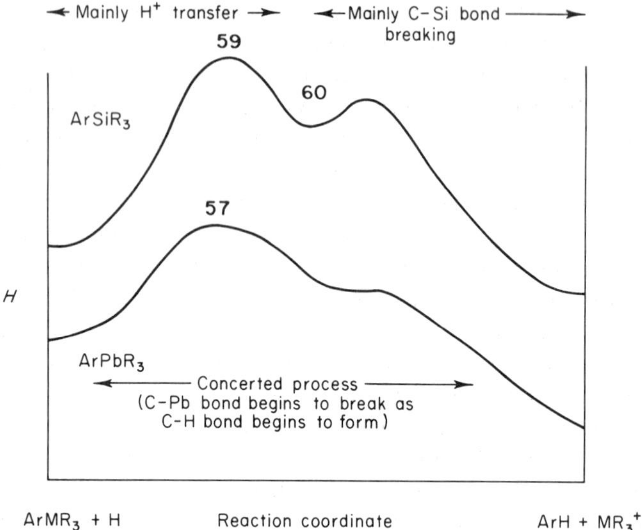

Fig. 3-4. Qualitative plots of H vs. reaction coordinate for the reactions of H_3O^+ with $ArSiR_3$ and $ArPbR_3$.

Cleavages of $ArSnMe_3$ by 1 M sodium hydroxide in MeOH–MeOD show isotope effects k_H/k_D in the range 3.4–4.6, clearly indicating that proton transfer is involved in the transition state and free Ar^- is not an intermediate.[133] Cleavages of the benzyl compounds $ArCH_2SiMe_3$ showed smaller isotope effects, 1.4–1.6, and $ArCH_2SnMe_3$ yielded k_H/k_D values of 2.0–2.8.

Nasielski and co-workers have reported $\rho = -2.24$ for the cleavage of $ArSnMe_3$ by acetic acid.[134] Chloride ion is more effective than perchlorate ion in accelerating this reaction.

The reaction of $ArSnMe_3$ with SO_2 to form $ArSO_2SnMe_3$ seems to have characteristics in common with protodestannation reactions and is included here even though it really belongs in a class by itself. Fong and Kitching found a good correlation with σ^+ which yielded $\rho = -1.87$.[135] A four-center transition state involving electrophilic attack of the sulfur on the benzene ring and nucleophilic attack of the oxygen on the tin seems likely.

Alkylboron compounds can also be cleaved by acids, but it turns out that the tendency of boron to complex with a nucleophile before it is displaced from carbon is the dominant influence on the rates. Strong acids do not react readily with trialkylboranes, carboxylic acids do react at $\sim$100°C. Toporcer, Dessy, and Green found that the rate of cleavage of triethylborane increases as the acid strength of RCO_2H decreases, that is, as the nucleophilic strength of RCO_2H increases.[136] The Taft ρ^* correlating variation of R in RCO_2H is -0.94 for this reaction. The deuterium isotope effect k_H/k_D is 3.3, indicating proton transfer in the transition state (**61**). Protodeboronation is a useful synthetic tool, since it can be used following hydroboration to achieve hydrogenation of olefins.[137] All three alkyl groups can be cleaved from the boron in refluxing propionic acid. The reaction has been shown to proceed with retention at carbon.

$$Et_3B + RCO_2H \rightleftharpoons Et_3B \leftarrow O{=}C(R){-}O{-}H \longrightarrow$$

$$[\text{61}] \longrightarrow H_3CCH_3 + Et_2B{-}O{-}C({=}O){-}R$$

61

Davies and Roberts have reported that deutero-deboronation of optically active 1-phenylethaneboronic acid in D_2O–NaOD proceeds with 54% net inversion.[138] A transition state having a considerable amount of carbanion character (**62**) seems reasonable for this reaction. The proposed mechanism resembles the inversions found by Cram and co-workers in carbon–carbon bond cleavages.[8]

$$H{-}C(Ph)(Me){-}B(OD)_3^- \xrightarrow{D_2O} \left[DO{\cdots}D{\cdots}C(H)(Ph)(Me){\cdots}B(OD)_3\right]^- \longrightarrow DO^- + D{-}C(Ph)(Me){-}H + B(OD)_3$$

62

Alkylsilicon compounds are generally inert to protolysis, at least under convenient experimental conditions. Of the other Group IV metal alkyls, tetraalkyllead compounds have proved to cleave readily near room temperature, and the reactivity series $Me_4Pb > Et_4Pb > Pr_4Pb \simeq Bu_4Pb$ in acetic acid with perchloric acid has been reported.[139] Tetraalkyltins are cleaved by HCl in benzene with the relative rates Me_4Sn, 1.0; Et_4Sn, 7.5; Pr_4Sn, 3; iso-Pr_4Sn, 3.[140]

D. Antimony as Electrofuge

McEwen and co-workers have studied the cleavage of tetraphenylstibonium methoxide by methanolic sodium methoxide to yield benzene and triphenylstibonium dimethoxide.[141] Rough kinetic data show that the rate is dependent on the methoxide concentration. A reasonable mechanism would involve addition of a methoxide ion to the antimony followed by proton transfer from the solvent methanol to the benzene ring in the transition state **(63)**. Although McEwen and co-workers suggested that the rate-determining step might involve cleavage to PhNa, the mechanism written here seems more reasonable, since MeOH is a much stronger acid than Na^+. The similarity of the proposed transition state **(63)** to that for alkaline cleavage of arylboron

$$\underset{\text{(crystalline covalent compound)}}{Ph_4SbOMe} + MeO^- \rightleftharpoons Ph_4Sb(OMe)_2 \xrightarrow{MeOH}$$

MeO···H···C_6H_5···$Ph_3Sb(OMe)_2$ $\longrightarrow$ $MeO^- + PhH + Ph_3Sb(OMe)_2$

63

compounds **(57)** might be noted, as well as the consistency with the k_H/k_D found in the base-catalyzed cleavage of $ArSnMe_3$ (Section IV,C). McEwen and co-workers also found that the relative rates of cleavage of aryl groups from $XC_6H_4Ph_3SbOMe$ are *m*- or *p*-NO_2 > *p*-Br > H > *p*-Me > *p*-MeO, the order typical of electrophilic displacements. No biphenyls were found, and it appears that free radicals are not involved in the mechanism.

E. Electropositive Metal Electrofuges

Protodemetallation of the reactive organometallic compounds is relatively difficult to study, and data are necessarily incomplete. Pocker and Exner have provided one of the more definitive studies on protodemetallation of PhM and $PhCH_2M$ compounds, where M is magnesium, sodium, or lithium.[142] The k_H/k_D values were determined by a competitive method with various partially deuterated proton sources in ether or tetrahydrofuran. With the oxygen-bound proton sources PhOH, H_2O, MeOH, and *t*-BuOH the k_H/k_D values are small, in the range 1.0–1.5, suggesting that proton transfer occurs after the rate-determining step, which appears to be displacement of one oxygen ligand by another to form a short-lived intermediate **(64)**.

$$Ph{-}Mg(\leftarrow OEt_2)_2{-}Br + H_2O \longrightarrow Ph{-}Mg(\leftarrow OEt_2)(\leftarrow OH_2){-}Br + Et_2O \longrightarrow PhH + HO{-}Mg(\leftarrow OEt_2){-}Br$$

64

In contrast, isotope effects are in the range 1.4–10.8 with the carbon-bound proton sources phenylacetylene, indene, fluorene, *p*-biphenylyldiphenylmethane, and triphenylmethane.[142] The maximum isotope effect should occur when the proton is just half-way transferred in the transition state, and acids stronger or weaker than optimum for a given organometallic substrate will place the transition state too early or too late along the reaction coordinate, yielding lower k_H/k_D values. Isotope effects observed in THF are summarized in Table 3-6.

TABLE 3-6

k_H/k_D VALUES FOR ORGANOMETALLIC COMPOUNDS AND CARBON ACIDS AT 15°C[a]

	Acid			
Organometallic	PhC≡CH	Indene	Fluorene	Ph_3CH
$PhCH_2MgCl$	3.4 ± 0.2	5.4	8.2	—
PhMgBr	3.0	—	—	—
$PhCH_2Li$	2.0	3.4	4.0	10.8
PhLi	1.4	6.5	8.1	4.6
$PhCH_2Na$	1.4	2.9	3.6	8.4

[a] From Y. Pocker and J. H. Exner, *J. Amer. Chem. Soc.* **91**, 6764 (1968).

In view of the range of isotope effects observed with the carbon acids, Pocker's suggested explanation for the low isotope effects found with oxygen acids may be oversimplified. Rates of reaction with the oxygen acids are generally very fast, meaning that there is inherently little energy barrier and little chance for discrimination between isotopes. The transition state might merely come very early along the reaction coordinate, before there is much C—H bond formation or much C—O bond breaking. The possibility of some sort of oxygen–metal coordination being present in the transition state is, of course, reasonable, but there is no reason to believe that replacement of one oxygen ligand by another would involve more of an energy barrier than the replacement of the metal by a proton.

The rates of reaction with the carbon acids were in easily measurable ranges. For PhMgCl with PhC≡CH in ether the second-order k is 1.2×10^{-4} at 37°C and k_H/k_D is 5.2. Under similar conditions, $PhCH_2MgCl$ yields $k = 1.2 \times 10^{-5}$ and $k_H/k_D = 6.2$. The more selective Grignard reagents generally yield higher isotope effects than the reactive lithium or sodium compounds.

With PhLi the maximum k_H/k_D is about 8 or 9 (Table 3-6) and occurs with acids in the pK range 22–26, indene (pK 22) providing the highest point actually observed. $PhCH_2Li$ and $PhCH_2Na$ do not have maximum isotope effects with acids of pK below 32 (Ph_3CH). Since the pK of $PhCH_3$ is about 35, this seems reasonable. However, the maximum found for PhLi occurs with a proton donor about 10 pK units stronger than benzene (pK 36–37), which does not coincide with the notion that the maximum isotope effect should occur when the proton is suspended half-way between two acids of equal strength in the transition state. There may be some difference between the mechanisms for the benzyl and the phenyl compounds causing this effect. For example, if the benzyl metal compounds tended to exchange the metal for a proton from a benzylic proton donor, the transition state would be fairly symmetrical and the isotope effect maximized by equal carbanion base strengths. The phenylmetal compound might tend to transfer the metal cation to the ether solvent rather than to the benzylic carbanion in the transition state, providing an asymmetric transition state in which the effective basicity of the metal cation-solvated phenyl anion is comparable to that of the unsolvated benzylic anion. Pocker has pointed out that there is delocalization of charge in the benzylic anions, not in the phenyl, which could produce the effect just described, since the benzyl anion would not solvate as effectively. Or one might merely remark that acid strengths are not independent of the medium.

Dessy and co-workers have studied the rates of reaction of Grignard reagents with the proton donor 1-hexyne.[143] For a series of XC_6H_4MgBr where X is *p*-Me, H, *p*-Cl, *m*-Cl, or *m*-CF_3 the Hammett $\rho = -2.5$. With ethylmagnesium bromide and butyne, k_H/k_D was found to be about 4.[144] P. West *et al.* have found that the rates of metalation of triphenylmethane by organolithium reagents in tetrahydrofuran show dependence on the degree of aggregation of the lithium reagent.[145] The kinetic order in "RLi" is about 1 for benzyllithium and allyllithium, which are apparently monomeric in THF. The order falls to 0.64 for phenyllithium (largely dimeric in ethereal solvents) and to around 0.25 for *n*-butyllithium, methyllithium, and vinyllithium (tetrameric). The active metalating agent appears to be organolithium monomer in each case. The reaction is first order in triphenylmethane if RLi is in excess, but if the Ph_3CH is in excess the 500 nm absorption of Ph_3C^- grows part way and then slows down drastically, as if some $Ph_2CLiC_6H_4Li$ or other extraneous species is formed and then metalates Ph_3CH much more slowly than does RLi.

Metalation of toluene with alkyllithium reagents does not proceed readily unless a tertiary amine is added to complex with the lithium cation and activate the reagent.[146,147] R. West and Jones have found that 1,2-bis-(dimethylamino)ethane, $Me_2NCH_2CH_2NMe_2$, and *n*-butyllithium lead to di- and trilithiation of toluene.[148] The second and third lithium atoms tend to attack the α and *para* positions. These dilithium and trilithium compounds were reacted with chlorotrimethylsilane to form trimethylsilyl derivatives.

Mechanistic details of the proton transfer in this lithiation are unknown, but it must come about as close to a free carbanion mechanism as is possible in hydrocarbons of this low acidity. Even so, the products would appear to be covalent organolithium compounds rather than ionic salts, since it seems unlikely that an alkyllithium would lithiate a benzyl carbanion to form a dianion.

Dialkylzincs are cleaved at convenient rates by very weak proton acids such as primary amines. Abraham and Hill have carried out a study of these reactions.[18] It was found that the cleavage of dipropylzinc by *p*-toluidine is ten times faster than by cyclohexylamine, indicating that acid strength outweighs nucleophilicity in determining the relative reactivities. A four-center transition state (**65**) was postulated. The probable similarity of this reaction

$$Pr_2Zn + ArNH_2 \longrightarrow \begin{array}{c} Pr{-}Zn \cdots\cdots CH_2CH_2CH_3 \\ \uparrow \qquad\quad \searrow \\ Ar{-}\underset{H}{N} \cdots\cdots H \end{array} \longrightarrow Pr{-}Zn{-}NHAr + CH_3CH_2CH_3$$

65

to those of Grignard reagents studied by Pocker and Exner[142] might be noted. It seems probable that nitrogen–zinc complexing precedes proton transfer, but that proton transfer is rate-determining in this case.

V. Group IV and V Electrophiles with Grignard and Related Reagents

A. Introduction

Reactions of Grignard reagents with carbonyl compounds are included here, as well as related reactions involving phosphinate esters. Scattered studies of other combinations of carbon or silicon electrophiles with reactive organometallic compounds have also been reported. The difficulties of working with the more reactive organometallic compounds have kept the number of publications in this field rather low, and the available coverage in the literature is spotty rather than comprehensive.

The structure of the Grignard reagent, essential to any understanding of its reaction mechanisms, has been discussed in Chapter 1, Section II,A, Fig. 1-1, and Chapter 2, Section IV,B.

B. Grignard Reagents and Ketones

Kinetic studies on the reaction of Grignard reagents with ketones had to await modern flow-system techniques. However, Swain was able to get good kinetics on the BuMgBr–PhCN reaction in 1947, analyzing for unreacted

BuMgBr by converting it to butane.[149] The reaction is first order in each reactant, $k = 3.7 \times 10^{-4}$ liter/mole second at 25°C, E_a about 12 kcal/mole.

Early kinetics on the reaction of methylmagnesium bromide with benzophenone leave quite a bit to be desired. Anteunis[150] and Bikales and Becker[151] did get similar activation energies, 11 kcal/mole, corresponding to $\Delta S^* = -25$ eu if the straightforward second-order rate law[151] is accepted, but the data could be interpreted as second order in Grignard and first order in ketone if one were so inclined.[150]

Holm used a flow reactor and followed the reaction of acetone with butylmagnesium halides by time–temperature plots, the reaction being exothermic.[152,153] Evidence was found for a 1:1 complex of RMgX with $R_2C{=}O$, which decomposes by a first-order rate law.[152] Equilibrium favored the complex over free RMgX and R_2CO under the conditions used. The rates were found to be sensitive to the halide in ether, Cl > Br > I, but in tetrahydrofuran there was little difference between the three halides and dibutylmagnesium.[153]

House and Traficante showed that 3-pentanone reacts faster with dialkylmagnesium than with alkylmagnesium bromide, and that added magnesium bromide slows the reaction of dialkylmagnesium in ether.[154] Even though magnesium bromide slows the reaction, it improves the yield of tertiary alcohol because it suppresses a side reaction, the reaction of the initial product, $RMgOCR_3$ or $BrMgOCR_3$, with remaining ketone to form enolate and RH. Alkylmagnesium alkoxides are much less active than Grignard reagents toward carbonyl groups.

In more recent work, House and Oliver have shown that the magnesium can be pentacoordinate in the transition state for the reaction of dimethylmagnesium with benzophenone.[155] The rate law is $k[Me_2Mg][Ph_2CO]$. Addition of 1,2-bis(dimethylamino)ethane, $Me_2NCH_2CH_2NMe_2$, in a 1:1 ratio with the dimethylmagnesium reduces the rate in ether by a factor of about 4, and additional diamine has little effect. 1-Dimethylamino-2-methoxyethane or 1,2-dimethoxyethane increases the rate by a factor of 2 or 3, and again little effect is produced by more than a 1:1 mole ratio. It is apparent that these bidentate ligands chelate with the magnesium, that the chelates do not dissociate extensively, and that the chelated form of the Grignard is the reacting species under these conditions. There is other evidence that the magnesium complexes with the carbonyl oxygen in the transition state (see the preceding and following paragraphs), and the sum of the ligands on the magnesium then comes out to be five. Stable pentacoordinate and hexacoordinate magnesium compounds are known (see Chapter 2, Section IV,B) and this is not surprising. It may also be noted that formation of a magnesium–oxygen bond in the product probably has much to do with the exothermic character of the reaction, and that magnesium–oxygen bond formation would therefore be expected to precede magnesium–carbon bond breaking (see Chapter 1, Section II,B; Chapter 2, Section IV,B; Chapter 3, Section II,D, Fig. 3-2).

Further evidence for solvation of the magnesium in the transition state by (at least one) ether is provided by the observation that an optically active ether induces asymmetry in the product from dimethylmagnesium and benzaldehyde.[156]

S. G. Smith and co-workers have obtained the first accurate and comprehensive kinetic data on the reactions of alkylmagnesium compounds with ketones.[157–159] A stopped-flow technique with rapid-scan ultraviolet spectrophotometry was used. The 4-(methylmercapto)phenylcarbonyl group was found to have good ultraviolet absorption properties. The ketones complex very rapidly and reversibly with the Grignard reagent, and the complexes absorb at a distinctly longer wavelength than the parent ketones. The equilibrium constants for the formation of the complexes were in an easily measurable range. The Grignard reagent was used in large excess, so that the fraction of ketone converted to complex (usually in the range 10–90%) remained constant throughout a given kinetic run as a consequence of the equilibrium relationship, $k[RMgX] = [complex]/[R_2CO]$. Reactions were pseudo-first-order in ketone out to 90% reaction.

The best kinetic data were obtained with dimethylmagnesium and 4-(methylmercapto)acetophenone.[158] The equilibrium constant $K = 6.2$ liters/mole for complex formation and the first-order rate constant, $k = 19.5\ sec^{-1}$, for rearrangement of the complex in diethyl ether at 25°C give satisfactory correlation of all the data. The concentration of $MgMe_2$ was varied from 0.01 to 0.6 M and the concentration of $MeSC_6H_4COCH_3$ from 5×10^{-4} to 10^{-2} M. The straightforward interpretation is that the $MgMe_2$–$ArCOCH_3$ complex **66** rearranges to the tertiary alcoholate product by way of transition state **67**. A small amount of proton transfer to yield enolate and methane occurs as a side reaction (4%).

Transition state structure **67** includes details considerably beyond Smith's actual conclusions, as any structural formula must. The two molecules of ether are postulated from House's work[155] and the known structures of Grignard reagents (Chapter 2, Section IV,B). The remainder of the stoichiometry of **67** is given by the kinetics, but the assumption that intermediate **66** rearranges directly to the transition state (**67**) cannot be proved, since any other mode of combination of $MgMe_2$ with $ArCOCH_3$ would yield the same kinetics. However, the ultimate need to form a magnesium–oxygen bond in the product and its probable partial formation by the time the transition state is reached should be kept in mind. If the magnesium is indeed

$$(Et_2O)_2Mg(CH_3)_2 + ArC(=O)CH_3 \overset{K}{\rightleftharpoons} (Et_2O)_2Mg(CH_3)_2 \leftarrow O{=}C(Ar)CH_3 \quad \mathbf{66}$$

$$\xrightarrow{k} \text{transition state } \mathbf{67} \longrightarrow (Et_2O)_2Mg(CH_3)\text{—O—}C(Ar)(CH_3)_2$$

$$\xrightarrow{k} (Et_2O)_2Mg(CH_3)\text{—O—}C(Ar){=}CH_2$$

pentacoordinate in the transition state, a choice of equatorial (tightly bound) and axial (more loosely bound) groups must be made in the drawing, and the choice is safe from being challenged by any imaginable experimental test.

It might be noted that the following factors would favor the postulated four-membered ring. Pentacoordinate magnesium would allow or even favor a 90° C—Mg—O bond angle. The Mg—C—C angle at the migrating CH_3 group should be less than 90°, since only one sp^3 orbital is used for the two-electron three-center bond. The C—C—O angle at the carbonyl carbon could easily be about 90°, since the initial interaction with the attacking methyl group utilizes the carbonyl carbon p orbital perpendicular to the main plane of the molecule. The cross-ring C---Mg attraction would be weak and have negligible effect. The only bond angle which is too large is the Mg—O—C angle, which might prefer 105°–120°. Even that need not be bent quite as far out of shape as it might seem, because the Mg—O and O—C bonds are relatively short (tending to widen the angle at O) and the Mg—CH_3 and CH_3—C distances are long (allowing the desired acute angle at CH_3) in the transition state (**67**). Thus, it may be concluded that there is not necessarily much angle strain in the four-membered ring of **67**. Organic chemists thinking along traditional lines have sometimes proposed involvement of two molecules of Grignard reagent in six-membered-ring transition states,[160] but in the $MgMe_2$–$MeSC_6H_4COCH_3$ system such a mechanism is (a) forbidden by the experimental evidence and (b) not favored by theoretical principles.

The experimental results are more complicated with methylmagnesium bromide. The ketone substrates used included 4-(methylmercapto)phenyl 2,4-dimethylphenyl ketone,[157] 4-(methylmercapto)acetophenone,[159] and benzophenone,[159] which all gave qualitatively similar results. Nuisance number one is that methylmagnesium bromide as normally prepared contains an unknown reactive impurity which causes an increase in the apparent rate constants at low ketone concentrations. One method of combatting this problem was to add acetone together with the aromatic ketone. The acetone reacts preferentially with the impurity and does not add any interfering ultraviolet absorption, and the methylmagnesium bromide is in excess so that its concentration is negligibly affected. A better solution was to recrystallize the methylmagnesium bromide, which removed the impurity and led to good kinetic data for 0.001–0.01 *M* $MeSC_6H_4COCH_3$. Benzophenone always misbehaved at low concentrations and yielded evidence of free radicals and ketyl coupling products.

The equilibrium constants *K* for the complexing of MeMgBr with the ketones were generally self-consistent over a wide range of concentrations. For example, $MeSC_6H_4COCH_3$ has $\lambda_{max} = 300$ nm, its complex with MeMgBr has $\lambda_{max} = 336$ nm, $\epsilon = 1.5 \times 10^4$, and the complex is formed within the 0.03 second time of mixing, with $K = 15.1 \pm 0.6$ liter/mole for MeMgBr in the range 0.04–0.4 *M* in ether at 25°C. The total of ketone plus complex was kept roughly constant, and the free ketone varied by a factor of 6.

The reaction is cleanly first order in MeMgBr at low concentrations. The actual observation is that the pseudo-first-order *k*'s increase linearly with MeMgBr, since the MeMgBr is in excess and the fraction of ketone converted to complex is small. Above about 0.1 *M* MeMgBr, where over half of the $MeSC_6H_4COCH_3$ is complexed, the pseudo-first-order *k* does not level off toward an upper limit (ketone 100% complexed) as fast as it should. By 0.5 *M* MeMgBr, *k* is about twice the "expected" value extrapolated from *K* and the value of *k* at 0.1 *M*. The deviation is linear in MeMgBr concentration.

Smith pointed out that no definite conclusion can be drawn from the deviations from first-order dependence on MeMgBr.[159] The highly polar MeMgBr may produce a large medium effect at 0.1 *M* in ether, or there might be a kinetic term second order in MeMgBr or, if one believes in the dimerization of MeMgBr (Chapter 2, Section IV,B), first order in $(MeMgBr)_2$. If the reaction is purely first order in MeMgBr, *k* is 0.70 sec^{-1} at 25°C, and if there is a second-order term, the first-order *k* may be as low as 0.14 sec^{-1}.

Ashby and co-workers once concluded that the reaction of MeMgBr with benzophenone has a prominent kinetic term second order in MeMgBr.[160,161] However, Billet and Smith showed that the observed *k* is sensitive to the ratio of MeMgBr to Ph_2CO and that free-radical side reactions are prominent at the concentrations of Ph_2CO used.[159] In summary, there is no strong evidence for any term second order in MeMgBr.

The measurements made by Smith and co-workers were confined to concentration ranges where MeMgBr was in large excess.[157–159] Ashby and co-workers have extended the measurements to solutions containing excess 2-methylbenzophenone and have shown that reaction by way of $MgMe_2$ competes appreciably with the direct reaction of MeMgBr.[162] The kinetic term in $MgMe_2$ cannot be detected from data involving MeMgBr alone because the Schlenk equilibrium [Eq. (3-7)] yields the same kinetic order in $MgMe_2$ as in MeMgBr [Eqs. (3-8) to (3-10)].

$$2\ \text{MeMgBr} \xrightleftharpoons{K_s} \text{MgMe}_2 + \text{MgBr}_2 \tag{3-7}$$

$$K_s = \frac{[\text{MgMe}_2][\text{MgBr}_2]}{[\text{MeMgBr}]^2} \tag{3-8}$$

$$[\text{MgMe}_2] = [\text{MgBr}_2] \tag{3-9}$$

$$[\text{MgMe}_2] = K_s^{1/2}[\text{MeMgBr}] \tag{3-10}$$

Where there is excess ketone, Eq. (3-9) is modified by a proportionality constant. Each equilibrium of the form $MgX_2 + R_2CO = R_2CO \cdot MgX_2$ maintains a constant ratio of complexed to uncomplexed MgX_2 because $[R_2CO \cdot MgX_2]/[MgX_2] = K[R_2CO]$ and the value of $[R_2CO]$ does not change appreciably in a given kinetic run. Thus, $[MgMe_2]$ remains a constant fraction of [MeMgBr].

Addition of $MgBr_2$ suppresses the disproportionation of MeMgBr and lowers the rate of reaction with 2-methylbenzophenone.[162] The amount of reduction of the initial rate reaches about 30–35%, where it levels off and is unaffected by further increase of the $MgBr_2$ concentration. Without added $MgBr_2$ the pseudo-first-order plots are curved because the product magnesium alcoholate, $BrMgOCMeAr_2$, complexes with MeMgBr and reduces its effective concentration, but in the presence of excess $MgBr_2$ the product complexes with $MgBr_2$ instead and the kinetic plots are linear.

Ashby and co-workers estimated the Schlenk constant K_s to be 0.0022 from variations in the zero-time ultraviolet absorption with $MgBr_2$ concentration.[162] They found the formation constant K for the MeMgBr–2-methylbenzophenone complex to be ~1.35 and K for the $MgBr_2$ complex to be ~4. They computed their rate constants as if K for the $MgMe_2$ complex were small, but a better estimate from the relationships in Smith's results[157–159] would be about 0.6. Using this value, the author has recalculated Ashby's data and found a modest improvement in the constancy of the apparent rate constants. The average value of k for reaction of the $Ar_2CO \cdot MeMgBr$ complex is 0.0230 sec^{-1}, based on the six most self-consistent data points, and the k for the $Ar_2CO \cdot MgMe_2$ complex is 0.345 sec^{-1}.

Ashby and co-workers have also found that the reaction of $MgMe_2$ with 2-methylbenzophenone is first order in each reactant, $k = 0.27$ liter/mole second, and have postulated a transition state which is analogous to **67** when revised to the same symbolism.[163,164] The recalculated first-order k for reaction of the $Ar_2CO \cdot MgMe_2$ complex, assuming K for its formation is 0.6, is 0.45 sec^{-1}, which is in reasonable agreement with the value 0.345 sec^{-1} estimated from the data on the Ar_2CO–MeMgBr system. It should be noted that given Ashby's value of 0.0022 for the Schlenk constant K_s of MeMgBr, the data of Smith and co-workers[157–159] would also require that a substantial fraction of the reaction of MeMgBr proceeds by way of $MgMe_2$, but in accord with the relationship of Eq. (3-10), the kinetic data do not reveal this extra pathway.

Smith and co-workers have studied the reaction of methyllithium with a ketone, 2,4-dimethyl-4′-methylmercaptobenzophenone.[165] In contrast to the magnesium reagents, methyllithium in ether does not perturb the ultraviolet spectrum of the ketone, and there is thus no evidence that it forms a complex, though it does seem likely that a short-lived complex is an intermediate. The reaction is one-fourth order in methyllithium, indicating that dissociation of the tetramer (Chapter 1, Section III,F and Chapter 2, Section V,A) to yield a small equilibrium concentration of monomer (either free or complexed with the ketone) occurs prior to the rate-determining step. The reaction is also first order in ketone. At 25°C, $k = 200 \pm 7$ $(\text{liters/mole})^{1/4}$ second^{-1}. The transition state for the rate-determining step probably resembles those in the reactions of magnesium reagents.

The foregoing mechanisms require retention of the configuration of the carbon in the Grignard or lithium reagent. Retention has been proved in the reactions of carbon dioxide with 1-methyl-2,2-diphenylcyclopropylmagnesium bromide[166] and with *endo*-norbornylmagnesium bromide[68] (preparation: Section II,E). Special methods were used to prepare these Grignard reagents, since reactions of alkyl halides with magnesium result in racemization (Chapter 8, Section IV). Lithium reagents having some optical activity can be prepared at low temperatures in pentane. Retention has been demonstrated for the reactions of carbon dioxide with 1-methyl-2,2-diphenylcyclopropyllithium,[100] norbornyllithium,[97] menthyllithium,[98] and 4-*t*-butylcyclohexyllithium.[98]

The transition state model **67** also appears to be compatible with the steric model used by Karabatsos for calculating diastereoisomer ratios to within 0.2 kcal/mole for the reaction of Grignard reagents with asymmetric ketones.[167]

C. A Phosphinate Ester and PhMgBr

The mechanism found with ketones also appears to apply to the reaction of Grignard reagents with phosphoryl compounds. Hays has found that phenylmagnesium bromide forms a 1:1 complex (**68**) with ethyl diphenylphosphinate in tetrahydrofuran.[168] From the kinetics at low concentrations, the estimated K for complex formation is greater than 6000, three powers of 10 larger than typical values with ketones. The complex equilibrates rapidly on the nmr time scale with excess ethyl diphenylphosphinate. The rearrangement of this complex is four or five orders of magnitude slower than the rates found with ketones, and the first-order rearrangement could be followed by gas chromatographic analysis of aliquots out to more than 85% completion, with some slowing at 90–95% completion. Rate constants found were 6.04×10^{-6} sec^{-1} at 14.2°C; 1.51×10^{-5} at 25°C; 8.59×10^{-5} at 50°C; and 2.55×10^{-4} at 67.8°C. From these $\Delta H^* = 12.8$ kcal/mole and $\Delta S^* = -33.2$ eu. The highly negative ΔS^* value is consistent with a cyclic transition state (**69**), which is analogous to that proposed for the ketone reaction.

$Ph_2P(=O)OEt$ + PhMgBr·2THF ⇌ **68** ⟶

69 ⟶ ⟶ ⟶ Ph_3PO + MgBrOEt + 2 THF

Much of the geometry of transition state **69** is necessarily conjectural. The notion that the transferring phenyl group ought to attack the phosphorus in the apical (more weakly bonded) position is supported by the observation by Sommer and Bauman that groups attacking or departing at silicon favor apical positions.[169]

D. Silicon Electrophiles

The major point of interest with silicon electrophiles is that the stereochemistry at the silicon atom can be established. This contrasts with such electrophiles as carbonyl or phosphoryl groups, where the cyclic transition states that have been proposed (**67** or **69**) require front-side attack on the carbon atom but stereochemical proof is inherently impossible. It turns out that the cyclic transition state model extends to some silicon compounds but not all.

Sommer and co-workers have shown that the stereochemistry at silicon depends on the nature of the departing nucleophile.[170] For example, retention at silicon is favored in the reaction of optically active α-naphthylphenylmethylmethoxysilane (R_3SiOMe) with ethylmagnesium bromide, which is compatible with cyclic transition state **70**.

$$R_3Si{-}OMe + EtMgBr \cdot 2Et_2O \longrightarrow \mathbf{70} \longrightarrow R_3SiEt + BrMgOEt + 2\,Et_2O$$

R, R, R on Si; Si—O(Me); O···Mg—Br; Mg···Et···Si; Et_2O, OEt_2 on Mg

70

In contrast, the silicon is mostly inverted when the corresponding chloride (R_3SiCl) reacts with Grignard or lithium reagents[170] or a wide variety of other nucleophiles.[171] This is in accord with the general principle that displacement of good leaving groups from silicon generally proceeds with inversion.[171] Good leaving groups generally include any X^- which has a conjugate acid HX with a pK below 6.[172]

Retention is not universal with methoxysilane reactions, but depends on the particular Grignard or lithium reagent. This nucleophile dependence is typical for poor leaving groups X^-, such as OR^-, SR^-, NR_2^-, or F^-.[172] With R_3SiOMe, retention predominates with alkyllithium, aryllithium, $PhC{\equiv}CLi$, and p-$MeOC_6H_4CH_2Li$, but inversion is favored with the more ionic reagents $PhCH_2Li$, $CH_2{=}CHCH_2Li$, and Ph_2CHLi.[173] The very basic hydride leaving group of R_3SiH led to predominant retention with every lithium reagent tested except Ph_2CHLi, which caused inversion.[173] Fluoride in R_3SiF behaves much like methoxide, simple alkyllithiums yielding retention and benzyllithium, allyllithium, benzylmagnesium bromide, and the like yielding net inversion.[170] It is characteristic of these reactions to proceed with a mixture of inversion and retention at silicon, with the preference approaching 100% either way only if all characteristics are highly favorable in a particular direction.

Glaze and co-workers have demonstrated retention at carbon in the reaction of menthyl- or 4-t-butylcyclohexyllithium with chlorotrimethylsilane, where the silicon cannot be observed directly but, from the foregoing, must be inverted.[98] The simple general picture of a cyclic transition state will not work here. Glaze has suggested that the transition state might be cyclic (**71**) if attack and departure were equatorial with a 120° R—Si—Cl angle,[98] but this requires some rather special and arbitrary motion of the alkyl groups to achieve inversion at silicon. Sommer's view that both attacking and leaving groups are axial and that inversion of silicon resembles the familiar Walden inversion of carbon is supported by much experimental evidence,[174] and an open transition state (**72**) can easily allow inversion at silicon and retention at carbon.

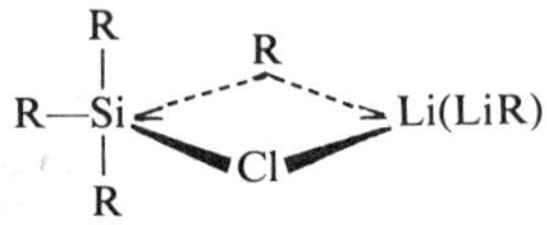

71
(unlikely)

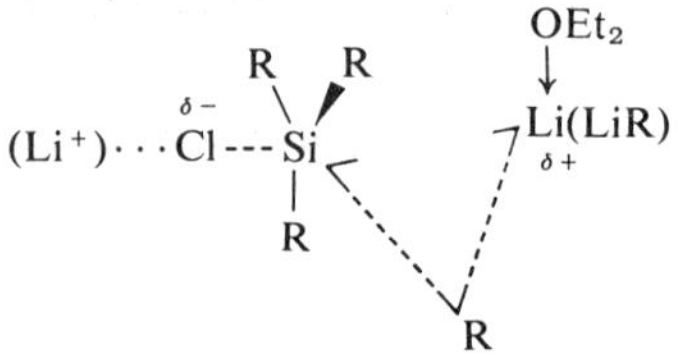

72

The evidence regarding the general mechanisms of retention and inversion at silicon is too extensive to review in detail in a chapter on electrophilic displacement at carbon, most of the nucleophiles used for such studies being more electronegative elements such as oxygen or nitrogen. The essential conclusion may be summarized by saying that retention at silicon implies a cyclic transition state analogous to **70**, inversion an open one analogous to **72**, except that the three-center Li—R—Si bond in **72** would be replaced by a simple linkage if R is replaced by a nitrogen or oxygen anion.

E. Aluminum Compounds as Analogs of Grignard Reagents

Dimeric aluminum alkyls react directly with ketones in benzene. Ashby and co-workers have found that the reaction of Al_2Me_6 with benzophenone is first order in each reactant.[175] The product is a stable oxygen-bridged dialuminum compound.[176,177] Jeffery and Mole have worked with several compounds of this type and shown that the Al—O—Al bridge does not dissociate measurably even at 110°C[178] (see Chapter 2, Section III,G). It seems reasonable to postulate an Al—O—Al bridge in the transition state (**73**) which would lead directly to the bridged product. Several details of the proposed transition state are arbitrary. The symmetry written is not the highest possible, but it appears that there are too many electron pairs to allow everything to be symmetrically bridged without increasing the energy. The six-membered ring transition state (**74**) proposed by Ashby and co-workers seems less compatible with the favored narrow angles of three-center bridge bonds and does not directly yield the right product.

Lundeen and Oehlschlager have found that the reaction of Al_2Et_6 with *cis*-2-butene epoxide inverts the carbon which becomes ethylated.[179] The transition state must obviously be an open one, not a cyclic analog of **73**,

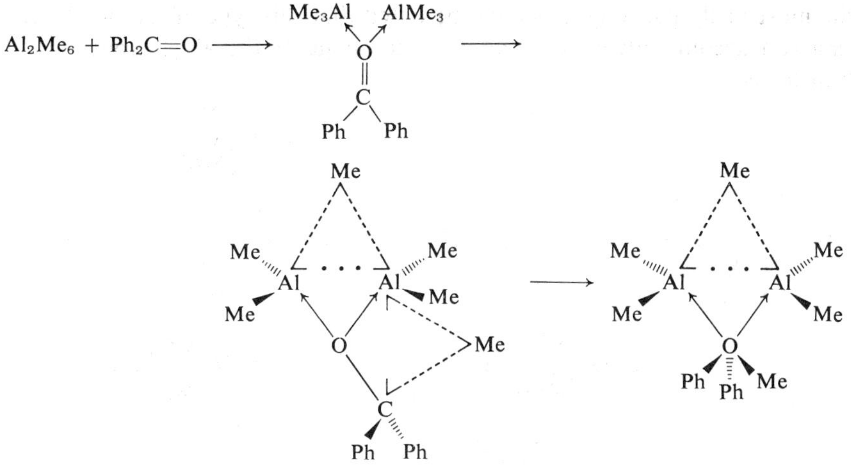

$Al_2Me_6 + Ph_2C{=}O \longrightarrow$ $Me_3Al \rightarrow O(=CPh_2) \leftarrow AlMe_3$ $\longrightarrow$

73 $\longrightarrow$

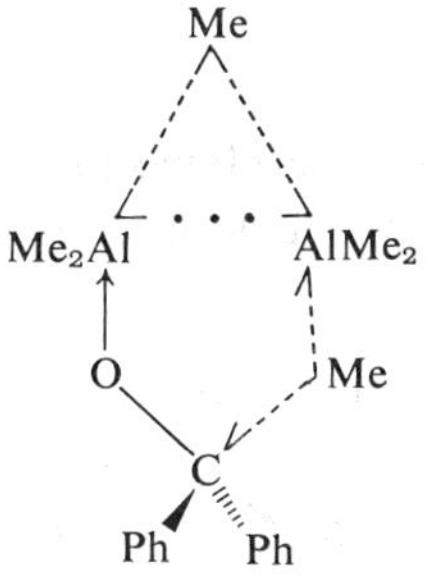

74

(less likely)

though it is the β-carbon instead of the α-carbon that is being attacked and there is no need to expect great similarity.

F. Conjugate Addition to α,β-Unsaturated Ketones

Grignard reagents normally yield mixtures of normal and conjugate addition products with α,β-unsaturated ketones. House and Traficante have shown that the ratio of normal to conjugate addition is unaffected by the R_2Mg concentration in reactions of Me_2Mg, Et_2Mg, or Ph_2Mg with *trans*-3-penten-2-one or *trans*-4-phenyl-3-buten-2-one.[180] Thus, the 1,2- and 1,4-addition both have the same kinetic order. The presence of magnesium bromide increases the proportion of 1,2-addition, but only slightly. There is no direct experimental proof for the popular notion that the transition state for 1,4-addition (**75**) contains a six-membered ring, but with the 1:1 reactant ratio and intermediate complex inferred from Smith's work[158] this seems the most likely pathway. It should be noted that this type of six-membered ring can accommodate the required acute angle at the alkyl group being transferred.

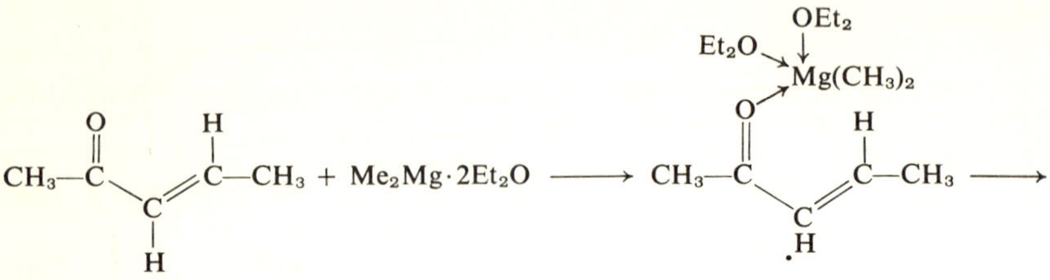

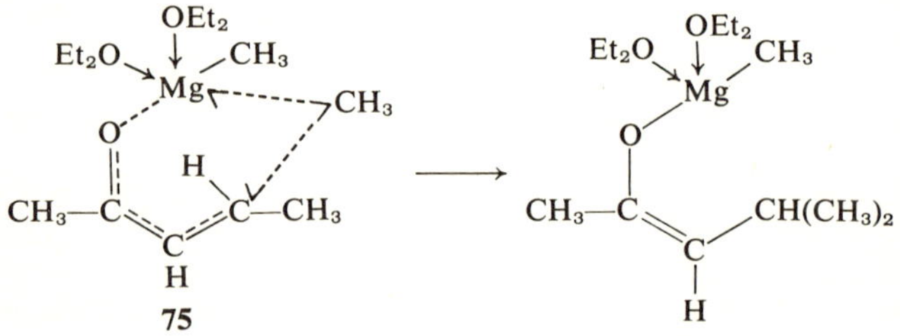

CH_3—C(=O)—CH=CH—CH_3 + $Me_2Mg \cdot 2Et_2O$ ⟶ [Et_2O, OEt_2 → $Mg(CH_3)_2$ ← O=C(CH_3)—CH=CH—CH_3] ⟶

75 ⟶ CH_3—C(O—Mg(CH_3)(OEt_2)(Et_2O))=CH—$CH(CH_3)_2$

A cyclic transition state with 5-methylcyclohexen-3-one (**76**) and dimethylmagnesium would be so strained that it is hard to visualize. It is easy to

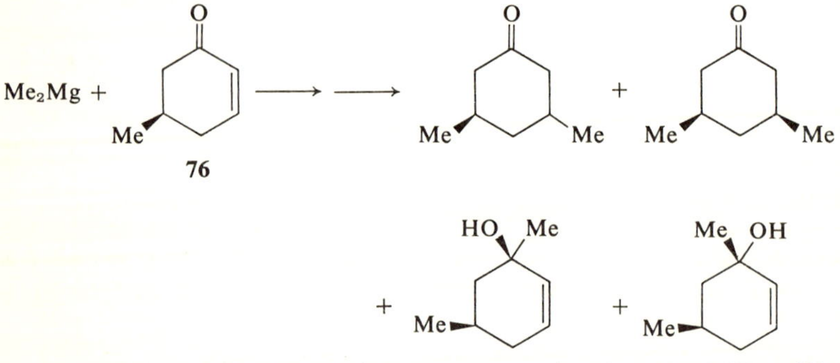

Me_2Mg + **76** ⟶ ⟶ (cis- and trans-3,5-dimethylcyclohexanone) + (1,5-dimethylcyclohex-2-en-1-ol, two isomers: HO, Me; Me, OH)

understand that 97% of the reaction was observed to be 1,2-addition,[181] but hard to see how the reported 3% 1,4-addition can occur, at least by an intramolecular process. The 1,4-addition product is 90% *trans*, the 1,2-product 87% *trans*.[181]

Copper salts have long been known to promote 1,4-addition of Grignard reagents to α,β-unsaturated ketones. The organocopper intermediates involved are outside the scope of this book as strictly defined, but a few comments will be made anyway in Chapter 8, Section III. For now, it will merely be noted that conjugate addition of copper reagents to α,β-unsaturated ketones definitely does not involve any sort of six-center cyclic transition state (analogous to **75**), as shown by stereochemical studies in which the predominant isomer formed was the one which could not be formed in a cyclic transition state.[182]

References

1. S. Winstein, T. G. Traylor, and C. S. Garner, *J. Amer. Chem. Soc.* **77**, 3741 (1955).
2. G. S. Hammond, *J. Amer. Chem. Soc.* **77**, 334 (1955).
3. S. Winstein and T. G. Traylor, *J. Amer. Chem. Soc.* **77**, 2747 (1955).
4. F. R. Jensen, L. D. Whipple, D. K. Wedegaertner, and J. A. Landgrebe, *J. Amer. Chem. Soc.* **82**, 2466 (1960).
5. F. R. Jensen and L. H. Gale, *J. Amer. Chem. Soc.* **82**, 148 (1960).
6. D. S. Matteson and J. O. Waldbillig, *J. Amer. Chem. Soc.* **86**, 3778 (1964).
7. F. R. Jensen and D. D. Davis, *J. Amer. Chem. Soc.* **93**, 4048 (1971); D. D. Davis, Ph.D. Thesis, Univ. of California (Berkeley), 1966.
8. D. J. Cram, "Fundamentals of Carbanion Chemistry," p. 137. Academic Press, New York, 1965.
9. J. N. Roitman and D. J. Cram, *J. Amer. Chem. Soc.* **93**, 2231 (1971).
10. H. B. Charman, E. D. Hughes, C. K. Ingold, and H. C. Volger, *J. Chem. Soc.* 1142 (1961).
11. E. D. Hughes, C. K. Ingold, and R. M. G. Roberts, *J. Chem. Soc.* 3900 (1964).
12. C. R. Hart and C. K. Ingold, *J. Chem. Soc.* 4372 (1964).
13. O. A. Reutov, B. Praisnar, I. P. Beletskaya, and V. I. Sokolov, *Bull. Acad. Sci. USSR, Div. Chem. Sci.* (*English transl.*) 884 (1963).
14. F. R. Jensen and D. Heyman, *J. Amer. Chem. Soc.* **88**, 3438 (1966).
15. F. R. Jensen and B. Rickborn, "Electrophilic Substitution of Organomercurials." McGraw-Hill, New York, 1968. (a) pp. 157–181. (b) p. 116. (c) pp. 75–99. (d) pp. 49–74.
16. I. P. Beletskaya, K. P. Butin, and O. A. Reutov, *Organometal. Chem. Rev. A* **7**, 51 (1971).
17. B. J. Gregory and C. K. Ingold, *J. Chem. Soc. B* 276 (1969).
18. M. H. Abraham and J. A. Hill, *J. Organometal. Chem.* **7**, 23 (1967).
19. C. G. Swain and E. E. Pegues, *J. Amer. Chem. Soc.* **80**, 812 (1958).
20. (a) R. Breslow, J. T. Groves, and G. Ryan, *J. Amer. Chem. Soc.* **89**, 5048 (1967); (b) D. G. Farnum, G. Mehta, and R. G. Silberman, *ibid.* **89**, 5048 (1967).
21. N. L. Allinger, J. C. Tai, and F. T. Wu, *J. Amer. Chem. Soc.* **92**, 579 (1970).
22. A. Gamba, G. Morosi, and M. Simonetta, *Chem. Phys. Lett.* **3**, 20 (1969).
23. G. A. Olah, G. Klopman, and R. H. Schlosberg, *J. Amer. Chem. Soc.* **91**, 3261 (1969). The calculation method used was that of M. J. S. Dewar and G. Klopman, *ibid.* **89**, 3089 (1967).
24. E. M. Kosower, "An Introduction to Physical Organic Chemistry," p. 31. Wiley, New York, 1968.
25. L. C. Allen, "Quantum Theory of Atoms, Molecules, and the Solid State" (P. O. Lowdin, ed.), p. 62. Academic Press, New York, 1966.
26. (a) F. H. Field and M. S. B. Munson, *J. Amer. Chem. Soc.* **87**, 3289 (1965); (b) A. Giardini-Guidoni and L. Friedman, *J. Chem. Phys.* **45**, 837 (1966).
27. G. A. Olah, Y. Halpern, J. Shen, and Y. K. Mo, *J. Amer. Chem. Soc.* **93**, 1251 (1971).
28. G. A. Olah and H. C. Lin, *J. Amer. Chem. Soc.* **93**, 1259 (1971).
29. M. M. Kreevoy and J. E. C. Hutchins, *J. Amer. Chem. Soc.* **91**, 4329 (1969).

30. R. E. Davis, E. Bromels, and C. L. Kibby, *J. Amer. Chem. Soc.* **84**, 885 (1962).
31. R. E. Davis, *J. Amer. Chem. Soc.* **84**, 892 (1962).
32. R. E. Davis and R. E. Kenson, *J. Amer. Chem. Soc.* **87**, 1384 (1967).
33. F. C. Whitmore, "Organic Compounds of Mercury." Van Nostrand Reinhold, Princeton, New Jersey, 1921.
34. O. A. Reutov and I. P. Beletskaya, "Reaction Mechanisms of Organometallic Compounds." North-Holland Publ. Amsterdam, 1968 (translated into English by A. M. A. Mincer).
35. G. F. Wright, *Can. J. Chem.* **30**, 268 (1952).
36. H. B. Charman, E. D. Hughes, and C. K. Ingold, *J. Chem. Soc.* 2523 (1959).
37. H. B. Charman, E. D. Hughes, and C. K. Ingold, *J. Chem. Soc.* 2530 (1959).
38. F. R. Jensen, *J. Amer. Chem. Soc.* **82**, 2469 (1960).
39. O. A. Reutov and E. V. Uglova, *Bull. Acad. Sci. USSR, Div. Chem. Sci.* 1628 (1959).
40. F. R. Jensen and J. A. Landgrebe, *J. Amer. Chem. Soc.* **82**, 1004 (1960).
41. H. B. Charman, E. D. Hughes, C. K. Ingold, and F. G. Thorpe, *J. Chem. Soc.* 1121 (1961).
42. E. D. Hughes, C. K. Ingold, F. G. Thorpe, and H. C. Volger, *J. Chem. Soc.* 1133 (1961).
43. D. S. Matteson, *Organometal. Chem. Rev. A* **4**, 263 (1969).
44. M. H. Abraham, D. Dodd, M. D. Johnson, E. S. Lewis, and R. A. More O'Ferrall, *J. Chem. Soc. B* 762 (1971).
45. D. S. Matteson and R. W. H. Mah, *J. Amer. Chem. Soc.* **85**, 2599 (1963), footnote 21.
46. R. L. Burwell and R. G. Pearson, *J. Phys. Chem.* **70**, 300 (1966).
47. R. E. Dessy and Y. K. Lee, *J. Amer. Chem. Soc.* **82**, 689 (1960).
48. D. Grdenic, *Quart. Rev.* **19**, 303 (1965).
49. E. Khotinsky and M. Melamed, *Chem. Ber.* **42**, 3090 (1909).
50. J. B. Honeycutt, Jr., and J. M. Riddle, *J. Amer. Chem. Soc.* **82**, 3051 (1960).
51. K. Torssell, *Acta Chem. Scand.* **13**, 115 (1959).
52. D. S. Matteson and J. O. Waldbillig, *J. Amer. Chem. Soc.* **86**, 3778 (1964).
53. D. S. Matteson and R. A. Bowie, *J. Amer. Chem. Soc.* **87**, 2587 (1965).
54. H. C. Brown and G. Zweifel, *J. Amer. Chem. Soc.* **83**, 2544 (1961).
55. D. S. Matteson and J. O. Waldbillig, *J. Org. Chem.* **28**, 366 (1963).
56. D. S. Matteson and E. Krämer, *J. Amer. Chem. Soc.* **90**, 7261 (1968).
57. D. S. Matteson, R. A. Bowie, and G. Srivastava, *J. Organometal. Chem.* **16**, 33 (1969).
58. H. G. Kuivila and T. C. Muller, *J. Amer. Chem. Soc.* **84**, 377 (1962).
59. D. S. Matteson and J. G. Shdo, *J. Org. Chem.* **29**, 2742 (1964).
60. (a) R. B. Castle and D. S. Matteson, *J. Amer. Chem. Soc.* **90**, 2194 (1968); (b) R. B. Castle and D. S. Matteson, *J. Organometal. Chem.* **20**, 19 (1969).
61. (a) D. S. Matteson and P. G. Allies, *J. Amer. Chem. Soc.* **92**, 1801 (1970); (b) D. S. Matteson and P. G. Allies, *J. Organometal. Chem.* **54**, 35 (1973).
62. D. S. Matteson and G. L. Larson, *J. Amer. Chem. Soc.* **91**, 6541 (1969).
63. D. S. Matteson and J. R. Thomas, *J. Organometal Chem.* **24**, 263 (1970).
64. W. Hanstein, H. J. Berwin, and T. G. Traylor, *J. Amer. Chem. Soc.* **92**, 829 (1970); W. Hanstein and T. G. Traylor, *Tetrahedron Lett.* 4451 (1967).
64a. H. C. Brown, E. Negishi, and S. K. Gupta, *J. Amer. Chem. Soc*, **92**, 2460 (1970).
65. D. S. Matteson, R. B. Castle, and G. L. Larson, *J. Amer. Chem. Soc.* **92**, 231 (1970).
66. A. Michaelis and P. Becker, *Ber.* **15**, 180 (1882).

67. D. S. Matteson, J. O. Waldbillig, and S. W. Peterson, *J. Amer. Chem. Soc.*, **86**, 3781 (1964).
68. F. R. Jensen and K. L. Nakamaye, *J. Amer. Chem. Soc.* **88**, 3437 (1966).
69. M. Gielen and J. Nasielski, *Ind. Chim. Belge* **26**, 1393 (1961).
70. M. Gielen and J. Nasielski, *Bull. Soc. Chim. Belges* **71**, 32 (1962).
71. M. Gielen and J. Nasielski, *Bull. Soc. Chim. Belges* **71**, 701 (1962).
72. M. Gielen and J. Nasielski, *J. Organometal. Chem.* **1**, 173 (1963).
73. M. Gielen, J. Nasielski, J. E. Dubois, and P. Fresnet, *Bull. Soc. Chim. Belges* **73**, 293 (1964).
74. M. Gielen and J. Nasielski, *Rec. Trav. Chim.* **82**, 228 (1963).
75. M. Gielen and J. Nasielski, *J. Organometal. Chem.* **7**, 273 (1967).
76. S. Boué, M. Gielen, and J. Nasielski, *J. Organometal. Chem.* **9**, 443 (1967).
77. S. Boué, M. Gielen, and J. Nasielski, *J. Organometal. Chem.* **9**, 461 (1967).
78. S. Boué, M. Gielen, and J. Nasielski, *J. Organometal. Chem.* **9**, 481 (1967).
79. P. Baekelmans, M. Gielen, and J. Nasielski, *Tetrahedron Lett.* 1149 (1967).
80. G. Pilloni and G. Tagliavini, *J. Organometal. Chem.* **11**, 557 (1968).
81. F. R. Jensen and D. D. Davis, *J. Amer. Chem. Soc.* **93**, 4047 (1971).
82. D. D. Davis and F. R. Jensen, *J. Org. Chem.* **35**, 3410 (1970).
83. A. Streitwieser, Jr., "Solvolytic Displacement Reactions," p. 13. McGraw-Hill, New York, 1962.
84. L. H. Sommer, K. W. Michael, and W. D. Korte, *J. Amer. Chem. Soc.* **85**, 3712 (1963).
85. C. Eaborn and O. W. Steward, *J. Chem. Soc.* 521 (1965).
86. G. M. Whitesides and D. J. Boschetto, *J. Amer. Chem. Soc.* **93**, 1529 (1971).
87. F. R. Jensen, V. Madan, and D. H. Buchanan, *J. Amer. Chem. Soc.* **93**, 5283 (1971).
88. H. C. Brown, M. W. Rathke, and M. M. Rogić, *J. Amer. Chem. Soc.* **90**, 5038 (1968).
89. H. C. Brown and C. F. Lane, *J. Amer. Chem. Soc.* **92**, 6660 (1970).
90. H. C. Brown and C. F. Lane, *J. Chem. Soc. D* 521 (1971).
91. H. G. Kuivila and E. K. Easterbrook, *J. Amer. Chem. Soc.* **73**, 4629 (1951).
92. H. G. Kuivila and A. R. Hendrickson, *J. Amer. Chem. Soc.* **74**, 5068 (1952).
93. H. G. Kuivila and E. J. Soboczenski, *J. Amer. Chem. Soc.* **76**, 2675 (1954).
94. H. G. Kuivila and R. M. Williams, *J. Amer. Chem. Soc.* **76**, 2679 (1954).
95. H. G. Kuivila and L. E. Benjamin, *J. Amer. Chem. Soc.* **77**, 4834 (1955).
96. H. G. Kuivila, L. E. Benjamin, C. J. Murphy, A. D. Price, and J. H. Polevy, *J. Org. Chem.* **27**, 825 (1962).
97. D. E. Applequist and G. N. Chmurny, *J. Amer. Chem. Soc.* **89**, 875 (1967).
98. (a) W. H. Glaze, C. M. Selman, A. L. Ball, Jr., and L. E. Bray, *J. Org. Chem.* **34**, 641 (1969); (b) W. H. Glaze and C. M. Selman, *J. Organometal. Chem.* **11**, P3 (1968).
99. W. H. Glaze and C. H. Freeman, *J. Amer. Chem. Soc.* **91**, 7198 (1969).
100. H. M. Walborsky, F. J. Impastato, and A. E. Young, *J. Amer. Chem. Soc.* **86**, 3283, 3286 (1964).
101. D. E. Applequist and A. H. Peterson, *J. Amer. Chem. Soc.* **83**, 862 (1961).
102. R. E. Dessy, G. F. Reynolds, and J. Y. Kim, *J. Amer. Chem. Soc.* **81**, 2683 (1959).
103. R. E. Dessy and J. Y. Kim, *J. Amer. Chem. Soc.* **83**, 1167 (1961).
104. H. Zimmer and S. Makower, *Naturwissenschaften* **41**, 551 (1954).
105. R. D. Brown, A. S. Buchanon, and A. A. Humffray, *Aust. J. Chem.* **18**, 1507, 1513 (1965).
106. W. Kitching, *Organometal. Chem. Rev. A* **3**, 35 (1968).

107. C. Perrin and F. H. Westheimer, *J. Amer. Chem. Soc.* **85**, 2773 (1963).
108. H. C. Brown and C. W. McGary, Jr., *J. Amer. Chem. Soc.* **77**, 2300, 2306, 2310 (1955).
109. C. W. McGary, Jr., Y. Okamoto, and H. C. Brown, *J. Amer. Chem. Soc.* **77**, 3037 (1955).
110. E. C. Taylor, F. Kienzle, R. L. Robey, and A. McKillop, *J. Amer. Chem. Soc.* **92**, 2175 (1970).
111. E. C. Taylor, F. Kienzle, R. L. Robey, A. McKillop, and J. D. Hunt, *J. Amer. Chem. Soc.* **93**, 4865 (1971).
112. S. Winstein and T. G. Traylor, *J. Amer. Chem. Soc.* **78**, 2597 (1956).
113. L. H. Gale, J. Landgrebe, and F. R. Jensen, *Chem. Ind.* (London) 118 (1960).
114. W. Kitching and P. R. Wells, *Aust. J. Chem.* **18**, 305 (1965).
115. I. P. Beletskaya, L. A. Fedorov, and O. A. Reutov, *Dokl. Akad. Nauk SSSR* **163**, 1381 (1965).
116. B. F. Hegarty, W. Kitching, and P. R. Wells, *J. Amer. Chem. Soc.* **89**, 4816 (1967).
117. F. Nerdel and S. Makower, *Naturwissenschaften* **45**, 490 (1958).
118. M. M. Kreevoy and R. L. Hansen, *J. Phys. Chem.* **65**, 1055 (1961).
119. J. R. Coad and M. D. Johnson, *J. Chem. Soc. B* 633 (1967).
120. (a) D. Dodd and M. D. Johnson, *J. Chem. Soc. B* 1071 (1969). (b) D. Dodd, C. K. Ingold, and M. D. Johnson, *J. Chem. Soc. B* 1076 (1969).
121. D. Dodd and M. D. Johnson, *J. Chem. Soc. B* 1337 (1970).
122. H. G. Kuivila and K. V. Nahabedian, *J. Amer. Chem. Soc.* **83**, 2159 (1961).
123. H. G. Kuivila and K. V. Nahabedian, *J. Amer. Chem. Soc.* **83**, 2164 (1961).
124. H. G. Kuivila and K. V. Nahabedian, *J. Amer. Chem. Soc.* **83**, 2167 (1961).
125. H. G. Kuivila, J. F. Reuwer, and J. A. Mangarity, *Can. J. Chem.* **41**, 3081 (1963).
126. C. Eaborn, *J. Chem. Soc.* 4858 (1956).
127. C. Eaborn and K. C. Pande, *J. Chem. Soc.* 297 (1961).
128. C. Eaborn and J. A. Waters, *J. Chem. Soc.* 542 (1961).
129. C. Eaborn and K. C. Pande, *J. Chem. Soc.* 3715 (1961).
130. C. Eaborn and K. C. Pande, *J. Chem. Soc.* 1566 (1960).
131. R. W. Bott, C. Eaborn, and P. M. Greasley, *J. Chem. Soc.* 4804 (1964).
132. Y. Yukawa and Y. Tsuno, *Bull. Chem. Soc. Japan* **32**, 971 (1959).
133. R. Alexander, C. Eaborn, and T. G. Traylor, *J. Organometal. Chem.* **21**, P65 (1970).
134. J. Nasielski, O. Buchman, M. Grosjean, and M. Jauquet, *J. Organometal. Chem.* **19**, 353 (1969).
135. C. W. Fong and W. Kitching, *J. Amer. Chem. Soc.* **93**, 3791 (1971).
136. L. H. Toporcer, R. E. Dessy, and F. I. E. Green, *J. Amer. Chem. Soc.* **87**, 1236 (1965).
137. H. C. Brown, "Hydroboration," p. 64. Benjamin, New York, 1962.
138. A. G. Davies and B. P. Roberts, *J. Chem. Soc. C* 1474 (1968).
139. G. C. Robinson, *J. Org. Chem.* **28**, 843 (1963).
140. R. Walraevens, Thesis, U. Libre de Bruxelles, 1962, quoted by M. Gielen in "Mecanismes de rupture de la liaison carbone–metal." Univ. Libre de Bruxelles, 1966.
141. W. E. McEwen, G. H. Briles, and B. E. Giddings, *J. Amer. Chem. Soc.* **91**, 7079 (1969).
142. Y. Pocker and J. H. Exner, *J. Amer. Chem. Soc.* **91**, 6764 (1968).
143. R. E. Dessy and R. M. Salinger, *J. Org. Chem.* **26**, 3519 (1961).
144. R. E. Dessy, J. Wotiz, and C. Hollingsworth, *J. Amer. Chem. Soc.* **79**, 358 (1957).
145. P. West, R. Waack, and J. I. Purmort, *J. Amer. Chem. Soc.* **92**, 840 (1970).

146. H. Gilman, *Org. Reactions* **8**, 258 (1957).
147. G. G. Eberhardt and W. A. Butte, *J. Org. Chem.* **29**, 2928 (1964).
148. R. West and P. C. Jones, *J. Amer. Chem. Soc.* **90**, 2656 (1968).
149. C. G. Swain, *J. Amer. Chem. Soc.* **69**, 2306 (1947).
150. M. Anteunis, *J. Org. Chem.* **26**, 4214 (1961).
151. N. M. Bikales and E. I. Becker, *Can. J. Chem.* **41**, 1329 (1963).
152. T. Holm, *Acta Chem. Scand.* **19**, 1819 (1965).
153. T. Holm, *Acta Chem. Scand.* **20**, 1139 (1966).
154. H. O. House and D. D. Traficante, *J. Org. Chem.* **28**, 355 (1963).
155. H. O. House and J. E. Oliver, *J. Org. Chem.* **33**, 929 (1968).
156. W. French and G. F. Wright, *Can. J. Chem.* **42**, 2474 (1964).
157. S. G. Smith and G. Su, *J. Amer. Chem. Soc.* **88**, 3995 (1966).
158. S. G. Smith and J. Billet, *J. Amer. Chem. Soc.* **89**, 6948 (1967).
159. J. Billet and S. G. Smith, *J. Amer. Chem. Soc.* **90**, 4108 (1968).
160. E. C. Ashby, *Quart. Rev. (London)* **21**, 259 (1967).
161. E. C. Ashby, R. B. Duke, and H. M. Neumann, *J. Amer. Chem. Soc.* **89**, 1964 (1967).
162. E. C. Ashby, J. Laemmle, and H. M. Neumann, *J. Amer. Chem. Soc.* **93**, 4601 (1971); **94**, 5421 (1972).
163. E. C. Ashby, J. Laemmle, and H. M. Neumann, *J. Amer. Chem. Soc.* **94**, 2158 (1972).
164. J. Laemmle, E. C. Ashby, and H. M. Neumann, *J. Amer. Chem. Soc.* **93**, 5120 (1971).
165. S. G. Smith, L. F. Charbonneau, D. P. Novak, and T. L. Brown, *J. Amer. Chem. Soc.* **94**, 7059 (1972).
166. H. M. Walborsky and A. E. Young, *J. Amer. Chem. Soc.* **86**, 3288 (1964).
167. G. J. Karabatsos, *J. Amer. Chem. Soc.* **89**, 1367 (1967).
168. H. R. Hays, *J. Amer. Chem. Soc.* **91**, 2736 (1969).
169. L. H. Sommer and D. L. Bauman, *J. Amer. Chem. Soc.* **91**, 7045 (1969).
170. L. H. Sommer, W. D. Korte, and P. G. Rodewald, *J. Amer. Chem. Soc.* **89**, 862 (1967).
171. L. H. Sommer, G. A. Parker, N. C. Lloyd, C. L. Frye, and K. W. Michael, *J. Amer. Chem. Soc.* **89**, 857 (1967).
172. L. H. Sommer and H. Fujimoto, *J. Amer. Chem. Soc.* **90**, 982 (1968).
173. L. H. Sommer and W. D. Korte, *J. Amer. Chem. Soc.* **89**, 5802 (1967).
174. L. H. Sommer and H. Fujimoto, *J. Amer. Chem. Soc.* **91**, 7040 (1969).
175. E. C. Ashby, J. Laemmle, and H. M. Neumann, *J. Amer. Chem. Soc.* **90**, 5179 (1968).
176. E. C. Ashby, J. Laemmle, and G. E. Parris, *J. Organometal. Chem.* **19**, P24 (1969).
177. E. C. Ashby and J. T. Laemmle, *J. Org. Chem.* **33**, 3398 (1968).
178. E. A. Jeffery and T. Mole, *Aust. J. Chem.* **23**, 715 (1970).
179. A. Lundeen and A. C. Oehlschlager, *J. Organometal. Chem.* **25**, 337 (1970).
180. H. O. House and D. D. Traficante, *J. Org. Chem.* **28**, 348 (1963).
181. H. O. House and W. F. Fischer, Jr., *J. Org. Chem.* **33**, 949 (1968).
182. H. O. House, W. L. Respess, and G. M. Whitesides, *J. Org. Chem.* **31**, 3128 (1966).

CHAPTER 4

Electrophilic Displacements Involving Neighboring Sites

I. Introduction

Most of the chemistry to be discussed in this chapter consists of intramolecular electrophilic displacements, often described as migration of carbon from an electron-rich to an electron-deficient atom. Typical examples from organic chemistry include Wagner-Meerwein rearrangements of carbonium ions, Hofmann hypohalite rearrangements of amides, Wolff rearrangements of diazoketones, Beckmann rearrangements of oximes, and many related reactions. In organometallic chemistry the electron-rich atom is a metal, the electron-deficient atom is carbon or a nonmetal, and the metal atom is oxidized as the alkyl or aryl group migrates away. A typical example is the reaction of benzeneboronic acid (**1**) with hydrogen peroxide.[1] The distinction

$$\underset{\mathbf{1}}{PhB(OH)_2} + OOH^- \rightleftharpoons HO{-}\overset{\overset{OH}{|}}{\underset{\underset{Ph}{|}}{B^-}}{-}O{-}OH \longrightarrow HO{-}\overset{\overset{OH}{|}}{\underset{\underset{Ph}{|}}{B}}{-}O + OH^-$$

between bimolecular and intramolecular electrophilic displacements is often arbitrary, as in the case of the hydrogen peroxide deboronation just described.

This chapter begins with migrations to electron-deficient oxygen and nitrogen, which are closely related to the material of the preceding chapter. Migrations to electron-deficient carbon follow. Finally, reactions where the electrophile attacks one carbon and the electrofuge leaves another, including cyclopropane openings and closures and displacements accompanied by allylic rearrangement, will be discussed.

Effects of neighboring metal atoms (mainly boron) on electrophilic displacements have been covered in Chapter 3, Section II,D.

II. Migrations to Electron-Deficient Oxygen or Nitrogen

A. Peroxides and Amine Oxides

Kuivila and co-workers have carried out a detailed study of the reaction of areneboronic acids with hydrogen peroxide.[1–3] In neutral or basic solution the reaction is first order in $PhB(OH)_2$ and first order in OOH^-. The first step of the reaction is undoubtedly coordination of the OOH^- ion with the boron, analogous to the coordination of boron with hydroxide ion which is characteristic of the acid behavior of boronic acids.[4,5] The transition state (**2**) must involve rearrangement of the intermediate $PhB(OH)_2OOH^-$.

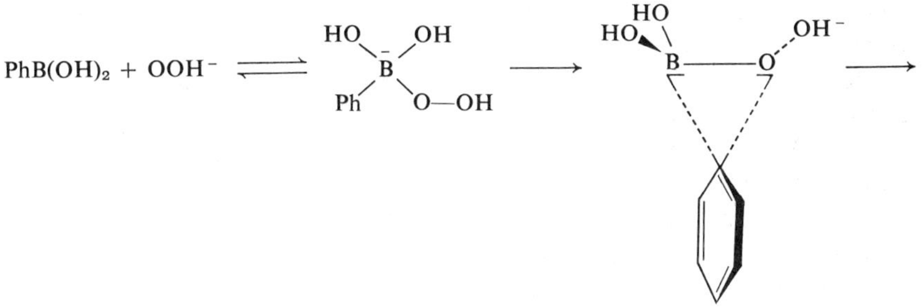

$$PhB(OH)_2 + OOH^- \rightleftharpoons PhB^-(OH)_2(O{-}OH) \longrightarrow \mathbf{2} \longrightarrow$$

2

$$(HO)_2B{-}O{-}Ph + OH^- \longrightarrow B(OH)_3 + PhO^-$$

Transition state **2** is structurally analogous to a phenonium ion (**3**),[6] or more properly to a transition state leading to formation of a phenonium ion, since **2** would have to undergo complete loss of the departing hydroxyl group to become strictly analogous to **3**. However, migration of the phenyl group from boron to oxygen in **2** is highly exothermic and undoubtedly complete by the time the oxygen–oxygen bond is broken, in contrast to the situation

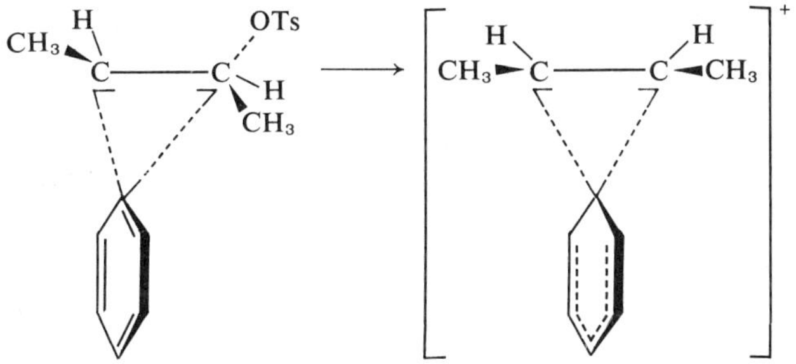

3

with **3**, where rearrangement in either direction requires equal energy and **3** itself is an energy minimum. Three-center electron-pair bonding supplemented by an antisymmetric component involving the π electrons of the benzene ring is appropriate for describing both **2** and **3**.

In addition to the $k[ArB(OH)_2][OOH^-]$ term in the rate law, there is also a term $k'[ArB(OH)_2]^2[OOH^-]$ (where the $[OOH^-]$ dependence is observed as a function of $[H_2O_2]$ and pH). This term evidently involves some form of boronic acid anion in place of hydroxide ion as the leaving group. The rearranging species must be $Ar\bar{B}(OH)_2O$—O—B(OH)Ar or a hydrate of it. In more acidic solutions an "uncatalyzed" process is observed, actually a pH-independent mechanism involving both H^+ and OH^- to form $Ar\bar{B}(OH)_2O$—OH_2^+, which rearranges by way of a transition state analogous to **2** but with water in place of hydroxide ion as the leaving group. In strong acids there are H^+-catalyzed mechanisms in which the degree of hydration or coordination of the boron atom is not definitely established. Hammett correlations are good with σ, not σ^+, and the rates are insensitive to substituents, unlike typical electrophilic aromatic substitutions but consistent with the intramolecular migration mechanism.[1] Specifically, ρ is $+0.007$ for the base-catalyzed reaction in 25% ethanol, $+0.3$ for the term $k'[ArB(OH)_2]^2[OOH^-]$, -0.04 for the pH-independent mechanism, -1.1 for phosphoric acid catalysis, and -1.3 for sulfuric acid catalysis. It may be concluded that there is little involvement of the π electrons of the benzene ring in transition state **2**, and that cyclic three-centered electron-pair bonding predominates, just as it would in an analogous alkyl migration.

Retention at carbon is obviously required in alkyl migrations from boron to oxygen by way of a transition state analogous to **2**. This stereochemistry has been verified in studies of hydroboration[7] and has been used as proof of stereochemistry in a number of studies of boron compounds.[8–10]

Minato, Ware, and Traylor have measured the rates of reaction of a variety of boronic acids with alkaline hydrogen peroxide.[11] In order to compare the actual rate constants for the rearrangement process, correction must be made for the differing acidities K_a of the boronic acids, so that the basis for the comparison is the anion $RB(OH)_2OOH^-$ rather than $RB(OH)_2$. The results are summarized in Table 4-1.

The K_a's of the first five alkaneboronic acids listed in Table 4-1 only vary by a factor of 2. The K_a's of the benzyl, phenyl, and vinyl compounds are an order of magnitude or more greater, accounting for most of the increase in reactivity compared to methyl. It appears that the alkyl migration from boron to oxygen is greatly aided by electron donation to the migratory site and is not much affected by steric hindrance. This is consistent with the usual three-center bond model for the transition state. It is also consistent with Kuivila's Hammett correlation, where electron-withdrawing groups must

TABLE 4-1

RATES OF REACTION OF BORONIC ACIDS WITH HYDROGEN PEROXIDE[a]

R in $RB(OH)_2$	k_2 at pH 5.23 (liter/mole second)	Relative k_2/K_a
Methyl	0.000127	1
1-Butyl	0.00480	52
2-Butyl	0.0233	185
t-Butyl	0.0718	330
1-Bicycloheptyl	0.11	680
Benzyl	0.0875	24
Phenyl	0.0168	2.3
Vinyl	0.0068	4.2

[a] From H. Minato, J. C. Ware, and T. G. Traylor, *J. Amer. Chem. Soc.* **85**, 3024 (1963).

increase the concentration of the active species $ArB(OH)_2OOH^-$ but slow its rate of rearrangement so that the net effect on the rate is very small in the aryl series.

Chromic acid oxidations proved more sensitive to structure,[12] but Lansbury and Nienhouse have shown that the mechanism is more complicated than the originally postulated alkyl migration from boron to oxygen.[13] In 0.114 *M* perchloric acid at 30°C the rate constants found included t-$BuB(OH)_2$, 7.5×10^{-2}; $EtB(OH)_2$, 6.6×10^{-4}; $MeB(OH)_2$, 2.4×10^{-7} liter/mole second.[12] These results imply a considerable degree of carbonium ion character in the transition state, and the rearrangement of norbornenylborane (**4**)

BR_2 $\xrightarrow{CrO_3}$

4

(49%) (51%)

on oxidation with chromic acid[13] requires either a free carbonium ion or radical intermediate. (Norbornenyl alcohol does not rearrange on oxidation with chromic acid.) The intramolecular migration mechanism might still apply to cases such as $MeB(OH)_2$, where carbonium ion formation would be difficult.

Chromic acid in acetic acid also oxidizes tetraalkyltins.[14] The relative rates are Me_4Sn, 10; Et_4Sn, 125; Pr_4Sn, 90; Bu_4Sn, 70; iso-Pr_4Sn, 210. The structural sensitivity is comparable to that for boronic acids with hydrogen peroxide.[11] The mechanism may involve an intramolecular alkyl migration from tin to oxygen or there could be carbonium ion intermediates. With an oxidizing agent as strong as chromium(VI) and the probable involvement of the intermediates chromium(V) and (IV) as oxidizing agents, the total mechanism of these reactions is probably complex. The product chromium(III) retards the reaction,[14] perhaps by reducing chromium(V) to chromium(IV), perhaps by acting as some sort of free-radical trap.

Although deboronations with hydrogen peroxide proceed by the intramolecular migration mechanism in most cases studied, even that has its exceptions. Pasto *et al.* have found that phenylethane-1,2-diboronic acid (**5**) yields a gross mixture of fragmentation products on treatment with alkaline hydrogen peroxide, probably by way of radical intermediates.[15]

$$\underset{\mathbf{5}}{\mathrm{PhCH(B(OH)_2)—CH_2—B(OH)_2}} + \mathrm{H_2O_2} \xrightarrow{\mathrm{OH^-}} \mathrm{PhCH_2CH_2OH,\ PhCH(OH)CH_3,\ PhCHO,\ PhCH(OH)CH_2OH,}\ \text{and others}$$

Trimethylamine oxide reacts as an analog of hydrogen peroxide with trialkylboranes.[16] Evidently the adduct $R_3B^-—ON^+Me_3$ undergoes intramolecular alkyl migration with expulsion of NMe_3 to yield $R_2B—OR$, which may be further converted to $RB(OR)_2$. Alkylboranes react much faster than arylboranes.

B. Azides, Chloramines, and Related Reagents

Migrations of carbon from boron to nitrogen analogous to the boron to oxygen migrations are known. Paetzold and co-workers have studied the thermal rearrangement of dialkyl- and diarylboron azides.[17,18] Dibutylboron azide decomposes at 230°–240°C to nitrogen and a mixture of hexabutylborazine (**6**), 1-butene, and borazines lacking one or more butyl groups at nitrogen or boron. Diphenylboron azide decomposes below 200°C to nitrogen, hexaphenylborazine (**7**), and an oily compound which was formulated as tetraphenyldiazadiboretidine (**8**).

The four-membered ring of **8** is isoelectronic with cyclobutadiene and the structure is therefore open to some question, even though molecular orbital calculations suggest that the lower symmetry of the boron–nitrogen ring may allow moderate stability.[19] A similar four-membered ring in a related compound is supported by X-ray evidence.[20]

More to the point with regard to the rearrangement mechanism, Paetzold

$$Bu_2B{-}N{=}\overset{+}{N}{=}\overset{-}{N} \longrightarrow [Bu{-}B{\cdots}N(Bu){\cdots}N_2] \longrightarrow N_2 + Bu{-}B{=}N{-}Bu \longrightarrow (BuBNBu)_3 \quad \mathbf{6}$$

$$Ph_2B{-}N{=}\overset{+}{N}{=}\overset{-}{N} \longrightarrow [Ph{-}B{\cdots}N(Ph){\cdots}N_2] \longrightarrow N_2 + Ph{-}B{=}N{-}Ph \longrightarrow (PhBNPh)_3 \; \mathbf{7} + (PhBNPh)_2 \; \mathbf{8}$$

and co-workers have generated the postulated borimide intermediates (**9**) by another route and proved their independent existence by trapping with 1,3-dipolar reagents.[21,22]

$$PhBCl_2NH_2C_6F_5 \xrightarrow{-HCl} \underset{\mathbf{9}}{PhB{=}NC_6F_5} \xrightarrow{Ph{-}C{\equiv}\overset{+}{N}{-}\overset{-}{O}} \text{3-Ph-4-}C_6F_5\text{-5-Ph-1,2,4,5-oxadiazaborole}$$

It should be expected that any alkylmetal azide which can be prepared will undergo a rearrangement analogous to those of boron compounds. For example, similar rearrangements have been reported for silyl azides such as **10**.[23]

$$\underset{\mathbf{10}}{\text{Me—Si(Ph)}_2\text{—N}_3} \longrightarrow \underset{\text{(unstable)}}{\text{Me—Si(Ph)=N—Ph}} + N_2 \longrightarrow [\text{MeSi(Ph)—NPh}]_2$$

Mechanistically related rearrangements can occur if an alkyl or aryl azide reacts with a trialkylborane.[24] For example, butyl azide (**11**) and triethylborane yield an intermediate which can be hydrolyzed to butylethylamine, and this type of reaction is a generally useful synthetic route to secondary amines.

$$\underset{\mathbf{11}}{\text{Bu—N=}\overset{+}{\text{N}}\text{=}\overset{-}{\text{N}}} + BEt_3 \xrightarrow[\text{xylene}]{135°C} \text{Bu—N(}N^{+}\equiv N\text{)—}\overset{-}{B}Et_3 \longrightarrow \text{Bu—N(Et)—BEt}_2 + N_2 \xrightarrow{H_2O} \text{BuNHEt} + \text{HOBEt}_2$$

Reactions involving similar migrations of alkyl groups from boron to nitrogen are found when trialkylboranes react with chloramine[25] or hydroxylaminesulfonic acid[26] to form amine boranes. In place of the N_2 of the azide, the departing nucleofuges are Cl^- and SO_4^{2-} or HSO_4^-. Detailed mechanistic studies of these reactions are not available, but the gross classification as intramolecular rearrangements is readily apparent from the products.

Dimethylchloramine (**12**) reacts with trialkylboranes by both radical and polar mechanisms.[27] The radical path can be inhibited by traces of galvinoxyl (a stable free radical).

$$\underset{\mathbf{12}}{Me_2NCl} + Bu_3B \longrightarrow Bu_3\overset{-}{B}\text{—}\overset{+}{N}Me_2(Cl) \longrightarrow Bu_2B\text{—}\overset{+}{N}(Bu)Me_2\ Cl^- \xrightarrow{H_2O} Bu_2BOH + BuNMe_2H^+Cl^-$$

$$\xrightarrow{X\cdot}$$

$$\left.\begin{array}{l} Me_2N\cdot + Bu_3B \longrightarrow Bu_2B\text{—}NMe_2 + Bu\cdot \\ Bu\cdot + Me_2NCl \longrightarrow BuCl + Me_2N\cdot \end{array}\right\}\text{chain reaction}$$

III. Migrations to Electron-Deficient Carbon

A. Diazo Compounds

This class of reactions might also be called "migrations away from an electron-rich metal atom," since such things are relative. Many examples have been found. Boron compounds have particularly convenient properties for this type of reaction for two reasons. First, boron–carbon bonds are generally compatible with the presence of other reactive functional groups. Second, the boron atom acts as a Lewis acid. The major mechanisms generalized and simplified most often involve combination of either $X^- +$ $X_2B—CR_2Y$ or $X_3B + {}^-CR_2Y$ to form $X_3B^-—CR_2Y$, which undergoes concerted migration of X^- and loss of Y^- to form $X_2B—CR_2X$. Compounds of other metals can undergo analogous reactions. The number of detailed mechanistic studies on this class of reactions is fairly small, but it is a rapidly expanding field of considerable current interest.

The classical example of this type of reaction is the formation of polymethylene from diazomethane (**13**) in the presence of borate esters and other

$$(MeO)_3B + CH_2{=}\overset{+}{N}{=}\overset{-}{N} \longrightarrow (MeO)_3\overset{-}{B}—CH_2—\overset{+}{N}{\equiv}N \longrightarrow$$
$$\textbf{13}$$

$$(MeO)_2B—CH_2—OMe + N_2 \xrightarrow{CH_2N_2} (MeO)_2B(—CH_2OMe)(—CH_2—\overset{+}{N}{\equiv}N) \longrightarrow$$

$$(MeO)_2B—CH_2—CH_2OMe + N_2 \longrightarrow \longrightarrow (MeO)_2B(CH_2)_nOMe$$

boron compounds.[28] With diazoacetone (**14**) and trialkylboranes polymerization is avoided and the reaction provides a useful synthetic route to ketones.[29]

$$CH_3\overset{O}{\overset{\|}{C}}—CH{=}\overset{+}{N}{=}\overset{-}{N} + BR_3 \longrightarrow CH_3\overset{O}{\overset{\|}{C}}—CH(—\overset{-}{B}R_3)(—N^+{\equiv}N) \longrightarrow CH_3\overset{O}{\overset{\|}{C}}—CH(R)(BR_2) + N_2 \longrightarrow$$
$$\textbf{14}$$

$$CH_3C(—O—BR_2){=}CH—R \xrightarrow{H_2O} CH_3\overset{O}{\overset{\|}{C}}—CH_2R + R_2BOH$$

(reasonable but not proved)

Analogous reactions have been observed with ethyl diazoacetate or diazoacetonitrile.[30] The analogy between these reactions of diazo compounds and those of alkyl azides described in the preceding section is readily apparent. The suggested rearrangement of the α-borylketone to an *O*-borylenol before hydrolysis has not been proved but is likely to be exothermic by ~25 kcal/mole or more. The basis for this estimate is bond energies, to be discussed later in Section III,B.

B. Halides

α-Haloalkylboron compounds provide many examples of boron-to-carbon migrations. The nucleophile which migrates and displaces the halide is an aryl or alkyl group if one is available, but if not alkoxy and other nucleophilic ligands will also migrate readily. Matteson and Mah happened onto one of the earlier examples as a sequel to the synthesis of α-bromoalkylboron compounds.[31] These were adducts of bromotrichloromethane with vinylboron compounds. Failure to obtain the expected adduct from *B*-vinyl-*B*-phenyl-*B*-butoxyborane led to the discovery that aqueous sodium bicarbonate used in the work-up was a strong enough base to cause the *B*-α-bromoalkyl-*B*-phenyl-*B*-butoxyborane (**15**) to rearrange to α-phenylalkaneboronic acid or its dibutyl ester (**19**). Proof of an intermediate anion (**17**) was obtained by treating the corresponding α-bromoalkaneboronic ester (**16**) with phenylmagnesium bromide to form **17**, which was converted by acid below −40°C to the unrearranged product **15** but at room temperature rearranged to **19**. Similar results were obtained when **16** was treated with ethylmagnesium bromide or other Grignard reagents. Transition state model **18** was postulated.

$$\underset{\mathbf{15}}{R{-}\underset{\displaystyle Br}{CH}{-}B(Ph)(OBu)} \underset{H^+}{\overset{BuO^-}{\rightleftharpoons}} \underset{\mathbf{17}}{R{-}\underset{\displaystyle Br}{CH}{-}\overset{\displaystyle Ph}{\underset{\displaystyle OBu}{B^-}}{-}OBu} \xleftarrow{PhMgBr} \underset{\mathbf{16}}{R{-}\underset{\displaystyle Br}{CH}{-}B(OBu)_2}$$

$$\mathbf{17} \xrightarrow{25^\circ C} \underset{\mathbf{18}}{\left[\,R{-}\underset{\displaystyle Br}{CH}{\cdots}Ph{\cdots}B(OBu)_2\,\right]^-} \longrightarrow \underset{\mathbf{19}}{R{-}\overset{\displaystyle Ph}{CH}{-}B(OBu)_2}$$

$(R = Cl_3CCH_2)$

Butoxide ion very rapidly displaces bromide ion from α-bromoalkaneboronic esters such as **16**, a reaction which will be discussed in Section IV. Treatment of the resulting α-butoxyalkaneboronic ester with ethylmagnesium bromide yielded *B*-ethyl-*B*-α-butoxyalkyl-*B*-butoxyborane (**20**),[32] the isomer of the product **21** obtained directly from the α-bromoalkaneboronic ester **16** and ethylmagnesium bromide. Infrared spectra of samples of **21** showed no evidence of the presence of any **20**.

$$\underset{\mathbf{20}}{R{-}\underset{\displaystyle OBu}{CH}{-}B(OBu)(C_2H_5)} \qquad \underset{\mathbf{21}}{R{-}\underset{\displaystyle C_2H_5}{CH}{-}B(OBu)_2}$$

Why should there be a strong preference for carbon rather than oxygen migration in anion **17** and its analogs? Rearrangement is irreversible and must be kinetically controlled, but inspection of the thermodynamics suggests that the entire potential energy surface may be tilted strongly in favor of **21** over **20**. The more stable isomer **21** contains a B—O bond and a C—C bond where the less stable isomer **20** has a B—C bond and a C—O bond. The difference in energy between a C—O and a C—C bond is generally only 1 or 2 kcal/mole in favor of the former,[33] but the B—O may be favored over the B—C bond by 30–50 kcal.[34] Estimated mean B—C bond dissociation energies are for Me_3B, 88.1 $\pm$ 2.5; Et_3B, 82.6; Ph_3B, 106.0 kcal/mole. Mean B—O bond dissociation energies are for $(HO)_3B$, 132.8; $(MeO)_3B$, 118.0; $(EtO)_3B$, 117.7 kcal/mole. The relatively high value for the B—Ph bond depends on B—C π-bonding, and it is likely to drop toward the boron–alkyl bond energy range if one strongly back-bonding oxygen ligand is present on boron. The set of B—O bond energies is self-consistent only if it is assumed that the O—H bond is stronger in water than in alcohols, since the reaction of $(RO)_3B$ with H_2O to form $(HO)_3B$ and ROH is easily reversed and not highly exothermic. Thus, tabulated bond energies cannot be applied blindly, but are consistent with the generalization that it is more exothermic to oxidize boron than carbon by a considerable margin whenever these are the alternatives.

An additional consequence of the steep general slope of the potential energy surface is that there can be no metastable zwitterion intermediate, $RC^+H\text{–}B^-(OBu)_2Ph$, formed by ionization of the bromide ion from **17** prior to rearrangement. Such a concept was introduced by Jäger and Hesse as a rationalization of an analogous oxidative rearrangement of tetraalkylborate anions.[35] However, if the zwitterion is an intermediate there must be some energy barrier to its rearrangement. There is no such barrier in an analogous all-carbon system, the solvolysis and rearrangement of 3-phenyl-2-butyl tosylate.[6] In this case the classical cations $MeC^+HCHMePh$ at either end of the reaction coordinate have equivalent energies and structures and the symmetrical phenyl-bridged structure **3** (see section on deboronation with hydrogen peroxide) in the middle of the reaction coordinate has a lower energy. The major pathway involves phenyl migration to form **3** concerted with tosylate ion expulsion. The minor pathway through the classical carbonium ion involves strong solvation. In the boron system, the additional conditions of (1) a highly exothermic rearrangement and (2) a solvent (ether) having slight polarity and nucleophilicity reduce the likelihood of any energy barrier still further. If $RC^+H\text{–}B^-(OBu)_2Ph$ should somehow be formed, there should be no barrier whatever to its rearrangement, and phenyl migration should therefore be in progress before departure of the bromide from **17** is completed. The arguments are less rigorous for rearrangements of tetraalkylborate anions, but there is still no reason to expect any stable R_2C^+—$B^-R'_3$ zwitterion in any known system. The sort of energy relationships involved are illustrated in Fig. 4-1.

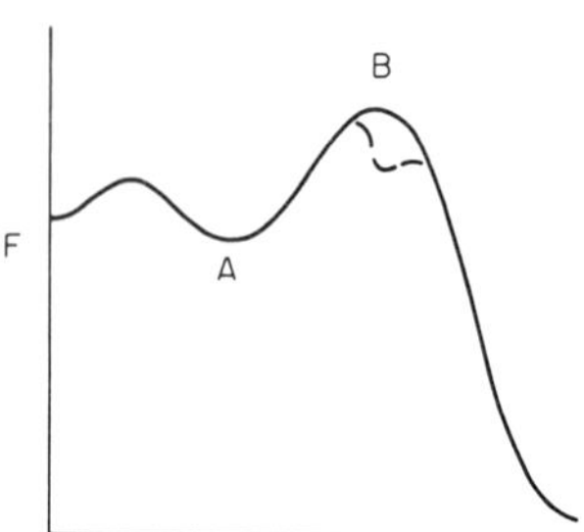

Fig. 4-1. Qualitative plot of *F* versus reaction coordinate for the reaction of RCHBrB(OBu)Ph (**15**) with BuO^- to form $RCHBrB(OBu)_2Ph^-$ (**17**) (point A on graph) and the rearrangement of **17** by way of the phenyl-bridged transition state **18** (point B) to the products $RCHPhB(OBu)_2$ (**19**) and Br^-. If there were a zwitterion intermediate $RC^+H\text{–}B^-(OBu)_2Ph$ it would have to lie in an energy minimum as illustrated by the broken curve under point B. [D. S. Matteson and R. W. H. Mah, *J. Amer. Chem. Soc.* **85**, 2599 (1963).]

Sound as these theoretical arguments may be, chemists would rather have experimental evidence. Stereochemistry is the crucial test in this case. Köbrich and Merkle showed that inversion of the carbon from which the halide is displaced occurs in the rearrangement of the anion derived from triphenylboron and the two geometrical isomers of α-methyl-β-chloro-β-styryllithium (**22** and **23**).[36] The stereospecificity precludes any free zwitterion.

Ph(Me)C=C(Li)Cl (**22**) + BPh_3 ⟶ Ph(Me)C=C($\bar{B}Ph_3$)Cl ⟶ [Ph(Me)C=C(Cl)–BPh_2 (Ph-bridged)]$^-$ ⟶

Ph(Me)C=C(Ph)BPh_2 + Cl^- $\xrightarrow{C_2H_5CO_2H}$ Ph(Me)C=C(Ph)H (isolated product)

Ph(Me)C=C(Cl)Li (**23**) + BPh_3 ⟶ ⟶ ⟶ Ph(Me)C=C(H)Ph

Concerted nucleophilic displacement at a vinyl carbon is usually very difficult to carry out, and the fact that this rather odd rearrangement proceeds efficiently underscores the great effect of the neighboring-group participation.

Zweifel and Arzoumanian have demonstrated similar stereospecificity in the rearrangement of the alkenylboranes (**24**) prepared by hydroboration of 1-bromo-1-hexyne with dicyclohexylborane or other dialkylboranes.[37]

Bu—C≡C—Br + R_2BH $\xrightarrow[\text{(cis addition known)}]{}$ (Bu)(H)C=C(Br)(BR_2) **24** $\xrightarrow{NaOCH_3}$

(R = cyclohexyl)

(Bu)(H)C=C(Br)($B^{-}R_2OCH_3$) $\xrightarrow{-Br^-}$ (Bu)(H)C=C(R)(B(R)OCH_3) $\xrightarrow[\text{(retention known)}]{HOAc}$ (Bu)(H)C=C(H)(R)

Using deuterium as the migrating atom, Pasto and Hickman have shown that (α-chloroalkyl)boranes rearrange to alkylboron chlorides with complete inversion of the α carbon.[38] In so doing, they have had to unravel systems of considerable complexity. One example is the deuteroboration of 1-chloro-*trans*-1-phenylpropene (**25**).

The reasoning which leads to the conclusion that inversion occurs is as follows. From other work, it is known that borane adds exclusively *cis* to double bonds. β-Chloroalkylboranes readily eliminate boron chloride *trans* to form an olefin which is in turn hydroborated under the reaction conditions. The percentages of the two directions of addition of borane to the olefin are known from other experiments. Configuration is retained on deboronation with hydrogen peroxide, and nmr spectra yield the identity and amounts of the diastereoisomers. Isomers are separated by gas chromatography. Specifically, the yields of alcohols **31**, **32**, and **33** are 54%, 25%, and 3.8% respectively, with 17% unaccounted for. Since hydroboration of *trans*-1-propenylbenzene, PhCH=$CHCH_3$, yields the isomers corresponding to **32** and **33** in the ratio 6.7:1 (25:3.8), all of the diastereoisomer **32** must have arisen from deuteroboration of PhCD=$CHCH_3$. That leaves rearrangement with inversion as the only path of reaction for the α-chloroborane **26**, which leads by way of **28** to **31**. Rearrangement with retention would yield **29** and then **32**, but all of the **32** has been accounted for by the elimination–rehydroboration route.

These conclusions are strengthened by further studies on 2-chloronorbornene and 2-chloro-1-methylcyclohexene, which yield similar results. The complexity of the data leaves some room for doubt, since the whole structure of the reasoning collapses if one product ratio or assumed stereospecificity

Ph(Cl)C=CH(CH$_3$) **25** + $BD_3 \cdot THF$ ⟶ **26** + **27**

26 ⟶ **28**; **28** $\xrightarrow{H_2O_2}$ **31**

27 ⟶ Ph(D)C=CH(CH$_3$) $\xrightarrow{BD_3 \cdot THF}$ **29** + **30**

29 $\xrightarrow{H_2O_2}$ **32**

30 $\xrightarrow{H_2O_2}$ $PhCD_2CHOH(CH_3)$ **33**

is moderately erroneous (see Section II,A),[15] and there is thus an opportunity for the operation of "Murphy's Law of Unfortunate Conclusions." However, the conclusions are the theoretically reasonable ones and are believable as much on that basis as on the grounds of the complicated evidence.

Pasto, Hickman, and Cheng studied the kinetics of the rearrangement of α-chloroisobutylborane (**34**), the product from hydroboration of 1-chloro-2-methylpropene in tetrahydrofuran,[39] to isobutylchloroborane (**35**). The rearrangement reaction is strongly catalyzed by Lewis acids such as $BH_3 \cdot THF$ or $BF_3 \cdot THF$ and is rapid until all the $BH_3 \cdot THF$ used in the hydroboration is consumed. Two different B—H couplings were seen in the ^{11}B nmr, and it was concluded that the borane **34** is dimeric in THF. After the $BH_3 \cdot THF$ is consumed the remainder of the α-chloroisobutylborane **34** rearranges by a first-order process (first order in dimer, assuming that is the major species present) over a period of a few hours and can be followed conveniently by nmr. The structure of **34** was proved to contain the α-chloroisobutylboron unit by hydrolysis to α-chloroisobutaneboronic acid (**36**).

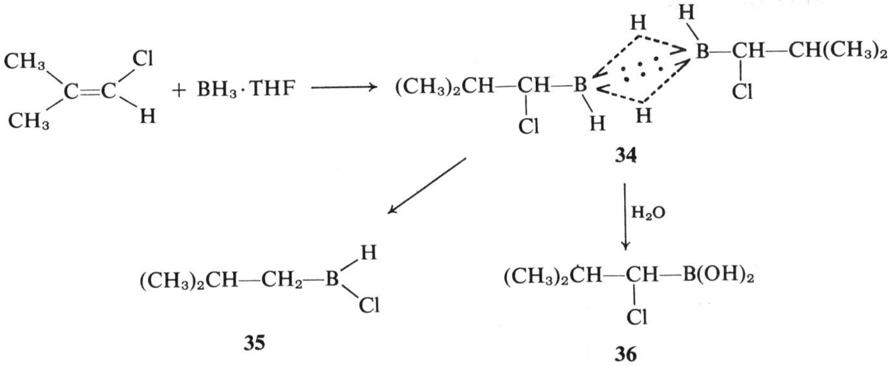

$$(CH_3)_2C{=}CHCl + BH_3\cdot THF \longrightarrow (CH_3)_2CH{-}CHCl{-}BH_2 \text{ dimer } (\mathbf{34}) \xrightarrow{H_2O} (CH_3)_2CH{-}CHCl{-}B(OH)_2\ (\mathbf{36})$$

$$\mathbf{34} \longrightarrow (CH_3)_2CH{-}CH_2{-}BHCl\ (\mathbf{35})$$

The evidence is insufficient to define the nature of the transition state in detail. In general, the transition state should contain the features of **37**.

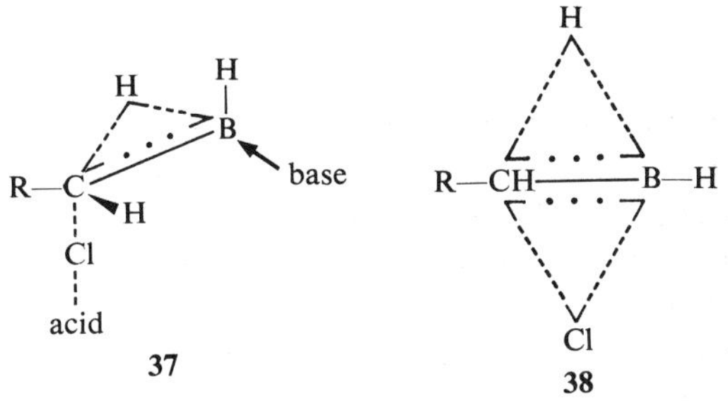

37

38

(violates bonding requirements)

The base would perhaps be a tetrahydrofuran molecule; the acid could be BH_3 or BF_3 in the case of the acid-catalyzed reaction or a second molecule of $RCHClBH_2$ in the uncatalyzed reaction. The kinetics would be the same whether intramolecular rearrangement of the dimer **34** or prior dissociation of **34** and recombination of the halves at a different site were the mechanism. One mechanism that can be ruled out *a priori* is simple migration of hydride from boron to carbon concurrent with chloride migration from carbon to boron (**38**). The boron atom of **38** has to be badly exposed, with insufficient ligands in the required directions. Complexing with THF could cure **38**, making it equivalent to **37** where "acid" becomes the boron atom itself. The same cure could be provided by hydrogen bridging to a second molecule of $RCHClBH_2$, and this rather than THF would be required if the predominant species in solution is in fact the dimer, as suggested by the nmr.

Returning to the question of stereospecificity, the intramolecular migration mechanism not only requires inversion of the carbon at which nucleophilic displacement occurs, but also requires retention of the configuration of the migrating group. Brown *et al.* have recorded observations which were thought to prove this point.[40]

CH_3 / BR_2 **39**

H_2O_2, NaOH → CH_3 / OH (retention known)

CO, $LiAlH(OMe)_3$ → CH_3 / CHO

$BrCH_2CO_2Et$, *t*-BuOK → CH_3 / CH_2CO_2Et

$CH_2{=}CH{-}CHO$ → CH_3 / CH_2CH_2CHO

The hidden catch to this scheme as a proof of stereochemistry is that only one of a pair of diastereoisomers (**39**) was available for testing. The fact that all the products were *trans*-1,2-disubstituted cyclopentanes provides assurance that retention can occur, but does not prove whether the *cis*-2-methylcyclopentylborane (not available from hydroboration) might not yield *trans* products in some cases, too. Subsequent investigation of the reaction with acrolein has shown it to be radical-catalyzed, involving in this case free 2-methylcyclopentyl radicals as chain carriers (see Chapter 7, Section III,B). Thus, the retention observed in the reaction of **39** with acrolein merely reflects a strong preference for reaction of the 2-methylcyclopentyl radical on the side *trans* to the methyl group. There is no great doubt that the other reactions listed are in fact stereospecific intramolecular rearrangements with retention, but this conclusion is based on other knowledge of these reactions and theoretical reasons, not the experiments cited.

Zweifel, Polston, and Whitney have reported complex but logical sequences converting *cis*-alkenylboranes (**40**) to *cis–trans* dienes.[41] These essentially synthetic experiments perhaps do not prove inversion at the displacement site and retention in the migrating group from a mechanistic purist's point of view, but it would be very hard to rationalize the stereospecificity on any other basis.

Zweifel and co-workers have demonstrated simultaneous retention of configuration of the migrating carbon and inversion of the carbon to which migration takes place in the reaction of sodium methoxide with *B*-bis(*trans*-2-

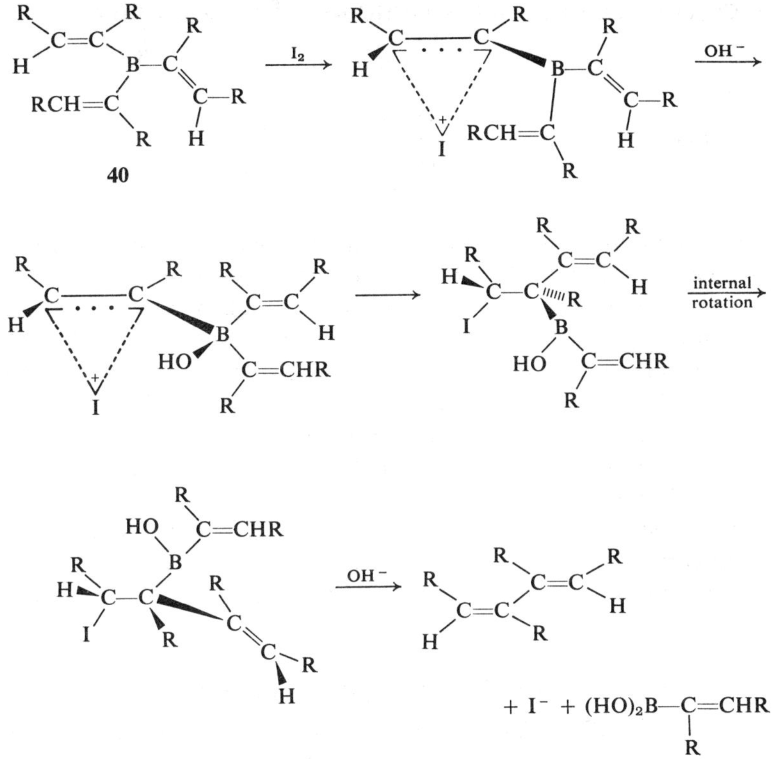

methylcyclohexyl)-*B*-1-bromo-1-hexenylborane (**41**).[42] The stereochemistry of the product was proved by conversion to the alkene with acetic acid (which cleaves the vinylic carbon–boron bond with retention).

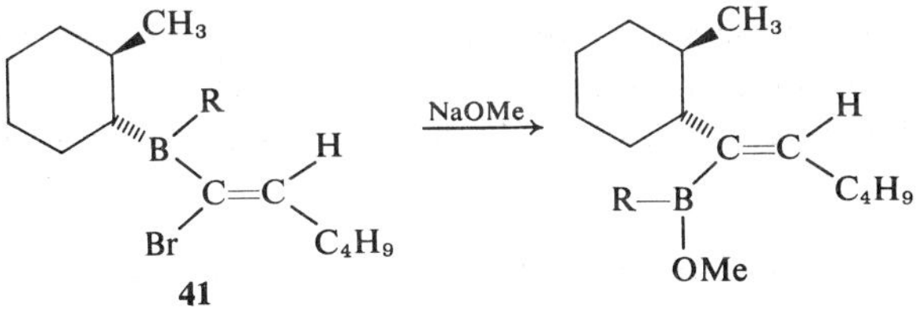

Since most of the other work published by Zweifel and co-workers is more synthetic than mechanistic, it will not be reviewed in detail, but this as well as H. C. Brown's hydroboration work provides excellent illustration of the utility of mechanistic thinking in devising useful synthetic schemes.

Alkyl migrations from boron to carbon have been exploited synthetically by Brown and co-workers, for example, in the reaction of trialkylboranes with ethyl bromoacetate and potassium *t*-butoxide to produce ethyl alkylacetates (**42**).[43] Again, the mechanism is complex and not really proved,

$$BrCH_2CO_2Et + {}^{-}OC(CH_3)_3 \longrightarrow Br\bar{C}HCO_2Et + HOC(CH_3)_3 \xrightarrow{R_3B} R_3\bar{B}{-}CH(Br)CO_2Et \longrightarrow$$

$$R_2B{-}CH(R)CO_2Et + Br^- \longrightarrow RCH{=}C(O{-}BR_2)(OEt)\ (?) \xrightarrow{HOC(CH_3)_3} RCH_2CO_2Et + R_2B{-}OC(CH_3)_3$$

42

though it would be hard to rationalize the product on any other basis. The *O*-boryl enolate immediately preceding **42** is conjecture by the author, though it is thermodynamically reasonable and would explain the extreme ease of deboronation. (Compare migrations of silicon atoms, Section V.) Many related examples of the formation and rearrangement of α-bromoalkylboron anions have been published by Brown and co-workers in the *Journal of the American Chemical Society*, but interesting as these synthetic applications are, they are outside the scope of this book. Another example is cited in Chapter 7, Section VI,A.

Groups other than halide can also be displaced in alkyl migrations from boron to carbon. For example, R_3B reacts with $Me_2S^+{-}C^-HCO_2Et$ in the same manner as with the anion from ethyl bromoacetate to yield RCH_2CO_2Et (**42**).[44]

C. Borane Adducts with Carbon Monoxide or Isonitriles

Another synthetically useful class of alkyl migrations from boron to carbon is based on borane–carbon monoxide adducts or their analogs, borane–isonitrile adducts. The original work in this field was reported by Hillman, who established that triethylborane reacts with carbon monoxide under pressure and the alkyl groups migrate, leading to a heterocyclic dimer (**43**) which rearranges in water at 140°C to triethylmethaneboronic acid (**44**).[45] No intermediate prior to **43** could be isolated, though monomeric

$$Et_3B + CO \longrightarrow Et_3\bar{B}{-}C{\equiv}\overset{+}{O} \longrightarrow Et_2B{-}C({=}O){-}Et \longrightarrow$$

$$[Et_2B{\leftarrow}O{=}C(Et){-}BEt_2{\leftarrow}O{=}C(Et)] \longrightarrow \underset{\mathbf{43}}{[Et{-}B{-}O{-}CEt_2{-}B(Et){-}O{-}CEt_2]} \xrightarrow{H_2O} \underset{\mathbf{44}}{Et_3C{-}B(OH)_2}$$

Et_2BCOEt is apparently trapped by acetaldehyde to form a heterocycle (**45**).

$$Et_2B{-}C({=}O){-}Et + CH_3CHO \longrightarrow [Et{-}B{-}O{-}CHCH_3{-}O{-}CEt_2] \quad \mathbf{45}$$

Ethyl migration from boron to carbon is required in the formation of **45**. Dimerization of Et_2BCOEt prior to the ethyl migrations required for formation of **43** seems likely, since the only conceivable rearrangement without prior dimerization would require formation of a three-membered B—O—C ring, and since the analogous dimer has been isolated in the series of compounds derived from boranes and isonitriles.

Casanova and Kiefer have studied the rearrangement of the triethylborane–phenyl isocyanide adduct.[46] No monomeric adduct could be isolated, but the initially isolated six-membered ring dimer (**46**) rearranged at 180°C to another (**47**), which rearranged to a five-membered ring heterocycle (**48**) at 300°C. The final step involves one more ethyl migration from boron to carbon.

$$\text{Ph—}\overset{+}{\text{N}}\equiv\overset{-}{\text{C}} + \text{BEt}_3 \longrightarrow \text{Ph—}\overset{+}{\text{N}}\equiv\text{C—}\overset{-}{\text{B}}\text{Et}_3 \longrightarrow \text{Ph—N=}\underset{\text{Et}}{\underset{|}{\text{C}}}\text{—BEt}_2 \longrightarrow$$

46 $\xrightarrow{180°\text{C}}$ **47** $\xrightarrow{300°\text{C}}$ **48**

46: six-membered ring Ph—N=C(Et)–BEt₂ ← N(Ph)=C(Et)–Et₂B ← N (Ph—N, C–Et, BEt₂, N—Ph, C–Et, Et₂B)

47: six-membered ring Ph—N–C(Et)(Et)–B—Et–N—Ph–C(Et)(Et)–B—Et

48: five-membered ring Ph—N—B(–C(Et)(Et)Et)–N—Ph–C(Et)(Et)–B—Et

Brown and Rathke have found that the high pressures used by Hillman for the carbon monoxide–borane reactions are not necessary, atmospheric pressure sufficing in diglyme at 125°C.[47] They added a number of synthetic innovations, but no basic knowledge about the alkyl migration mechanisms. However, they did find that the uptake of carbon monoxide is greatly accelerated by borohydrides[48] or trimethoxyaluminohydride.[49] Their postulated intermediate (**49**) is doubtful because all steps must be rapid and reversible up to the point where the reducing agent enters the reaction in order to account for acceleration by the reducing agent, and it seems unlikely that **48** would undergo a rapid alkyl migration in the wrong direction, from carbon to boron. On this basis, the intermediate is more likely something like **50**.

$$R_3B + CO \rightleftharpoons R_3\overset{-}{B}\text{—}\overset{+}{CO} \rightleftharpoons \underset{\textbf{49 (doubtful)}}{R_2B\text{—}\overset{\overset{\displaystyle O}{\|}}{C}\text{—}R} \xrightarrow{\text{LiAlH(OMe)}_3} R_2B\text{—}\underset{\underset{\displaystyle H}{|}}{\overset{\overset{\displaystyle O\bar{A}l(OMe)_3}{|}}{C}}\text{—}R$$

$$R_3\overset{-}{B}\text{—}\overset{+}{CO} \xrightarrow{\text{LiAlH(OMe)}_3} \underset{\textbf{50}}{R_3\overset{-}{B}\text{—}C(\text{=O}\rightarrow \text{Al(OMe)}_3)\text{H}} \xrightarrow{\text{R migration}} R_2B\text{—}CH(O\bar{A}l(OMe)_3)\text{—}R$$

D. An Alkylmanganese Carbonyl

This section on migrations of alkyl groups from metal atoms to carbon will be concluded with a reminder that a number of alkyl transition metal carbonyls undergo rearrangements analogous to those of trialkylborane carbonyls. The reverse of such a rearrangement, migration of a methyl group from carbon to manganese, has been observed to convert *trans*-acetyl(triphenylphosphine)manganese tetracarbonyl (**51**) to *cis*-methyl(triphenylphosphine)manganese tetracarbonyl (**52**).[50] Some aspects of the mechanism illustrated are speculative, but the stereochemistry indicates that the carbon monoxide expelled is not the same as the one from which the methyl group migrates, that is, carbonylation and decarbonylation are true migration reactions, not insertion of CO into the C—Mn bond and its reverse.

$CH_3C(=O)Mn(CO)_4PPh_3$ (**51**) $\xrightarrow{-CO}$ $CH_3C(=O)Mn(CO)_3PPh_3$ $\longrightarrow$ [H_3C bridging C(=O)–Mn, $Mn(CO)_3PPh_3$] $\longrightarrow$ $CH_3Mn(CO)_4PPh_3$ (**52**)

IV. Migrations of Group VI and VII Anions from Boron to Carbon

As pointed out in Section III,B, addition of an alkoxide ion to an α-haloalkylborane, $RCHXBR'_2$, results in alkyl migration from boron to carbon to form $RR'CHBROR''$. However, an α-haloalkaneboronic ester, $RCHXB(OR')_2$, has no alkyl group that can migrate, and an alkoxide ion migrates instead. The net result is merely a nucleophilic displacement, but the behavior of these compounds indicates that the alkoxide adds to the boron atom first, and the mechanism of the displacement is similar to that of an alkyl migration. It may be noted that a migration of hydride from boron to carbon has already been discussed in Section III,B.

Matteson and Mah noted the particularly facile displacement of bromide ion from dibutyl 1-bromo-3,3,3-trichloropropaneboronate (**16**) by butoxide ion to form dibutyl 1-butoxy-3,3,3-trichloropropaneboronate (**55**).[31] This contrasts with the characteristic elimination which occurs on base treatment of the analogous carboxylic ester (**56**). Accordingly, assistance to the displacement process by the neighboring boron is evident, and intermediate **53** and transition state **54** were postulated.

$$\underset{\mathbf{16}}{Cl_3CCH_2\underset{|}{C}H(Br){-}B(OBu)_2} + BuO^- \longrightarrow \underset{\mathbf{53}}{Cl_3CCH_2CH(Br){-}B(OBu)_3^-} \longrightarrow$$

$$\underset{\mathbf{54}}{\left[\text{H, }Cl_3CCH_2\text{, Br on C} \cdots \text{OBu} \cdots B(OBu)_2 \right]^-} \longrightarrow \underset{\mathbf{55}}{Cl_3CCH_2CH(OBu){-}B(OBu)_2}$$

$$\underset{\mathbf{56}}{Cl_3CCH_2CH(Br)CO_2Et} + \text{base} \longrightarrow Cl_3CCH{=}CHCO_2Et$$

Other evidence of participation by the neighboring boron atom appears in the reaction of the α-bromoalkaneboronic ester **16** with sodium iodide, which is 1.5–2.0 times faster than the reaction of allyl bromide with sodium iodide. Also, the reaction of the bromo compound **16** with butyl mercaptide ion readily yields the expected substitution product, $Cl_3CCH_2CH(SBu)B(OBu)_2$, but this is accompanied by small amounts of the α-butoxy compound **55**. This result is unusual because butyl mercaptan is a stronger acid than butanol, which requires that BuS^- be present in higher concentration than BuO^-, and furthermore BuS^- normally reacts much faster than BuO^- in nucleophilic displacements on carbon. However, the *B*-butoxy anion **53** would probably be present in substantial competition with the butylthio analog, $Cl_3CCH_2CHBrB(OBu)_2SBu^-$, because BuO^- would be more basic than BuS^- toward boron.

This competition between alkoxide and mercaptide ions has subsequently been verified with simpler boronic esters, which had to await development of a synthetic method. The reaction of dibutyl 1-bromo-1-phenylethaneboronate (**57**) with sodium thiophenolate in butanol yields only the butoxy substitution product and none of the expected sulfide.[51] Similarly,

$$\underset{\mathbf{57}}{Ph{-}C(CH_3)(Br){-}B(OBu)_2} + NaSPh + BuOH \longrightarrow Ph{-}C(OBu)(CH_3){-}B(OBu)_2 + HSPh + NaBr$$

$(CH_3)_2CBrB(OBu)_2$ and NaSPh in butanol yield the butoxy substitution product, $(CH_3)_2C(OBu)B(OBu)_2$, but $CH_3CHIB(OBu)_2$ yields the phenylthio-substitution product, $CH_3CH(SPh)B(OBu)_2$.[52] Evidently the outcome of the competition between basicity toward boron (BuO^-) and nucleophilicity toward carbon (PhS^-) is determined by the amount of branching at the α carbon. In accord with this trend, dibutyl iodomethaneboronate, $ICH_2B(OBu)_2$, alkylates anions from malonic ester and related compounds fairly efficiently, but its higher homologs give poor yields or fail altogether.[53]

The solvolysis of dibutyl 2-bromopropane-2-boronate, $(CH_3)_2CBrB(OBu)_2$, in aqueous ethanol was found by Matteson and Schaumberg to have a considerable degree of carbonium ion character in the transition state as shown by the Grunwald-Winstein m value, 0.7, but probably to involve a tightly bound solvent molecule, as shown by the rather negative ΔS^*, -15.5 eu.[52] Transition state **58** was postulated.

$$(CH_3)_2C(Br)\text{—}B(OR)_2 + H_2O \longrightarrow (CH_3)_2\overset{\delta+}{C}(\cdots OH_2)(\cdots Br^{\delta-})\text{—}B(OR)_2 \longrightarrow (CH_3)_2C(OH)\text{—}B(OR)_2$$

(R = H, Et, or Bu) **58**

The reaction of α-bromoalkaneboronic esters with sodium iodide in acetone to form the α-iodo compounds was found to be accelerated by the boron atom.[52] From competition experiments with allyl bromide, relative rates were found to be $(CH_3)_2CBrB(OBu)_2$, 1600; allyl bromide, 4000; $CH_3CHBrB(OBu)_2$, 6000; which may be compared to the relative values from the literature,[54] $(CH_3)_2CHBr$, 1.0; CH_3CH_2Br, 40. The boron-assisted transition states for these halide displacements probably resemble **58** postulated for the reaction with water. Where the nucleophile is not very basic, it is unlikely that any stable intermediate anion $R_2CBrB(OBu)_2I^-$ is formed, but probable that the vacant p orbital of the boron atom facilitates approach of the attacking or departure of the leaving halide ion, or both. The principle of microscopic reversibility must be considered in formulating a detailed path.[31,55] The boronic ester group is analogous to a carbonyl group in accelerating these displacements.

A facile S_N2' displacement, evidently assisted by neighboring boron since it would not be expected to proceed at all without assistance, has been found in the reaction of $Cl_3CCH{=}CHB(OBu)_2$ with PhMgBr, which yields $Cl_2C{=}CH\text{—}CHPhB(OBu)_2$.[56]

V. Migrations of Silicon from Carbon to Oxygen

Most of the familiar rearrangements involve migration of a carbon atom from an electron-rich to an electron-poor site. However, it is possible to imagine metal atoms migrating, and in view of the bonding properties of metals, migration toward an electron-rich site might be expected. Few mechanistically studied examples are available. One is the reaction of acylsilanes with diazomethane,[57] in which silicon migrates to carbon or oxygen with roughly equal probability. The *O*-silyl enol (**60**) is more stable than the isomeric α-silyl ketone (**59**), and the latter rearranges at 110°C, a second type of migration of silicon.[58] By the use of an optically active silyl group, Brook and co-workers showed that the silicon atom retains its configuration throughout these migrations.

(Np = 1-naphthyl)

Ph—Si(Me)(Np)—C(=O)—Ph $\xrightarrow{CH_2N_2}$ [Ph—Si(Me)(Np)—C(O⁻)(CH₂N₂⁺)—Ph] ⟶ Ph—Si(Me)(Np)—CH₂—C(=O)—Ph (**59**)

[intermediate] ⟶ Ph—Si(Me)(Np)—O—C(Ph)=CH₂ (**60**); **59** $\xrightarrow{110°C}$ **60**

60 $\xrightarrow{LiAlH_4}$ Ph—Si(Me)(Np)—H (retention known)

Biernbaum and Mosher have shown that the rearrangement of silylcarbinols to silyl ethers proceeds with inversion at carbon.[59] Treatment of (+)-phenyltriphenylsilylcarbinol-*O*-*d* (**61**) with triethylamine in chloroform yielded (+)-benzyl-α-*d* triphenylsilyl ether (**63**), which was cleaved to benzyl-α-*d* alcohol with lithium aluminum hydride. The absolute configurations of the starting material and product are both known. Ion pair **62** seems a reasonable structure for the transition state.

Ph_3Si—C(H)(OD)—Ph (**61**) $\xrightarrow{NEt_3}$ Ph_3Si—C(H)(O⁻)—Ph $D\overset{+}{N}Et_3$ ⟶ [Ph_3Si bridging O and C, C(H)(Ph)] $D\overset{+}{N}Et_3$ (**62**) ⟶

Ph_3Si—O—C(H)(Ph)(D) ≡ Ph_3SiO—C(Ph)—D (**63**)

Transition state **62** does not include a cyclic three-center bond symbol but indicates a pentacovalent silicon because there are two electron pairs involved in the bonding. The antisymmetric silicon *d* orbital fits correctly into this proposed bonding scheme. If the amount of electron density accepted by the silicon *d* orbital is small enough, the semi-open three-center bond symbol suggested for phenyl bridging between two aluminum atoms in Chapter 2, Section III,D, would be appropriate. This problem with the symbolism merely illustrates once more the difficulty of representing a continuum of possibilities with a small number of discrete symbols, especially when there is no information about just where the bonding lies along the continuum in a particular case.

Brook and Pascoe have found evidence in support of Biernbaum and Mosher's conclusion that the carbon atom undergoes inversion in this rearrangement.[60] From X-ray studies, the reduction of an asymmetric silyl ketone, MePhNpSiCOPh, with EtMgBr yields the opposite diastereoisomeric alcohol, MePhNpSiCHOHPh, from that predicted by Cram's rules,[60] reversing the basis for an earlier conclusion that the carbon retained its configuration in the silylcarbinol rearrangement.[61] It has been shown unequivocally that the silicon atom retains its configuration,[61] as required by transition state **62**.

VI. Cyclopropane Openings and Closings

A. General Mechanism

These reactions are closely related to alkyl migrations. The relationship is easily seen by considering the transition state (or intermediate) (**64**) for methyl migration in *n*-propyl cation. If **64** were to lose a proton from the methyl

$$CH_3CH_2\overset{+}{C}{}^{*}H_2 \longrightarrow H_2C\overset{+}{\cdots}C^{*}H_2 \;(\text{bridged by } CH_3) \longrightarrow \overset{+}{C}H_2C^{*}H_2CH_3$$

64

group it would become cyclopropane. If the methyl group were substituted with a metal in the first place, $MCH_2CH_2CH_2^+$, loss of the metal cation M^+ would become likely as closure toward the analog of **64** was approached. Not surprisingly, several examples of cyclopropane ring openings and closures involving compounds $MCH_2CH_2CH_2X$ (where X is halogen or the like) are known, and the generalized transition state expected is **65**.

$$MCH_2CH_2CH_2X \longrightarrow \underset{\overset{\delta+}{M}\cdots CH_2}{H_2C\text{—}CH_2\cdots X^{\delta-}} \longrightarrow \underset{CH_2}{H_2C\text{—}CH_2} + M^+ + X^-$$

65

Some structural details have been omitted from **65**. Where the nucleophile X^- departs, that carbon will probably be inverted. There is no *a priori* way to predict the stereochemistry of the simultaneous electrophilic displacement of the metal cation M^+, but experimental evidence indicates that inversion of the carbon is preferred and retention is possible if inversion is sterically prohibited. Finally, it should be noted that where the metal cation M^+ is a fairly strong electrophile and the anion X^- a good nucleofuge (leaving group), there may be an intermediate edge-metalated cyclopropane. For the analogous protonated cyclopropane, accurate SCF–MO calculations indicate that the edge-protonated form is the optimum geometry.[62] Bonding is due principally to an interaction of the proton with in-plane carbon 2*p* atomic orbitals of the $3e' \rightarrow 6a_1$ molecular orbital of cyclopropane, illustrated by structure **66**. The cyclopropane ring opens out so that the bond angle opposite

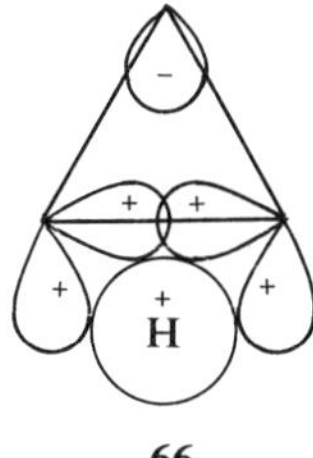

66

(+ and − represent symmetry, not charge; lobes of carbon *p* orbitals are shown)

the H^+ is 80° to give maximum stability, mainly because of the H^+—C nuclear repulsion. (The electrons would favor a 60° angle.) In a metal analog, for example with Hg^{2+} in place of H^+, there could also be interaction of *p* and *d* orbitals of the metal with antisymmetric components of the cyclopropane wave function, further strengthening the bonding. With those electrophiles such as mercury(II) which can open cyclopropane rings, edge metalation seems a likely first step. However, the ring opens along an edge away from the mercury atom in the case examined experimentally, inverting the carbon to which the mercury finally bonds. This may seem contrary to naive theoretical principles, but the lowest unoccupied cyclopropane molecular orbital to which a mercury *d* orbital might donate electrons is antisymmetric and thus antibonding along all three edges of the cyclopropane ring.

B. Cyclopropane Ring Closures

One of the pioneering examples of the sort of cyclopropane ring closure under discussion is that of 3-chloropropylboranes (**67**) reported by Hawthorne and Dupont in 1958.[63] Analogous tin compounds have been studied

$$R_2BCH_2CH_2CH_2Cl \xrightarrow{OH^-} R_2\bar{B}(OH){-}CH_2CH_2CH_2{-}Cl \longrightarrow \text{cyclo-}(CH_2)_3 + R_2BOH + Cl^-$$

67

by Kuivila and Scarpa.[64] Base catalysis is not required in order to make an effective electrofuge of the tin, and (3-tosyloxypropyl)trimethyltin (**68**) yields cyclopropane under solvolytic conditions. The tin compound solvolyzes 585

$$Me_3Sn{-}CH_2CH_2CH_2{-}OSO_2Ar \xrightarrow{HOAc}$$

68

$$[Me_3Sn\cdots CH_2(CH_2)\cdots CH_2\cdots OSO_2Ar] \longrightarrow \text{cyclo-}(CH_2)_3 + Me_3Sn^+ + ArSO_3^-$$

times faster than ethyl tosylate (data extrapolated to 120°C), indicating that loss of the tosylate and trimethyltin groups are concerted to some degree.

The stereochemistry of metal displacement in this type of cyclopropane ring closure was first examined in a norbornenyl system by Matteson and Waldbillig.[65] Reaction of either *exo*- or *endo*-5-norbornene-2-boronic acid (**69** or **70**) with mercuric chloride yields nortricyclylmercuric chloride (**71**). The *exo* isomer (**69**), in which the ring closure inverts the carbon from which boron is displaced, reacts much faster than the *endo* isomer (**70**), in which the ring closure results in retention of configuration.

B(OH)₂ — H — inversion — **69** — or — H — B(OH)₂ — retention — **70** — $\xrightarrow{HgCl_2}$ — ClHg — **71**

Since all the electrophilic displacements studied up to that time had shown a strong preference for retention, special care was taken to establish that the displacement of boron occurs during the rate-determining step and that the reaction is a direct ring closure, not an attack by mercury on the boron-bearing carbon followed by rearrangement. (However, claims for a "first" must be modified by noting that Cristol and LaLonde had previously encountered an analogous inversion in the bromination of quadricycloheptanedicarboxylic acid, though without advertising its significance to electrophilic displacement and without proving the mechanism.[66]) Evidence supporting carbon–boron bond breaking in the transition state was obtained by measurement of the $^{10}B/^{11}B$ isotope effects.[67] The measurements of isotope ratios were made by determining the ^{10}B neutron absorption of boric acid samples from reactions run to half completion, comparing the boric acid generated in the reaction with that derived by degradation of the unreacted boronic acid. The $^{10}B/^{11}B$ rate ratio for the *exo* isomer (**69**) was found to be 1.033 $\pm$ 0.004 and that for the *endo* isomer (**70**) to be 1.027 $\pm$ 0.003.

Kinetic measurements established that the rate law for the *exo* isomer (**69**) in aqueous acetone buffered with phthalic acid and potassium acid phthalate is[68]

$$\frac{-d[RB(OH)_2]}{dt} = \frac{k[RB(OH)_2][HgCl_3^-][RCO_2^-]}{[Cl^-]}$$

The *endo* isomer (**70**) was not tested with varying initial concentrations of reagents, but it did yield good straight lines when plots were made according to the integrated rate law.

An integrated rate law would not normally be used with a rate expression of such complexity, since it is usually possible to arrange reaction conditions to yield pseudo-first-order kinetics, but this simplification was precluded in this reaction by the limitations of the analytical method available at that time (treatment of the boronic acid with hydrogen peroxide followed by gas chromatography of the resulting alcohol). It might be noted that in cases of necessity such complex rate expressions can be integrated by partial fractions. For the equation

$$\frac{dx}{dt} = \frac{k(a - x)(b - x)(c - x)}{(d + x)}$$

the integrated rate equation is

$$kt = A \ln\left[\frac{a}{(a - x)}\right] + B \ln\left[\frac{b}{(b - x)}\right] + C \ln\left[\frac{c}{(c - x)}\right]$$

where

$$A = \left[\frac{(a + d)}{(a - b)(a - c)}\right]$$

$$B = \left[\frac{(b + d)}{(b - a)(b - c)}\right]$$

$$C = \left[\frac{(c + d)}{(c - a)(c - b)}\right]$$

The reader will be comforted by the thought that he will be most unlikely to ever need this particular equation in his own work, but it is included here to point out that such problems can be solved.

Returning to the chemistry itself, the *exo* isomer (**69**) reacted 400 ± 40 times faster than the *endo* isomer (**70**) at 25°C and 270 times faster at 45°C.[68] The difference in ΔG^* values between the two reactions is 3.55 kcal/mole at both temperatures and the difference in ΔH^* is 4.2 ± 1.0 kcal/mole, or in other words, the two are the same within experimental error. Thus, the slow reaction of the *endo* isomer (**70**) is not the result of some strange entropy effect or change in mechanism.

A mechanism consistent with the rate law is illustrated for the *exo* isomer **69** going to intermediate **72** and transition state **73**. Intermediate **72** is a

$HgCl_2$ + [**69**, $B(OH)_2$, H] ⟶ [**72**, $\overset{\delta-}{Cl_2Hg}$, $\delta+$, $B(OH)_2$, H] $\xrightarrow{RCO_2^-}$

[**73**, Cl_2Hg, OH, B---^-O_2CR, OH] ⟶ $\xrightarrow{H_2O}$ Cl^- + ClHg [**71**, H]

+ $B(OH)_3$ + RCO_2H

mercury-olefin π complex related to those thought to be intermediates in oxymercuration (see next chapter). A reversible addition of ClHg–Cl across the double bond would be equally consistent with the kinetics and would not change the essential nature of the ring-closure step. Two details of the kinetics differ from other mercuri-deboronations subsequently studied (see Chapter 3, Section II,D). In aqueous acetone the $HgCl_2$ is largely complexed to $HgCl_3^-$ by excess Cl^-, which does not happen in methanol, and in this reaction the carboxylate anion appears to be the effective base rather than the hydroxide or methoxide ion found in direct displacements in alcoholic media, though the amount of kinetic data available is too small to establish this beyond question.

The evidence so far cited does not rule out an alternative explanation not involving displacement with inversion in the rate-determining step but direct displacement to form the norbornenylmercuric chloride (**74**) with subsequent rapid rearrangement to the stable nortricyclyl compound (**71**). If such a route were involved, the *exo*-boronic acid (**69**) could undergo moderately rapid displacement with retention to form *exo*-5-norbornenyl-2-mercuric chloride (**74**), which might then rearrange very rapidly by some mechanism other than electrophilic displacement to **71**, leaving no proof that electrophilic

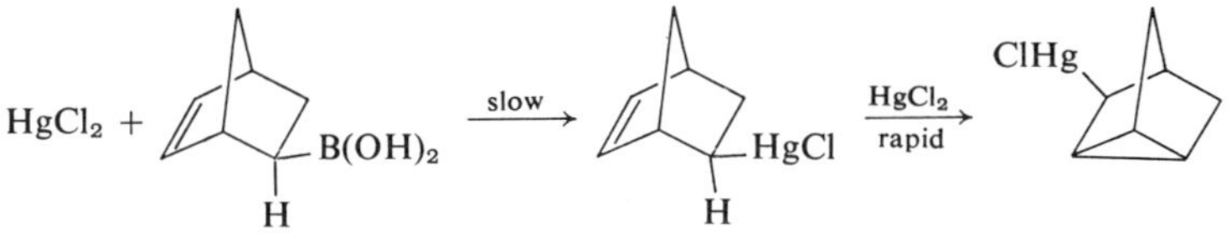

$HgCl_2$ + [69: B(OH)$_2$, H] —slow→ [74: HgCl, H] (unlikely) —$HgCl_2$, rapid→ [71: ClHg]

displacement with inversion has occurred at all. The slower reaction of the *endo*-boronic acid (**70**) would then be attributed to the steric hindrance typical of reactions at *endo* sites.

The initial argument against such a mechanism was that the reaction is many orders of magnitude too fast to be attributed to direct displacement, since the saturated analog norbornaneboronic acid could not be induced to react with mercuric chloride at all. This argument has gained considerable support from the more recent studies of direct mercuri-deboration (see Chapter 3, Section II,D) in which the benzylboronic esters appear to react more slowly than *exo*-norborneneboronic ester (**69**), though neither the conditions nor the rate laws are the same.

A very strong argument against **74** being an intermediate is provided by the analogous bicyclooctyl series of compounds.[69] In this case, *exo*-bicyclo-[2.2.2]oct-5-ene-2-boronic acid (**75**) reacts with mercuric chloride in the presence of sodium chloride to form tricyclo[2.2.2.0^{2,6}]octyl-3-mercuric chloride (**76**), which in the presence of mercuric chloride isomerizes to *exo*-bicyclo[2.2.2]oct-5-enyl-2-mercuric chloride (**77**). Thus, the stable direct displacement product (**77**) cannot be an intermediate in the formation of the unstable transannular displacement product (**76**) in this case, and since the reactions in the norbornenyl series are considerably faster than in the bicyclooctyl series, direct displacement can be dismissed as extremely improbable in the norbornenyl series as well.

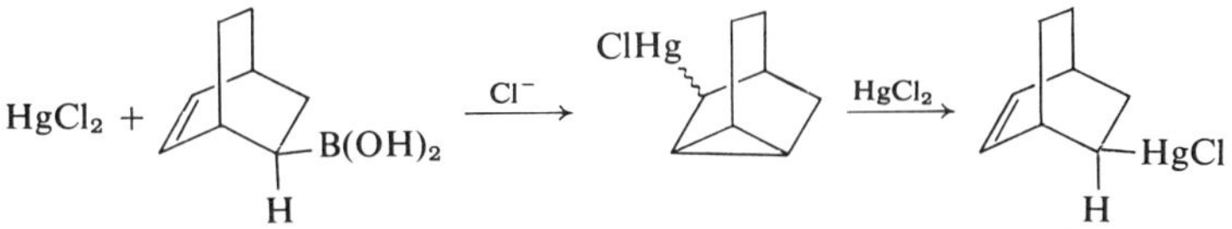

$HgCl_2$ + [75: B(OH)$_2$, H] —Cl^-→ [76: ClHg] —$HgCl_2$→ [77: HgCl, H]

The reaction of the *endo*-bicycloocteneboronic acid (**78**) with mercuric chloride proved so sluggish that it could not be run under conditions which would preserve the tricyclic mercury compound (**76**) and only rather small conversions and yields of the bicyclooctenylmercuric chloride (**77**) could be obtained. These reactions, especially of the *endo*-boronic acid (**78**), were not clean enough to justify serious attempts at kinetic measurements. However, it did appear that the *exo* isomer (**75**) reacts hundreds or perhaps thousands of

$HgCl_2$ + —H, $B(OH)_2$ $\xrightarrow{\text{slow}}$ —HgCl, H

78 **77**

times faster than the *endo* (**78**), thus providing another example of preferred inversion at the electrophilic displacement site in cyclopropane ring closure.

The role of the mercuric chloride in the isomerization of **76** to **77** was not established, but it might be either a transannular electrophilic displacement or some sort of carbonium ion or radical process catalyzed by $HgCl^+$. The fact that sodium chloride was fairly effective in suppressing the reaction does imply involvement of $HgCl^+$ or RHg^+.

The stereochemistry of the bicyclic mercury compound (**77**) was supported by bromodemercuration (retention) to the known corresponding bromide. However, it proved impossible to do anything to the tricyclic isomer (**76**) without rearranging it, and the configuration of the mercury is unknown.

A really determined referee could (and once did) find a couple more points to doubt. First, the Lewis acidity of the *exo*-boronic acid (**69**) should exceed that of the *endo* (**70**) for steric reasons, and the relative acidities should be determined to find out how much of the rate difference could be the result of mere interaction with the base (carboxylate ion) required in the rate-determining step. It was found that the difference in K_a's in aqueous glycerol is by a factor of 2.5–3,[65] which lacks a factor of about 150 of accounting for the 400-fold difference in rates. Also, such steric factors should be much less important in accounting for the behavior of the bicyclooctyl compounds (**75** and **78**). The second point is that the *endo* compound (**70**) might complex with the mercuric chloride and be less available for reaction because of that. However, when the *exo*- and *endo*-norborneneboronic acids (**69** and **70**) were allowed to compete for a limited amount of mercuric chloride, no gross change in the rate of reaction of the *exo* isomer resulted,[65] and in the bicyclooctyl series the *exo* isomer (**75**) reacted selectively with one equivalent of mercuric chloride in the presence of about four equivalents of *endo* isomer (**78**).[68] Therefore, complexing is not involved.

C. Cyclopropane Ring Openings

Inversion at the site of electrophilic displacement in a simple cyclopropane ring opening with mercuric acetate has been demonstrated by DeBoer and DePuy.[70] Mercuric acetate opens cyclopropanols or cyclopropyl methyl ethers to β-acetoxymercuri ketones in acetic acid at room temperature. Inversion was demonstrated with two diastereoisomers, 1-phenyl-*cis,trans*-2,3-dimethylcyclopropanol (**79**) yielding the *erythro* ketone **80** and 1-phenyl-

cis,cis-2,3-dimethylcyclopropanol (**81**) the *threo* ketone (**82**). The structures of the β-acetoxymercuri ketones (**80** and **82**) were proved by reaction with bromine in pyridine, which replaces mercury by bromine with retention

79 $\xrightarrow[\text{inversion}]{Hg(OAc)_2}$ **80**

81 $\xrightarrow[\text{inversion}]{Hg(OAc)_2}$ **82**

retention

(Chapter 3, Section III, B) and in this case yielded β-bromo ketones having known steric configurations. The ring-opening reaction was also found to be first order in the cyclopropanol and first order in mercuric acetate. The stereochemistry was additionally demonstrated to be inversion with the methyl ether of **79**. The hydroxycyclopropanes open about 10^4 times faster than the hydrocarbon analogs, methoxycyclopropanes about 0.05–0.1 as fast as hydroxycyclopropanes, and acetoxycyclopropanes about 10^{-4}–10^{-3} as fast, as if there is substantial carbonium ion character at the carbon of the carbonyl group being generated.

DePuy and co-workers had previously demonstrated that protons open cyclopropane rings with retention,[71] bromine with inversion,[72] at the site of electrophilic attack.

Ouellette and co-workers have carried out extensive kinetic studies of arylcyclopropane ring opening by mercury(II),[73] thallium(III),[74] and lead(IV)[75] acetates in acetic acid. The simple arylcyclopropanes used did not allow proof of stereochemistry, which was assumed to be retention in the interpretations but in view of DePuy's work is more likely inversion, or may even vary with the electrophile.

The cleavage of arylcyclopropanes by mercuric acetate in acetic acid at 25–75° is first order in each reactant. Electron-releasing substituents facilitate the reaction. From the series *p*-MeO, *p*-Me, *m*-Me, H, *p*-Cl, and *m*-Cl with σ^+ constants, ρ is -3.2 at 50°C. Activation parameters are shown in Table 4-2. From the near constancy of the ΔS^* values and the steady trend in ΔH^* values, it is apparent that the mechanism remains constant and the rates are

TABLE 4-2

ACTIVATION PARAMETERS FOR THE CLEAVAGE OF ARYLCYCLOPROPANES BY MERCURIC ACETATE[a]

Substituent	ΔH^* (kcal/mole)	ΔS^* (eu)
p-MeO	16.0	−10.6
p-Me	17.0	−13.6
m-Me	18.5	−12.4
H	19.1	−11.7
p-Cl	19.9	−11.4.
m-Cl	21.6	−9.8

[a] From R. J. Ouellette, R. D. Robins, and A. South, Jr., *J. Amer. Chem. Soc.* **90,** 1619 (1968).

controlled mainly by ΔH^*. The very negative ρ approaches that for the solvolysis of 2-phenyl-2-propyl chlorides, −4.54 in 90% acetone,[76] indicating substantial benzyl cation character in the transition state. The only moderately negative entropy of activation suggests loss of a ligand from the mercury in the transition state.[73] A mechanism consistent with the experimental observations is illustrated for phenylcyclopropane (**83**). Edge mercuration of the

$$\underset{\textbf{83}}{\text{PhCH}\langle(CH_2)_2\rangle} + Hg(OAc)_2 \cdot 2\,HOAc \rightleftharpoons \underset{\substack{\textbf{84}\\ \text{(conjectural)}}}{\text{PhCH}\langle(CH_2)_2\rangle \cdots Hg(OAc)} + 2\,HOAc \longrightarrow$$

$$\underset{\textbf{85}}{\overset{\delta+}{\text{PhCH}}(\cdots HOAc)\text{—}CH_2\text{—}CH_2\cdots Hg(OAc)\cdots \overset{\delta-}{OAc}} \longrightarrow \underset{\textbf{86}}{Ph\text{—}CH(OAc)\text{—}CH_2\text{—}CH_2HgOAc}$$

cyclopropane as in conjectural intermediate **84** would not be easy to prove or disprove. The transition state **85** might be approached directly from $Hg(OAc)_2$ (with one or both acetic acid molecules of solvation lost) and phenylcyclopropane (**83**), though the intermediate complex (**84**) might provide an energy trough for the loss of the ligands. The ligands lost would have to be acetic acid, not acetate ion, to be consistent with the kinetics. It is doubtful whether mere loosening of the Hg–OAc bond as depicted in **85** could account for the

ΔS^*, especially since incipient carbonium ion formation is likely to tie up another solvent molecule as shown. Formation of the major product **86** does not necessarily require a free carbonium ion intermediate.

Cleavage of the same set of six arylcyclopropanes with thallium(III) acetate goes by way of a transition state having more carbonium ion character, $\rho = -4.3$ at 50°C.[74] The presumed 3-aryl-3-acetoxypropylthallium(III) diacetate (**87**) and 1-phenylpropenyl-3-thallium(III) diacetate (**88**) intermediates are unstable and decompose to the corresponding acetoxy compounds (**89** and **90**) and thallium(I) acetate. The reaction is first order in

$$\text{Ar—CH}\langle(\text{CH}_2)_2\rangle + \text{Tl(OAc)}_3 \longrightarrow$$

$$\underset{\mathbf{87}}{\text{Ar—}\underset{\text{OAc}}{\underset{|}{\text{C}}}\text{HCH}_2\text{CH}_2\text{Tl(OAc)}_2} + \underset{\mathbf{88}}{\text{Ar—CH=CH—CH}_2\text{Tl(OAc)}_2} \longrightarrow$$

$$\underset{\mathbf{89}}{\text{Ar—}\underset{\text{OAc}}{\underset{|}{\text{C}}}\text{HCH}_2\text{CH}_2\text{OAc}} + \underset{\mathbf{90}}{\text{Ar—CH=CH—CH}_2\text{OAc}} + \text{TlOAc}$$

arylcyclopropane and first order in thallium(III) acetate, with some kinetic complication arising because TlOAc and $Tl(OAc)_3$ form a complex $Tl_2(OAc)_4$ (which has been isolated) which is less reactive than $Tl(OAc)_3$. Activation parameters for the cleavage of phenylcyclopropane by thallium acetate are $\Delta H^* = 12.4$ kcal/mole and $\Delta S^* = -29.2$ eu. The very negative ΔS^* is typical of bimolecular reactions, perhaps with some additional ring formation (such as bonding of thallium to the edge of the cyclopropane ring) and no net loss of acetic acid ligands, with solvent perhaps providing nucleophilic assistance to ring opening. It is hard to say whether the stereochemistry of thallium attack would necessarily be the same as that of mercury attack (inversion) and in view of the complexity of the reaction, experimental proof of stereochemistry would be difficult at best.

Lead(IV) acetate is qualitatively similar to thallium(III) acetate, the probable organolead intermediates (**91** and **92**) being likewise unstable and yielding the acetoxy products (**89** and **90**).[75] However, all the quantitative aspects of the reaction are quite different. The same series of six arylcyclopropanes yields $\rho = -1.75$, indicating much less carbonium ion character in the transition state. The reaction is greatly accelerated by perchloric acid, suggesting that $Pb(OAc)_3^+$ is a much more reactive electrophile than $Pb(OAc)_4$. The ρ value for the reaction of $Pb(OAc)_3^+$ is -1.3, indicating even less carbonium ion character than with $Pb(OAc)_4$. This anomaly was

$$\text{Ar—CH}\begin{matrix}\diagup CH_2 \\ | \\ \diagdown CH_2\end{matrix} + Pb(OAc)_4 \longrightarrow$$

$$\underset{\substack{|\\ \text{OAc}\\ \mathbf{91}}}{\text{Ar—CHCH}_2\text{CH}_2\text{Pb(OAc)}_3} + \underset{\mathbf{92}}{\text{Ar—CH=CH—CH}_2\text{Pb(OAc)}_3} \longrightarrow$$

$$\underset{\substack{|\\ \text{OAc}\\ \mathbf{89}}}{\text{Ar—CHCH}_2\text{CH}_2\text{OAc}} + \underset{\mathbf{90}}{\text{Ar—CH=CH—CH}_2\text{OAc}} + Pb(OAc)_2$$

explained away by referring to $Pb(OAc)_3^+$ as "less discriminating,"[75] which could be a contributing factor, but the overall result still seems a bit odd.

It is clear that $Pb(OAc)_4$ does not dissociate to $Pb(OAc)_3^+$ and OAc^- because the ρ values differ and the direction of the difference cannot be explained by assuming that a nucleophile is required in the transition state, OAc^- in the reaction with more carbonium ion character and HOAc in the one with less. Also, the reaction of $Pb(OAc)_4$ does not involve prior dissociation to $Pb(OAc)_3^+$ without further participation by OAc^- because lithium acetate does not retard the reaction, in addition to the evidence already mentioned.

At this point, about the only reasonable explanation of the behavior of the lead(IV) reactions is that the transition state occurs at a considerably earlier point along the reaction coordinate than it does with the mercury(II) or thallium(III) acetates. This would not be a satisfactory interpretation for a one-step reaction because lead(IV) acetate is the least reactive electrophile of the series, second-order k's for attack on phenylcyclopropane at 25°C being $Tl(OAc)_3$, 2.0×10^{-3}; $Hg(OAc)_2$, 1.7×10^{-4}; $Pb(OAc)_4$, 3.1×10^{-6}; or relative k's 12, 1, and 0.0018, respectively. The Hammond postulate would require lead(IV) acetate to have its transition state later along a one-step reaction coordinate, and therefore to be more discriminating and to show more carbonium ion character in the transition state (more negative ρ), contrary to fact.

A two-step mechanism accommodates the facts easily. For $Pb(OAc)_4$, the highest energy barrier could be the transition state **93** for formation of a lead(IV)–cyclopropane complex (**94**), which once formed might go irreversibly on to products. There would, of course, be little carbonium ion character at the benzyl carbon atom in this stage, and the less negative ρ for $Pb(OAc)_3^+$ attack could well be attributed to the transition state **95** occurring earlier along the reaction coordinate for formation of the complex (**93**). The thallium(III) and mercury(II) complexes analogous to **94** would have to be assumed to be formed reversibly, with less of an energy barrier to the displacement of acetic acid ligands from the divalent or trivalent metal than for displacement of acetate from the lead. The rate-determining step would then be the ring opening to a benzyl cation, probably assisted by the acetic acid solvent. The activation parameters for the lead(IV) reactions are consistent with this hypothesis. The ΔH^* values are nearly constant: p-CH_3, 19.2; H, 19.6; m-Cl, 19.8 kcal/mole. The ΔS^* values are, for p-CH_3, -16.4; H, -17.8; m-Cl, -20.4 eu; values which are in an appropriate range for some sort of simple bimolecular displacement without too much involvement of solvent.

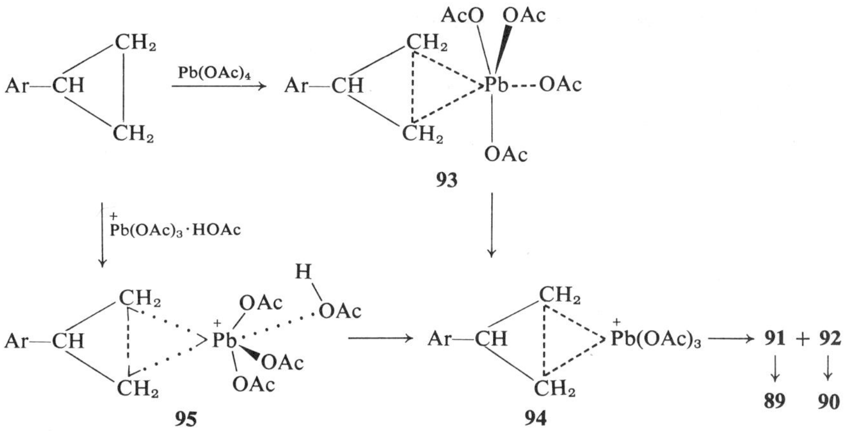

This interpretation is, of course, speculative, and differs in many details from Ouellette's suggestions,[73–75] which attributed most of the ΔS^* variations to differences in solvation of the metal atom. The trouble is that $\Delta S^* = -11.7$ for Hg(II), -29.2 for Tl(III), and -17.8 for Pb(IV) with phenylcyclopropane does not make a coherent three-point series, as there ought to be if the cause were a progressive change in the number of solvent molecules in the transition state as the metal valence increased. In summary, all the quantitative data suggest a different sort of transition state for the lead(IV) reaction, a rational explanation can be invented to fit the available data in terms of one type of two-step mechanism for all the reactions, but the data are far from sufficient to define the mechanism uniquely.

D. Carbanion Rearrangement by way of a Cyclopropane

Grovenstein and co-workers have investigated the migration of biphenylyl groups in alkali metal compounds.[77,78] 1,1-Dideutero-2-p-biphenylylethylpotassium (**96**) and the analogous cesium compound undergo rearrangement in refluxing tetrahydrofuran to randomize the CD_2 group between the 1- and

2-positions.[77] The postulated transition state or intermediate **97** is a cyclopropane. The corresponding lithium compound fails to rearrange in tetrahydrofuran at 0°C, conditions under which most of it decomposes.[77] However, the more highly substituted analog $(PhC_6H_4)_3CCH_2Li$ rearranges to $(PhC_6H_4)_2CLiCH_2C_6H_4Ph$.[78] The biphenylyl group evidently bears a high

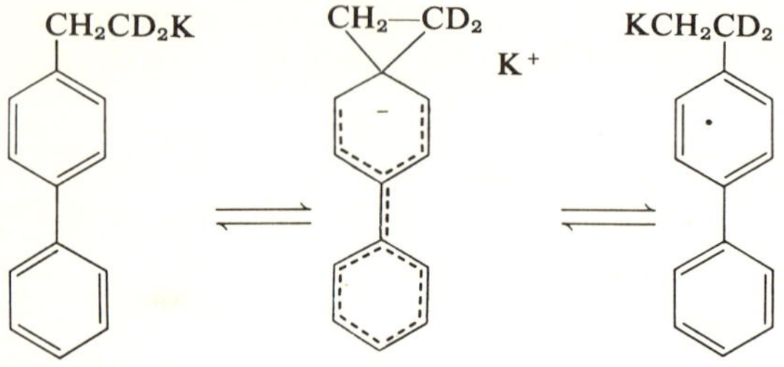

96 **97**

negative charge in the transition state, as shown by the 24.5-fold faster migration of a *p*-biphenylyl than a *m*-biphenylyl group where both are present in the starting compound to compete.

It may be noted that these aryl migrations in the carbanions are symmetry-allowed, as indicated by the cyclopropane structure **97**, but that migrations of alkyl groups in carbanions are symmetry-forbidden.[77] Benzyl groups can migrate, but evidently only by dissociation of the carbanion to olefin and benzyl anion (Chapter 5, Section VIII,C). Cyclic three-center bonding is impossible where two electron pairs are strongly involved.

VII. Electrophilic Displacements with Allylic Rearrangement

A. General Considerations

It is fairly easy to outline possible mechanisms for electrophilic displacement accompanied by allylic rearrangement, or S_E2' reactions, but there is not much detailed mechanistic information about such reactions in the literature. Possibilities for attack of an electrophile halide, EX, on an allylmetal compound $CH_2{=}CH{-}CH_2M$ include the transition state for direct S_E2' displacement, (**98**), intermediates involving π complexes (**99** and **100**), and the possible intermediate adduct of EX with the double bond (**101**), in addition to normal S_E2 displacement without rearrangement.

Because double bonds are readily susceptible to electrophilic attack, S_E2' and related reactions are expected to be more competitive with direct displacements than are their nucleophilic counterparts, S_N2' reactions.

$$EX + CH_2{=}CH{-}CH_2M \longrightarrow X{\cdots}E{\cdots}CH_2{\overset{\cdots}{=}}CH{\overset{\cdots}{=}}CH_2{\cdots}M$$

98 (stereochemistry unknown)

E^+ → $H_2C{-}CH{-}CH_2{-}M$ (bridged by E, +) **99** ⟶ $E{-}CH_2{-}CH{-}CH_2$ (bridged by M, +) **100**

X^- ⇌ $ECH_2CH(X)CH_2M$ **101**

101 $\xrightarrow{-MX}$ $E{-}CH_2{-}CH{=}CH_2$ ← **99**, **100**

B. Allylmercuric Salts

Winstein *et al.* observed electrophilic displacement with allylic rearrangement in the protolysis of crotylmercuric bromide, $CH_3CH{=}CH{-}CH_2HgBr$, with hydrogen chloride or perchloric acid.[79] The product is essentially pure 1-butene, $CH_3CH_2{-}CH{=}CH_2$, which clearly requires some sort of concerted displacement with rearrangement. The crotyl compound reacts with HCl about 10^7 times faster than does *n*-butylmercuric bromide. A simple S_E2' mechanism was suggested for the reaction of crotylmercuric bromide with perchloric acid, and a cyclic S_E1' path for the reaction with HCl in ethyl acetate or ether.

The cleavage of allylmercuric iodide by acid is probably also an S_E2' reaction. Kreevoy and co-workers found that the reaction follows the Brønsted catalysis law with $\alpha = 0.67$ with a series of acids such as acetic, formic, fluoroacetic, and others.[80]

C. Allyltins

The degree to which allylic rearrangement competes with direct displacement depends strongly on the electrophile and to some degree on the substrate. Kuivila and Cochran have found that allenyltin compounds such as **102** react with halogens at the allylic position.[81] However, protons (HCl in

$$(CH_3)(H)C{=}C{=}C(CH_3)(SnMe_3) + Br_2 \longrightarrow CH_3{-}CH(Br){-}C{\equiv}C{-}CH_3 + BrSnMe_3$$

102

moist methanol) attack both by direct and by allylic rearrangement paths. The relative amounts of direct (S_E2) and allylic (S_E2') displacement vary with substrate structure in ways that would be expected. A phenyl substituent

$$\underset{\mathbf{102}}{(CH_3)(H)C{=}C{=}C(CH_3)(SnMe_3)} \xrightarrow{H^+} (CH_3)(H)C{=}C{=}C(CH_3)(H) + CH_3{-}CH_2{-}C{\equiv}C{-}CH_3$$

(retention expected, not proved)

tends to end up conjugated with multiple bonds and steric hindrance tends to retard attack at a hindered site. These trends are summarized in Table 4-3.

TABLE 4-3

PROTOLYSIS RATES OF ALLENYLTIN COMPOUNDS[a]

Compound	k_2 for S_E2	k_2 for S_E2'
Me(H)C=C=C(Me)(SnMe$_3$)	0.296	0.223
Ph(H)C=C=C(Me)(SnMe$_3$)	0.0441	0.0670
Me(H)C=C=C(Ph)(SnMe$_3$)	0.106	0.0188
Me(H)C=C=C(Me)(SnEt$_3$)	0.100	0.199
Me(H)C=C=C(Me)(SnPh$_3$)	0.00082	0.00657

[a] From H. G. Kuivila and J. C. Cochran, *J. Amer. Chem. Soc.* **89**, 7152 (1967).

The cleavage of $Et_3SnCH_2CH{=}CHCH_3$ with mercuric iodide has been reported by R. M. G. Roberts to give the unrearranged product, $IHgCH_2CH{=}CHCH_3$.[82] It was not proved that the less stable (and unavailable) methallyl isomer, $CH_2{=}CHCH(HgI)CH_3$, would not rearrange under the reaction conditions. It was argued that since $Et_3SnCH_2CH{=}CH_2$ reacts faster than $Et_3SnCH_2CH{=}CHCH_3$, the displacement is probably of the direct S_N2 type, but this argument ignores steric factors. In view of the uncertainty of the interpretation, there seems no point in reviewing further details of the rate data.

Kitching *et al.* have found that allylic rearrangement accompanies the reaction of sulfur dioxide with allenyltrimethyltin (**103**), propargyltriphenyltin (**104**), and crotyltrimethyltin (**105**).[83] The reaction rates appeared qualitatively to be higher than usual for replacement of tin by SO_2.

$$\underset{\mathbf{103}}{Me_3Sn{-}CH{=}C{=}CH_2} + SO_2 \longrightarrow HC{\equiv}C{-}CH_2{-}\overset{O}{\overset{\|}{S}}{-}OSnMe_3$$

$$\underset{\mathbf{104}}{Ph_3SnCH_2C{\equiv}CH} + SO_2 \longrightarrow CH_2{=}C{=}CH{-}\overset{O}{\overset{\|}{S}}{-}OSnPh_3$$

$$\underset{\mathbf{105}}{Me_3SnCH_2{-}CH{=}CH{-}CH_3} + SO_2 \longrightarrow CH_2{=}CH{-}\underset{CH_3}{\underset{|}{CH}}{-}\overset{O}{\overset{\|}{S}}{-}OSnMe_3$$

D. Benzylmetal Compounds

Benzylmetal systems can show an appreciable amount of allylic rearrangement even though this path costs the aromatic stabilization energy of the benzene ring. Hanstein and Traylor found competition between direct displacement, S_E2' displacement, and electrophilic substitution in the ring during deuterodeboronation of a benzylboronic ester (**106**).[84] The ring substitution is greatly facilitated by hyperconjugative electron donation from the C—B bond (Chapter 5, Section II), which amounts to a loosening of the C—B bond and therefore is in a sense an S_E2' displacement which

$$\underset{\mathbf{106}}{C_6H_5{-}CH_2B(OR)_2} \xrightarrow{D^+} p\text{-}D{-}C_6H_4{-}CH_2B(OR)_2 + o\text{-}D{-}C_6H_4{-}CH_2B(OR)_2$$

$$+\ C_6H_5CH_2D + o{-}D{-}C_6H_4CH_2D + p{-}DC_6H_4CH_2D$$

stops short of going all the way. The σ^+ constant of the $-CH_2B(OR)_2$ substituent was estimated to be about -1.1.

Reutov and co-workers similarly found that the reaction of benzylmercuric chloride with DCl yields mostly α,*o*-dideuterotoluene, *o*-deuterated benzylmercuric chloride, and little if any α-deuterotoluene without ring deuteration.[85] Acetylation of benzylmercuric chloride with acetyl chloride and aluminum bromide yielded α,*p*-diacetyltoluene (**107**) as the major product.[86]

Another example of S_E2' displacement in a benzylmetal system is the reaction of $PhCH_2MgCl$ with $ClCH_2OCH_3$, which was found by Benkeser and de Talvo to yield a mixture of $PhCH_2CH_2OCH_3$, *o*- and *p*-$CH_3OCH_2C_6H_4CH_3$, and *o*- and *p*-$CH_3OCH_2C_6H_4CH_2CH_2OCH_3$.[87] The

CH_2HgCl $\xrightarrow[AlBr_3]{CH_3COCl}$ (CH_2; H, $COCH_3$) $\longrightarrow$ (CH_2COCH_3; $COCH_3$)

107

so-called S_E1 protodemercuration of γ-pyridylmethylmercuric chloride (Chapter 3, Section IV,B) is closely related to these S_E2' displacements.

E. A Norbornyltin Rearrangement and Elimination

Davis and co-workers have reported the rearrangement and elimination reaction of *anti*-7-trimethylstannyl-*exo*-2-norbornanol (**108**),[88] which probably passes through an intermediate (**109**) resembling model **99** in Section VII,A.

Me_3Sn, H, OH, H (**108**) $\xrightarrow[CCl_4]{PCl_3}$ Me_3Sn, H, + (**109**) $\longrightarrow$ Me_3Sn^+ + (norbornene) ≡ (norbornene)

This is a good example of a reaction which does not fit into any simple set of preconceived mechanistic labels, yet is easily understood with the aid of structural formulas.

References

1. H. G. Kuivila and A. G. Armour, *J. Amer. Chem. Soc.* **79**, 5659 (1957).
2. H. G. Kuivila, *J. Amer. Chem. Soc.* **76**, 870 (1954); **77**, 4014 (1955).
3. H. G. Kuivila and R. A. Wiles, *J. Amer. Chem. Soc.* **77**, 4830 (1955).
4. D. H. McDaniel and H. C. Brown, *J. Amer. Chem. Soc.* **77**, 3756 (1955).
5. M. J. S. Dewar and R. Jones, *J. Amer. Chem. Soc.* **89**, 2408 (1967).
6. D. J. Cram, *J. Amer. Chem. Soc.* **71**, 3863 (1949).
7. H. C. Brown and G. Zweifel, *J. Amer. Chem. Soc.* **83**, 2544 (1961).
8. H. C. Brown, M. M. Rogić, M. W. Rathke, and G. W. Kabalka, *J. Amer. Chem. Soc.* **91**, 2150 (1969).
9. D. S. Matteson and J. O. Waldbillig, *J. Org. Chem.* **28**, 366 (1963).
10. D. S. Matteson and R. A. Bowie, *J. Amer. Chem. Soc.* **87**, 2587 (1965).
11. H. Minato, J. C. Ware, and T. G. Traylor, *J. Amer. Chem. Soc.* **85**, 3024 (1963).
12. J. C. Ware and T. G. Traylor, *J. Amer. Chem. Soc.* **85**, 3026 (1963).
13. P. T. Lansbury and E. J. Nienhouse, *Chem. Commun.* 273 (1966).
14. C. Deblandre, M. Gielen, and J. Nasielski, *Bull. Soc. Chim. Belg.* **73**, 214 (1964).
15. D. J. Pasto, J. Chow, and S. K. Arora, *Tetrahedron Lett.* 723 (1967).
16. R. Köster and Y. Morita, *Angew. Chem.* **78**, 589 (1966).

17. P. I. Paetzold, P. P. Habereder, and R. Müllbauer, *J. Organometal. Chem.*, **7**, 51 (1967).
18. P. I. Paetzold and P. P. Habereder, *J. Organometal. Chem.* **7**, 61 (1967).
19. D. S. Matteson, *Progr. Boron Chem.* **3**, 117 (1970).
20. H. Hess, *Angew. Chem., Int. Ed. English* **6**, 975 (1967).
21. P. I. Paetzold and H. Maisch, *Chem. Ber.* **101**, 2870 (1968).
22. P. I. Paetzold and G. Stohr, *Chem. Ber.* **101**, 2874 (1968).
23. E. Ettenhuber and K. Rühlmann, *Chem. Ber.* **101**, 743 (1968).
24. A. Suzuki, S. Sono, M. Itoh, H. C. Brown, and M. M. Midland, *J. Amer. Chem. Soc.* **93**, 4329 (1971).
25. H. C. Brown, W. R. Heydkamp, E. Breuer, and W. S. Murphy, *J. Amer. Chem. Soc.* **86**, 3568 (1964).
26. M. W. Rathke, N. Inoue, K. R. Varma, and H. C. Brown, *J. Amer. Chem. Soc.* **88**, 2870 (1966).
27. A. G. Davies, S. C. W. Hook, and B. P. Roberts, *J. Organometal. Chem.* **23**, C11 (1970).
28. C. E. H. Bawn and A. Ledwith, *Progr. Boron Chem.* **1**, 345 (1964).
29. J. Hooz and S. Linke, *J. Amer. Chem. Soc.* **90**, 5936 (1968).
30. J. Hooz and S. Linke, *J. Amer. Chem. Soc.* **90**, 6891 (1968).
31. D. S. Matteson and R. W. H. Mah, *J. Amer. Chem. Soc.* **85**, 2599 (1963).
32. D. S. Matteson and R. W. H. Mah, *J. Org. Chem.* **28**, 2171 (1963).
33. J. Hine, "Physical Organic Chemistry," 2nd ed., p. 32. McGraw-Hill, New York, 1962.
34. A. Finch and P. J. Gardner, *Progr. Boron Chem.* **3**, 177 (1970).
35. H. Jäger and G. Hesse, *Chem. Ber.* **95**, 345 (1962).
36. G. Köbrich and H. R. Merkle, *Chem. Ber.* **100**, 3371 (1967).
37. G. Zweifel and H. Arzoumanian, *J. Amer. Chem. Soc.* **89**, 5086 (1967).
38. D. J. Pasto and J. Hickman, *J. Amer. Chem. Soc.* **89**, 5608 (1967).
39. D. J. Pasto, J. Hickman, and T. C. Cheng, *J. Amer. Chem. Soc.* **90**, 6259 (1968).
40. H. C. Brown, M. M. Rogić, M. W. Rathke, and G. W. Kabalka, *J. Amer. Chem. Soc.* **91**, 2150 (1969).
41. G. Zweifel, N. L. Polston, and C. C. Whitney, *J. Amer. Chem. Soc.* **90**, 6243 (1968).
42. G. Zweifel, R. P. Fisher, J. T. Snow, and C. C. Whitney, *J. Amer. Chem. Soc.* **93**, 6309 (1971).
43. H. C. Brown, M. M. Rogić, M. W. Rathke, and G. W. Kabalka, *J. Amer. Chem. Soc.* **90**, 819, 1911 (1968).
44. J. J. Tufariello, L. T. C. Lee, and P. Wojtkowski, *J. Amer. Chem. Soc.* **89**, 6804 (1967).
45. M. E. D. Hillman, *J. Amer. Chem. Soc.* **84**, 4715 (1962); **85**, 982, 1626 (1963).
46. J. Casanova, Jr., and H. R. Kiefer, *J. Org. Chem.* **34**, 2579 (1969).
47. H. C. Brown and M. W. Rathke, *J. Amer. Chem. Soc.* **89**, 2737 (1967).
48. M. W. Rathke and H. C. Brown, *J. Amer. Chem. Soc.* **89**, 2740 (1967).
49. H. C. Brown, R. A. Coleman, and M. W. Rathke, *J. Amer. Chem. Soc.* **90**, 499 (1968).
50. R. J. Mawby, F. Basolo, and R. G. Pearson, *J. Amer. Chem. Soc.* **86**, 5043 (1964).
51. D. S. Matteson, R. A. Bowie, and G. Srivastava, *J. Organometal. Chem.* **16**, 33 (1969).
52. D. S. Matteson and G. D. Schaumberg, *J. Org. Chem.* **31**, 726 (1966).
53. D. S. Matteson and T. C. Cheng, *J. Org. Chem.* **33**, 3055 (1968).
54. A. Streitwieser, Jr., "Solvolytic Displacement Reactions," pp. 12–13. McGraw-Hill, New York, 1962.

55. P. D. Bartlett and E. N. Trachtenberg, *J. Amer. Chem. Soc.* **80**, 5808 (1958).
56. D. S. Matteson and R. W. H. Mah, *J. Org. Chem.* **28**, 2174 (1963).
57. A. G. Brook, W. W. Limburg, D. M. MacRae, and S. A. Fieldhouse, *J. Amer. Chem. Soc.* **89**, 704 (1967).
58. A. G. Brook, D. M. MacRae, and W. W. Limburg, *J. Amer. Chem. Soc.* **89**, 5493 (1967).
59. M. S. Biernbaum and H. S. Mosher, *J. Amer. Chem. Soc.* **93**, 6221 (1971).
60. A. G. Brook and J. D. Pascoe, *J. Amer. Chem. Soc.* **93**, 6224 (1971).
61. A. G. Brook, C. M. Warner, and W. W. Limburg, *Can. J. Chem.* **45**, 1231 (1967).
62. J. D. Petke and J. L. Whitten, *J. Amer. Chem. Soc.* **90**, 3338 (1968).
63. M. F. Hawthorne and J. A. Dupont, *J. Amer. Chem. Soc.* **80**, 5830 (1958).
64. H. G. Kuivila and N. M. Scarpa, *J. Amer. Chem. Soc.* **92**, 6990 (1970).
65. D. S. Matteson and J. O. Waldbillig, *J. Amer. Chem. Soc.* **85**, 1019 (1963); **86**, 3778 (1964).
66. S. J. Cristol and R. T. LaLonde, *J. Amer. Chem. Soc.* **80**, 4355 (1958).
67. D. S. Matteson, J. O. Waldbillig, and S. W. Peterson, *J. Amer. Chem. Soc.* **86**, 3781 (1964).
68. D. S. Matteson and M. L. Talbot, *J. Amer. Chem. Soc.* **89**, 1119 (1967).
69. D. S. Matteson and M. L. Talbot, *J. Amer. Chem. Soc.* **89**, 1123 (1967).
70. A. DeBoer and C. H. DePuy, *J. Amer. Chem. Soc.* **92**, 4008 (1970).
71. C. H. DePuy, F. W. Breitbeil, and K. R. de Bruin, *J. Amer. Chem. Soc.* **88**, 3347 (1966).
72. C. H. DePuy, W. C. Arney, Jr., and D. H. Gibson, *J. Amer. Chem. Soc.* **90**, 1830 (1968).
73. R. J. Ouellette, R. D. Robins, and A. South, Jr., *J. Amer. Chem. Soc.* **90**, 1619 (1968).
74. A. South, Jr., and R. J. Ouellette, *J. Amer. Chem. Soc.* **90**, 7064 (1968).
75. R. J. Ouellette, D. Miller, A. South, Jr., and R. D. Robins, *J. Amer. Chem. Soc.* **91**, 971 (1969).
76. Y. Okamoto and H. C. Brown, *J. Org. Chem.* **22**, 485 (1957); *J. Amer. Chem. Soc.* **80**, 4979 (1958).
77. E. Grovenstein, Jr., and Y.-M. Cheng, *J. Amer. Chem. Soc.* **94**, 4971 (1972).
78. E. Grovenstein, Jr., and G. Wentworth, *J. Amer. Chem. Soc.* **89**, 2348 (1967).
79. P. D. Sleezer, S. Winstein, and W. G. Young, *J. Amer. Chem. Soc.* **85**, 1890 (1963).
80. M. M. Kreevoy, T. S. Straub, W. V. Kayser, and J. L. Melquist, *J. Amer. Chem. Soc.* **89**, 1201 (1967).
81. H. G. Kuivila and J. C. Cochran, *J. Amer. Chem. Soc.* **89**, 7152 (1967).
82. R. M. G. Roberts, *J. Organometal. Chem.* **18**, 307 (1969).
83. W. Kitching, C. W. Fong, and A. J. Smith, *J. Amer. Chem. Soc.* **91**, 767 (1969).
84. W. Hanstein and T. G. Traylor, *Tetrahedron Lett.* 4451 (1967).
85. Yu. G. Bundel, N. D. Antonova, and O. A. Reutov, *Dokl. Akad. Nauk SSSR* **166**, 1103 (1966).
86. Yu. G. Bundel, V. I. Rozenberg, A. L. Kurts, N. D. Antonova, and O. A. Reutov, *J. Organometal. Chem.* **18**, 209 (1969).
87. R. A. Benkeser and W. de Talvo, *J. Amer. Chem. Soc.* **89**, 2141 (1967).
88. D. D. Davis, R. L. Chambers, and H. T. Johnson, *J. Organometal. Chem.* **25**, C13 (1970).

CHAPTER 5

Polar 1, 2-Additions and Eliminations

I. Introduction

Many metal salts, hydrides, and alkyls add by polar or four-center mechanisms to carbon–carbon double bonds, and the reverse reaction, β elimination, is likely to occur whenever a metal atom is β to a potential nucleofuge such as halide or alkoxide. The additions include such synthetically useful reactions as oxymercuration, hydroboration, and homologation of aluminum alkyls with ethylene. The β eliminations generally convert interesting organometallic compounds to ordinary olefins, but there are people who want to make olefins, and there are also opportunistic mechanism chemists who study β eliminations because they often yield good data easily.

This chapter will begin with a brief discussion of hyperconjugation involving carbon–metal bonds, which is fundamental to understanding β eliminations. Hyperconjugation is also relevant to S_E2' mechanisms, discussed in the preceding chapter, which are themselves as closely related to eliminations as to displacements. Oxymercuration and deoxymercuration will be discussed next, since these reactions have been studied extensively and are well understood. Other β eliminations will then be discussed. The chapter will conclude with discussion of hydroboration and of additions of lithium and other alkyls to olefins.

II. Carbon–Metal Hyperconjugation

A. Evidence in Support

Carbon–metal hyperconjugation is basic to an understanding of β eliminations involving metals. Hanstein, Berwin, and Traylor have found definitive evidence that this phenomenon is real.[1,2] Charge-transfer spectra of dichloromaleic anhydride and tetracyanoethylene complexes indicate that $RHgCH_2$—,

Me_3SiCH_2—, and Ph_3PbCH_2— substituents on benzene rings are strong electron donors, lowering the energy of the excited state relative to the ground state.[1] A plot of charge-transfer absorption frequencies versus σ^+ of the MCH_2— substituent shows good correlation (Fig. 5-1).

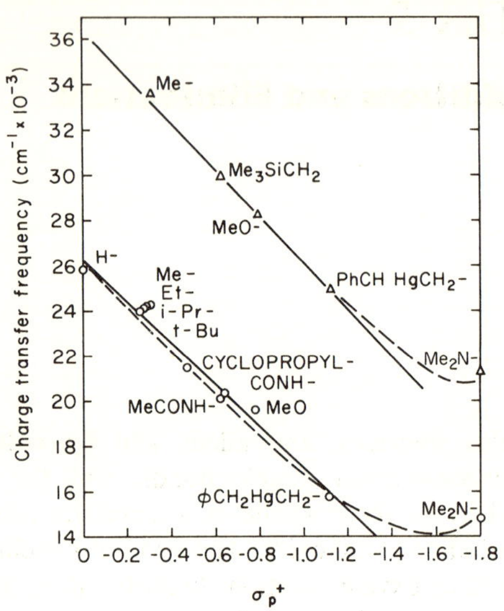

Fig. 5-1. Plots of charge-transfer frequencies of Y—C_6H_5–acceptor complexes against $\sigma_p{}^+$ of Y. [From W. Hanstein, H. J. Berwin, and T. G. Traylor, *J. Amer. Chem. Soc.* **92**, 829 (1970).]

The charge-transfer spectra particularly show that the electron-donating effect is purely electronic. The Franck-Condon principle requires that the absorption of visible light be rapid compared to vibrational motions of the molecule. Therefore, the stabilization of the excited state by electron donation occurs with the same configuration of atoms that was present in the ground state. This clearly rules out any argument that the electron donation could arise from some sort of neighboring-group interaction between the metal atom and the benzene ring, where the metal atom would have to move toward the ring to form some sort of bond. Such an argument could not be ruled out on the basis of chemical evidence such as relative reaction rates.

Hyperconjugation requires a particular molecular geometry in order for the carbon–metal bonding orbital to overlap with the π-bond system, as illustrated in Fig. 5-2. In molecules which have free internal rotation, part of the population will have geometries within a range which allows hyperconjugation, and that part will give rise to the observed effect. However, in rigid cyclic molecules where the metal atom is held in the plane of the benzene ring there

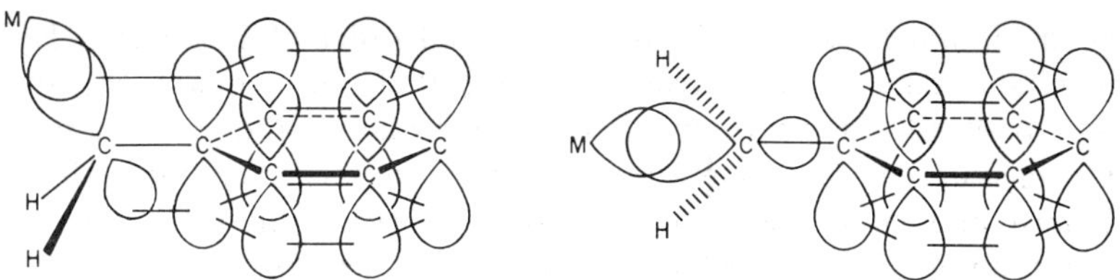

Fig. 5-2. Illustration of possible and impossible molecular geometries for hyperconjugation of the carbon–metal bond with the benzene ring.

can be no hyperconjugation, and if the interpretation is correct the observed effect on the spectrum should then become zero. This has been found true in a series of tin compounds.[2] Benzyltin compounds such as **1** form charge-transfer complexes with tetracyanoethylene which absorb at longer wavelengths than those of toluene or xylene, but rigid molecules such as **2** and **3** in which the carbon–tin bond cannot hyperconjugate with the benzene ring form charge-transfer complexes which absorb in the same range as those of hydrocarbons. The data are summarized in Chart 5-1.

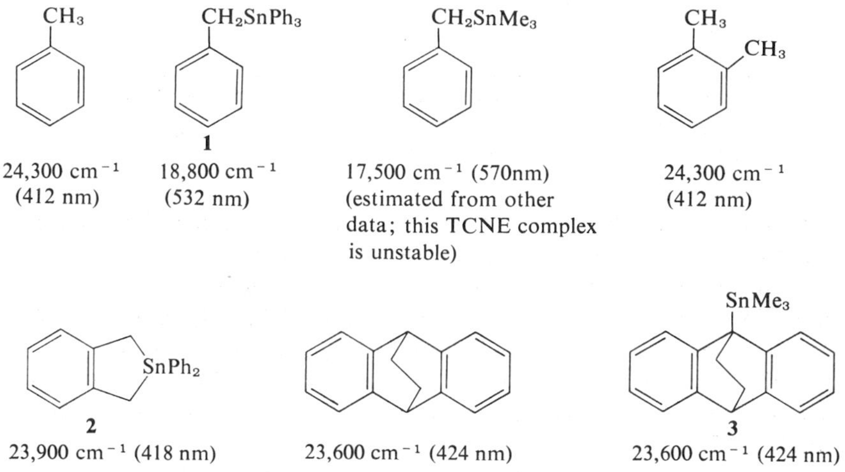

CHART 5–1. Frequencies (cm^{-1}) and wavelengths (nm) of the absorption maxima of charge-transfer complexes of tetracyanoethylene with benzyltin compounds and hydrocarbons. [From W. Hanstein, H. J. Berwin, and T. G. Traylor, *J. Amer. Chem. Soc.* **92,** 7476 (1970).]

There is much other evidence for carbon–metal bond hyperconjugation, and although each item by itself is not as definitive as the evidence just cited, the total evidence is highly convincing. For example, Waters and Kiefer found that the carbonyl absorption of $IHgCH_2COCH_2HgI$ appears at 1600 cm^{-1}.[3] This is comparable to the carbonyl band of the sodium salt of tropolone,

1606 cm^{-1}, which has a carbon–oxygen bond order of 1.5 or less. Hyperconjugation seems the only reasonable way that the mercury substitutents could grossly affect the carbonyl absorption frequency, though perhaps some sort of mercury–oxygen complex has not been entirely ruled out in this case.

Eaborn and co-workers have found evidence for hyperconjugation in silicon compounds.[4,5] Rates of perchloric acid cleavage of the aryl–silicon bond as influenced by *para* or *meta* Me_3SiCH_2—, $(Me_3Si)_2CH$—, and $(Me_3Si)_3C$— groups[4] are summarized in Table 5-1. From the tabulated rate

TABLE 5-1

RELATIVE RATES OF PROTOLYSIS OF THE ARYL–SILICON BOND OF XC_6H_4–$SiMe_3$[a,b]

Compound (*m* or *p*)	Relative *k*	
	Para isomer	Meta isomer
$Me_3SiCH_2C_6H_4SiMe_3$	670	6.5
$(Me_3Si)_2CHC_6H_4SiMe_3$	280	8.4
$(Me_3Si)_3CC_6H_4SiMe_3$	200	3.4
$CH_3C_6H_4SiMe_3$	21	2.4
$C_6H_5SiMe_3$	1	1

[a] Where X is H, CH_3, Me_3SiCH_2—, $(Me_3Si)_2CH$—, or $(Me_3Si)_3C$—.
[b] From A. R. Bassindale, C. Eaborn, D. R. M. Walton, and D. J. Young, *J. Organometal. Chem.* **20**, 49 (1969).

constants it is readily apparent that some sort of conjugative effect is operating, since the substituents are much less effective in promoting the reaction if they are in the *meta*-position than they are in the *para*-position. Also, the results are contrary to what would be expected if the inductive effect of the Me_3Si group were the primary influence. Only one carbon–silicon bond can be pointed in the best direction for hyperconjugation with the benzene ring, and the failure of a second Me_3Si group on the substituent carbon to accelerate the reaction agrees with the hyperconjugation model. In addition to these kinetic results, ^{19}F chemical shifts in *p*-$Me_3SiCH_2C_6H_4F$ and related compounds likewise indicate hyperconjugative electron release.[4]

Further evidence for hyperconjugation is provided by the σ and σ^+ constants of Me_3SiCH_2— and related groups.[5] From the dissociation constants of the *p*-substituted carboxylic acids, the σ constants are Me_3SiCH_2—, -0.29; $(Me_3Si)_2CH$—, -0.33; and $(Me_3Si)_3C$—, -0.27. From rates of solvolysis of the *p*-substituted cumyl chlorides (such as $Me_3SiCH_2C_6H_4CMe_2Cl$), the σ^+ constants are Me_3SiCH_2—, -0.54, $(Me_3Si)_2CH$—, -0.62, and $(Me_3Si)_3C$—, -0.52. The more negative values of the σ^+ constants compared to the σ constants support the hyperconjugation hypothesis.

The Me_3Si— group in *p*-$Me_3SiC_6H_4F$ shifts the ^{19}F resonance 0.5 ppm downfield relative to fluorobenzene, indicating slight π-electron withdrawal, and Me_3Ge, Me_3Sn, or Me_3Pb groups directly on the benzene ring shift the *p*-fluorine resonance slightly upfield by 0.2–0.55 ppm. These shifts contrast with the 7–8 ppm upfield shifts caused by the metal–CH_2— groups Me_3SiCH_2—, Me_3GeCH_2—, Me_3SnCH_2—, and Ph_3PbCH_2— *para* to the fluorine.[6] The *meta* isomers have upfield shifts of only about 1 ppm. For comparison, a *p-t*-butyl group causes an upfield shift of 5.6 ppm. These results are again consistent with hyperconjugation, assuming that the carbon–carbon bond of the *t*-butyl group can enter into hyperconjugation to a reasonably high fraction of the degree shown by the carbon–metal bonds of Me_3SiCH_2 and related groups.

Bock and Alt have pointed out that there is a carbon–silicon bond in the right position for hyperconjugation in $PhSiMe_3$, but the silicon atom transmits the effect only weakly because silicon–carbon π-bonding is weak.[7] The charge-transfer complex of $PhSiMe_3$ with tetracyanoethylene shows only a slight shift in the absorption maximum from that of benzene, in contrast to the large shift shown by the complex of $PhCH_2SiMe_3$.

In an earlier volume in this series, Ramsey has reviewed the ultraviolet spectral evidence for hyperconjugation in allylmetal compounds.[8]

The strong evidence in favor of hyperconjugation involving carbon–metal bonds is not to be confused with some of the historical arguments surrounding carbon–hydrogen bond hyperconjugation. Although carbon–hydrogen bond hyperconjugation is a significant factor in the stability of carbonium ions, it is among the weakest of the common hyperconjugation interactions. The Baker-Nathan effect, which led to the idea of hyperconjugation in the first place, has turned out to be opposite the true order of hyperconjugation effects (C—C is a better donor than C—H) and is in fact nothing much more than a solvation effect.[9]

B. Chemical Consequences

Some of the chemical consequences of carbon–metal bond hyperconjugation have already been noted in Chapter 4, Section VII,D. Much ring deuteration occurs in the *ortho* and *para* positions of benzylboron and benzylmercury compounds when they are cleaved with D^+, and some of the deuteration occurs without carbon–metal bond cleavage, though the latter is the main reaction. The generally strong tendency of allylmetal compounds to undergo S_E2' displacement in preference to direct S_E2 displacement is probably a consequence of the favorable hyperconjugation which results when an electrophile attacks the allylic double bond, summarized in the generalized structure **4**.

$$M{-}CH_2{-}CH{=}CH_2 + E^+ \longrightarrow$$

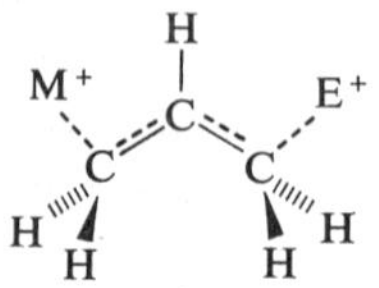

$$\mathbf{4} \equiv M{-}CH_2{-}CH{=}CH_2\ E^+ \longleftrightarrow M^+\ CH_2{=}CH{-}CH_2{-}E$$

Another chemical consequence of carbon–metal bond hyperconjugation is that hydrogen atoms β to a metal are surprisingly labile in several types of reactions. For example, Jerkunica and Traylor have found that triphenylmethyl fluoroborate readily abstracts hydride from the β position of ethyltrimethyltin (**5**).[10] The rate is first order in each reactant. Relative rates for

$$Ph_3C^+ + \underset{\mathbf{5}}{H_3CCH_2SnMe_3} \longrightarrow Ph_3CH + H_2C{=}CH_2 + Me_3Sn^+$$

Si, Ge, Sn, Pb, and Hg compounds parallel relative hyperconjugation effects. In a 1,3-ditin compound the 2-hydrogens are activated by nearly two powers of 10 compared to those of a monotin compound, which is consistent with an additive electronic effect and inconsistent with any sort of neighboring-group effect which could involve only one tin atom at a time. Rate constants statistically corrected for the number of β-hydrogens are shown in Table 5-2.

TABLE 5-2

RATE CONSTANTS PER β-HYDROGEN FOR REACTION OF ORGANOMETALLIC COMPOUNDS WITH $Ph_3C^+BF_4^-$ IN ACETONITRILE AT 29.8°C[a]

Compound	k_2 per β-hydrogen
Et_4Si	5.7×10^{-8}
Et_4Ge	$4\,0 \times 10^{-6}$
Et_4Sn	8.3×10^{-4}
$Me_3SnCH_2CH_2CH_2SnMe_3$	5.8×10^{-2}
$Me_3SnCH_2CH_2Ph$	4.8×10^{-2}
Et_4Pb	0.49
Et_2Hg	0.15
Cycloheptatriene	0.21

[a] From J. M. Jerkunica and T. G. Traylor, *J. Amer. Chem. Soc.* **93**, 6278 (1971).

The reactivity of these ethylmetal compounds may be fully appreciated by considering that tetraethyllead reacts faster than cycloheptatriene with the triphenylmethyl cation, and diethylmercury reacts almost as fast (Table 5-2). If it were not for the special reactivity of $Me_3SnCH_2CH_2CH_2SnMe_3$, it would not be necessary to apply the label "hyperconjugation" to explain these observations, since the carbon–metal bond is after all being broken in the transition state for concerted elimination, by definition. The transition state may be represented by the broken-line structure **6** or the set of resonance structures **7**, which are merely different representations of the same wave

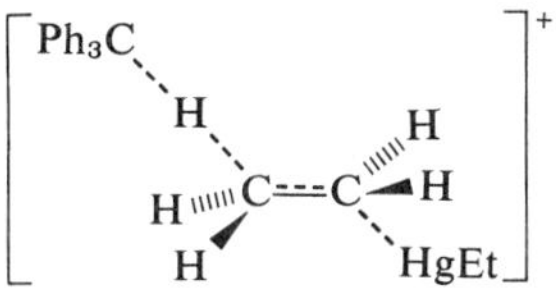

$$\left[\begin{array}{c} Ph_3C \cdots H \cdots \\ H_2C{=}{-}{-}CH_2 \cdots HgEt \end{array}\right]^+ \quad \textbf{6}$$

$$Ph_3C^+ \; H{-}CH_2{-}CH_2{-}HgEt \;\leftrightarrow\; Ph_3C{-}H \; {}^+CH_2{-}CH_2{-}HgEt \;\leftrightarrow\; Ph_3C{-}H \;\; CH_2{=}CH_2 \;\; {}^+HgEt \quad \textbf{7}$$

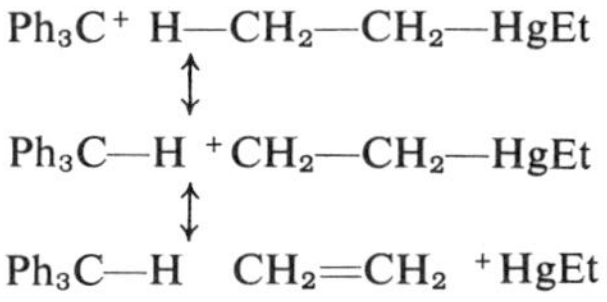

function. Structure **6** is immediately recognizable as a typical transition state for β elimination, and the resonance structures **7** immediately force recognition that the partial bond making and breaking or electron delocalization in **6** is identical with hyperconjugation.

Jerkunica and Traylor also pointed out the relationship of hyperconjugation as illustrated by these eliminations to the mechanism of oxymercuration, which will be discussed in Section III,E.

Another chemical consequence of carbon–metal bond hyperconjugation is the insertion of dichlorocarbene (CCl_2) into the carbon–hydrogen bond β to silicon, tin,[11] or mercury.[12] Only the β-hydrogens are activated toward dichlorocarbene insertion, though insertion into the carbon–metal bond also occurs. Landgrebe and Thurman have studied the stereochemistry of the reaction of dichlorocarbene with optically active bis[(*S*)-2-methylbutyl]-mercury (**8**).[12] Insertion into the β—C—H bond is 90% or more of the reaction. The net retention of configuration found was 23–32%, the remainder of the product being racemized. An ion pair intermediate (**9**) was postulated. Structure **9** and the transition state which precedes it must be related to the β elimination process discussed in the preceding paragraphs, except that the carbon–mercury bond probably never breaks completely (there being no evidence of elimination or rearrangement), and the stability of ion pair **9** may be attributed to hyperconjugation.

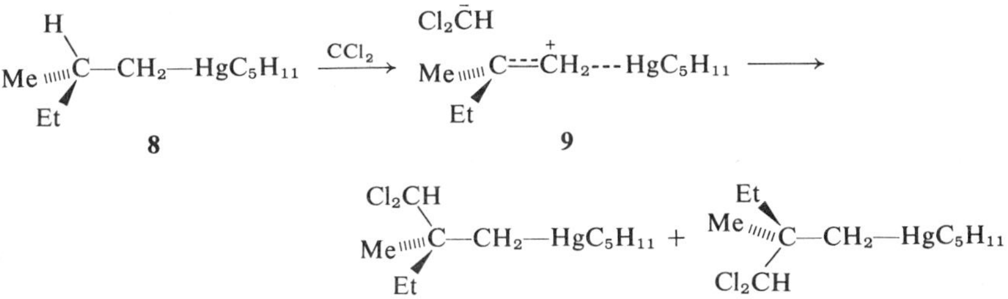

$$\underset{\textbf{8}}{(H)(Me)(Et)C{-}CH_2{-}HgC_5H_{11}} \xrightarrow{CCl_2} \underset{\textbf{9}}{(Cl_2\bar{C}H)(Me)(Et)C{=}{-}{-}\overset{+}{C}H_2{-}{-}{-}HgC_5H_{11}} \longrightarrow$$

$$\underset{\substack{\textbf{10}\\ \text{(predominant isomer)}}}{(Cl_2CH)(Me)(Et)C{-}CH_2{-}HgC_5H_{11}} + (Et)(Me)(Cl_2CH)C{-}CH_2{-}HgC_5H_{11}$$

The proof of configuration of the predominant enantiomer (**10**), though a digression from hyperconjugation, is an essential part of the investigation.

Cl_2CH, Me, Et C—CH_2HgR (**10**) $\xrightarrow{Br_2}$ Cl_2CH, Me, Et C—CH_2Br $\xrightarrow{NaCN}$ Cl_2CH, Me, Et C—CH_2CN $\xrightarrow[2.\ H_3O^+]{1.\ Ag_2O}$ HO_2C, Me, Et C—CH_2CO_2H

11
(known)

The innate perversity of chemicals was illustrated once more by the fact that the sign of rotation of the diacid **11** is concentration-dependent in chloroform. Naturally, Landgrebe and Thurman first chose a concentration sufficiently different from the previously reported conditions so that they unknowingly got the opposite rotation and assigned the wrong absolute configuration.[13] Inversion seemed an exceedingly strange result for the dichlorocarbene insertion and the error was soon found and corrected.[13] No doubt it is mean to bring up mistakes that people would rather forget. However, if this bit of history persuades one organometallic chemist somewhere to wait, do that control experiment, and catch his own error before parading it naked before the world, it will have been worth mentioning.

III. Oxymercuration and Oxythallation

A. Deoxymercuration

The place to start this topic is with the reverse reaction, deoxymercuration, which was studied extensively by Kreevoy and co-workers in the early 1960s and before. Kreevoy's major conclusions have been strongly supported by numerous recent studies of oxymercuration. H. C. Brown has registered some degree of dissent, but his data actually do fit in with the generally accepted view of the structure of the transition state and merely sharpen and refine the picture a bit.

The normally preferred stereochemistry of deoxymercuration is *trans*. The acid-catalyzed deoxymercuration of *trans*-1-methoxy-2-iodomercuricyclohexane (**12**) proceeds readily with $\Delta H^* = 17.7$ kcal/mole and $\Delta S^* = +4.6$ eu in aqueous perchloric acid.[14] The reaction is much slower with *cis*-1-methoxy-2-iodomercuricyclohexane (**13**), with $\Delta H^* = 26.2$ kcal/mole and $\Delta S^* = +4.5$ eu. These results immediately suggest some sort of neighboring-group participation with backside attack favored, and the mercurinium ion **14** was proposed as an intermediate. It should be emphasized that **14** is an intermediate and that the transition state implicit in this reaction pathway is **15**.

MeO H HgI H **12** ⇌ MeO H H HgI **12** $\xrightarrow{H^+}$

MeO H H HgI → [MeO H H H HgI]$^+$ **15**

MeOH + H H HgI **14** → H H + HgI^+

$\xrightarrow{MeOC_6H_{10}HgI}$ $HgI_2 + MeOC_6H_{10}Hg^+$

14 $\xleftarrow{H^+}$ H IHg H OMe **13**

The secondary deuterium isotope effect k_H/k_D found on comparison of $MeOCH_2CH_2HgI$ with $MeOCD_2CD_2HgI$ is only 1.06 ± 0.02.[15] If the transition state resembled a carbonium ion, k_H/k_D should be much greater than unity, and if it resembled the mercurinium ion (**16**), k_H/k_D would be less than 1. A structure somewhere between these two extremes (**17**, analogous to **15**) is consistent with the observed isotope effect.

For a series of ten X—$CH_2CH(OMe)CH_2HgI$, Taft's $\rho^* = -2.77$.[16] If the transition state closely resembled the protonated ether, ρ^* should be much less negative, and if it resembled a carbonium ion, ρ^* should be more negative, about -3.3 to -4.3. Again, a transition state resembling **17** or **15** is consistent with the evidence.

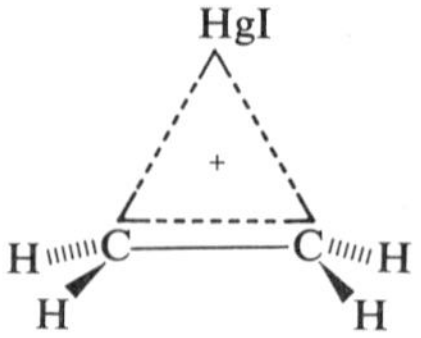

HgI, +, H, C, C, H, H, H

16

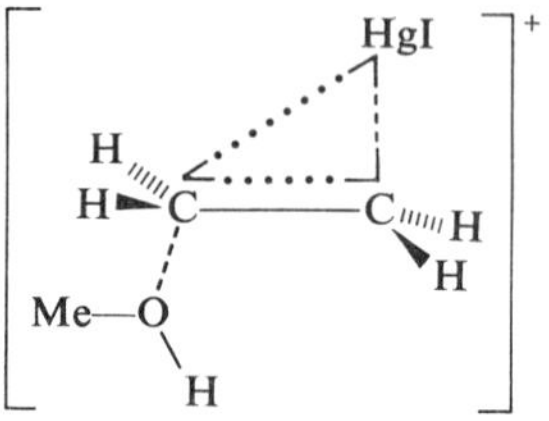

[HgI, H, H, C, C, H, H, Me—O, H]$^+$

17

B. Mercurinium Ions

Olah and Clifford have verified the existence of the ethylenemercurinium ion (**18**) in FSO_3H–SbF_5–SO_2 by observing its proton nmr spectrum.[17] The singlet of the mercurinium ion (**18**) appears at δ 7.68 and shows ^{199}Hg

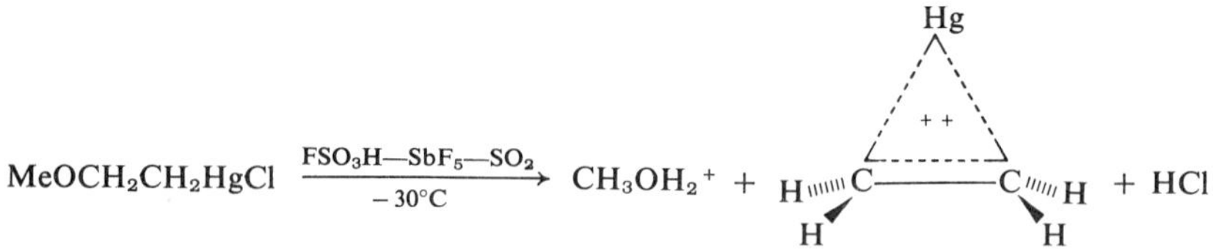

$$MeOCH_2CH_2HgCl \xrightarrow[-30^\circ C]{FSO_3H—SbF_5—SO_2} CH_3OH_2^+ + \mathbf{18} + HCl$$

Hg, ++, H, C, C, H, H, H

18

satellites with $J = 190$ Hz. The only difference between **18** and **16** is in the ligands other than ethylene on the mercury, and the actual number and types are not known in either case. The protonated methanol formed in the production of **18** shows up in the nmr spectrum as a CH_3 triplet at δ4.80 and an OH_2^+ quartet at δ9.77. The same mercurinium ion (**18**) was obtained from $EtOCH_2CH_2HgCl$.

The norbornenemercurinium ion was also observed at $-70°C$.[17] At $-30°C$ the protons equilibrate as if the ion is rearranging rapidly, and cooling back to $-70°C$ restores the original spectrum.

Olah and Clifford have made the cyclohexenemercurinium ion by mercuration of cyclohexene with mercuric trifluoroacetate in "super acid."[18] The two protons adjacent to the cationic site appear as a singlet at δ8.35, the next set of four appears as a multiplet at δ3.10, and the remaining four as a multiplet at δ2.12. Methylmercuric acetate mercurated cyclohexene to form the analogous cyclohexene(methylmercurinium) ion.

While it is clear that mercurinium ions are stable species in the absence of effective nucleophiles, it is also clear that mercurinium ions are not long-lived species in the presence of water, alcohols, or even weakly basic anions. Brown, Rei, and Liu have shown by nmr that the adducts formed from mercuric trifluoroacetate and norbornene or related olefins in benzene are covalent Hg—C—C—O_2CCF_3 structures and do not resemble the stable silver–olefin π complexes.[19] Parker and Roberts have also shown by ^{13}C nmr that mercuric acetate adducts do not resemble silver–olefin complexes of cyclopentene and cyclohexene.[20]

The observed stability of mercurinium ions in the absence of nucleophiles and instability in their presence is consistent with extended Hückel calculations carried out by Bach and Henneike.[21] The bonding of ethylene with Hg^{2+} was calculated to be intermediate in character between that with Cl^+ and that with Ag^+. The Hg^{2+} withdraws a considerable amount of electron density from the carbon atoms, in accord with the high reactivity of these complexes. The ethylene–Ag^+ complex was calculated to have little positive charge on carbon, in accord with its lack of chemical reactivity.

Studies of oxymercuration are generally consistent with the rapid and reversible formation of mercurinium ions in low concentrations prior to the rate-determining step. The evidence points toward transition states resembling those for deoxymercuration (**15**, **17**) as required by the principle of microscopic reversibility, which applies here because the conditions for the forward and reverse reactions differ mainly in the concentration of Hg^{2+} ion or related factors such as the presence or absence of halide ligands on mercury. Because mercurinium ions precede the rate-determining step, they cannot be detected by kinetic measurements, and they also need not be the only route to the transition state or even necessarily intermediates at all in some cases.

C. Oxymercuration of Olefins

Halpern and Tinker have measured the kinetics of hydroxymercuration of 20 olefins in aqueous perchloric acid.[22] The rate law in each case is first order in Hg^{2+} and first order in olefin. Several cases of the expected *trans* addition were confirmed, including the reactions of *cis*- and *trans*-2-butene (**19** and **20**). The rate is insensitive to oxygen or hydrogen peroxide and is independent

$$CH_3CH{=}CHCH_3\ (\textbf{19}) \xrightarrow[H_2O]{Hg^{2+}} CH_3CH(Hg^+)\text{—}CH(OH)CH_3$$

19

$$CH_3CH{=}CHCH_3\ (\textbf{20}) \xrightarrow[H_2O]{Hg^{2+}} CH_3CH(Hg^+)\text{—}CH(OH)CH_3$$

20

of H^+ in the pH 1–3 range. Mercurous ion does not catalyze the reaction, but does yield an oxymercuration rate equal to that expected from the amount of Hg^{2+} in equilibrium with $Hg_2{}^{2+}$ and Hg°.

These kinetic results require that any mercurinium ions present must be formed rapidly and reversibly and only in low concentrations. The lack of any fall-off in rate dependence on olefin concentration at high olefin concentrations implies that the upper limits for the complexing constants for olefins with Hg^{2+} are between 1 and 100, depending on the concentrations actually tested, and, of course, the actual values may be much less. These upper limits are in a range comparable to the complexing constants for olefins with Ag^+ and in no way conflict with the possibility of mercurinium ion formation.[22]

Application of the Taft ρ^* treatment to the series $CH_2{=}CHX$ where X is CH_2CN, CH_2Cl, CH_2OH, H, CH_2CH_2OH, CH_3, and C_2H_5 yielded an excellent correlation with $\rho^* = -3.3$. This negative ρ^* value implies a high degree of positive charge localization, approaching that of a free carbonium ion, in the transition state at the site where the water attacks.[22] The ρ^* found for this series of oxymercurations is more negative than that found for deoxymercurations (-2.77)[16] because the ligands on the mercury are different, the iodide ligand present in the deoxymercuration reducing the electron-attracting power of the mercury(II) cation.

The reaction of $CH_2{=}CHCH_2CH_2CH_2OH$ is exceptionally rapid as a result of hydroxyl group participation and leads to a cyclic product (**21**).[22]

$$CH_2{=}CHCH_2CH_2CH_2OH + Hg^{2+} \longrightarrow \overset{+}{Hg}{-}CH_2{-}\text{(tetrahydrofuran-2-yl)}$$

21

Because oxymercuration is reversible, exchange of Hg^{2+} and OH^- between two different olefins is possible. The reaction of $HOCHXCH_2Hg^+$ with $CH_2{=}CHY$ to form $CH_2{=}CHX$ and $HOCHYCH_2Hg^+$ generally yields a two-term rate law,[23]

$$\frac{-d[HOCHXCH_2Hg^+]}{dt} = k[HOCHXCH_2Hg^+][H^+] + k'[HOCHXCH_2Hg^+][CH_2{=}CHY].$$

The first term corresponds to rate-determining deoxymercuration of $HOCHXCH_2Hg^+$ with H^+ to yield H_2O, $CH_2{=}CHX$, and Hg^{2+} followed by rapid reaction of Hg^{2+} with the other olefin $CH_2{=}CHY$. It is thus a measurement of the reverse reaction under conditions where oxymercuration occurs and serves to verify and round out a complete description of the oxymercuration mechanism. The second term corresponds to formation of a transition state involving $HOCHXCH_2Hg^+$ and $CH_2{=}CHY$, probably leading to $HOCHXCH_2HgCH_2CHYOH$, which then deoxymercurates to $CH_2{=}CHX$ and $HOCHYCH_2Hg^+$. This second term amounts to just one more variation of the ligands on mercury in an oxymercuration.

D. Oxymercuration of Allenes

The oxymercuration of allenes has supplied some of the strongest evidence in favor of mercurinium ion intermediates. Here the choice is between a relatively stable allylic carbonium ion (**22**) and a cyclic mercurinium ion (**23**). The stereochemistry and regioselectivity expected of these two types of ions (**22** and **23**) are different. Depending on the ligand X on the mercury(II)

$$R_2C{=}C{=}CR_2 + \overset{+}{Hg}X \longrightarrow \underset{\mathbf{22}}{R_2C\overset{+}{\cdots}C(HgX)\cdots CR_2} \quad \text{or} \quad \underset{\mathbf{23}}{R_2C{=}C{-}CR_2 \text{ bridged by } \overset{+}{Hg}X}$$

cation, both types of ions have been found to be intermediates, and the contrasting behavior strengthens the evidence.

Waters and Kiefer found that the products from reactions of mercuric acetate with allenes are all consistent with intermediates of the mercurinium ion type (**23**).[3] For example, the mercury attacks the terminal carbons of allene itself, yielding 1,3-bis(acetoxymercuri)acetone (**24**), which is inconsistent with the allylic cation intermediate (**22**). An open vinyl cation,

$$H_2C{=}C{=}CH_2 + Hg(OAc)_2 \xrightarrow{MeOH} AcOHgCH_2{-}C(OMe)_2{-}CH_2HgOAc \xrightarrow{H_2O} \underset{\mathbf{24}}{AcOHgCH_2C({=}O)CH_2HgOAc}$$

$AcOHgCH_2C^+{=}CH_2$, would have been unlikely to compete with the allylic alternative (**22**), and some sort of stabilized cyclic intermediate (**23**) is indicated in order to account for the product.

Waters and Kiefer also found that the mercury(II) cation attacks the central carbon of 1,1-dimethylallene and that the methoxy group in the product (**25**) is exclusively on the tertiary carbon. They argued that an open allylic cation

$$CH_2{=}C{=}C(CH_3)_2 + Hg(OAc)_2 \xrightarrow{MeOH} \underset{\mathbf{25}}{CH_2{=}C(HgOAc){-}C(CH_3)_2OMe} \quad \text{and not} \quad \underset{\mathbf{26}}{MeOCH_2C(HgOAc){=}C(CH_3)_2}$$

(**22**) would in this case be attacked at the terminal CH_2 group to yield the more stable olefinic product (**26**), which was not found. Other evidence favoring the mercurinium ion intermediate (**23**) was also discussed, but the crucial evidence is more recent stereochemical studies.

Caserio *et al.* accomplished partial resolution of 1,3-dimethylallene by preferential destruction of one isomer by hydroboration with optically active

$Hg(OAc)_2$ + MeOH

27

28 + **29**

MeOH

MeOH

$-H^+$

$-H^+$

30

31

1. $NaBH_4$
2. H_2, Pt

32

Configuration known

33

Not an intermediate

diisopinocampheylborane.[24] By analogy to the behavior of diphenylallene, which has known configuration, they assigned the *R* configuration to (−)-1,3-dimethylallene (**27**). Both oxymercuration and halogenation were then shown to be stereospecific *trans* additions. In oxymercuration two diastereoisomeric mercurinium ions (**28** and **29**) are formed, presumably rapidly and reversibly. (These are the only two possibilities allowed by symmetry. A 180° rotation of the allene **27** about the vertical axis and a 90° rotation about the horizontal axis interchanges the positions of the two equivalent terminal methyl groups.) One mercurinium ion (**28**) is opened by methanol to form the oxymercuration product (**30**) having the carbon chain *trans* at the remaining double bond, and the other (**29**) yields the *cis* isomer (**31**). The transition states which determine the product ratio involve the attack of methanol on the mercurinium ions (**28** and **29**), and the ratio 83% of **30** to 17% of **31** seems consistent with the degree of steric shielding to methanol attack provided by the respective groups (H versus CH_3) at the other end of the original allene moiety. The absolute configuration of the product was found by reducing one of the oxymercuration products (**30**) to 2-methoxypentane (**32**), which has a known configuration. The stereospecificity of the oxymercuration is high, and the planar allylic cation (**33**) cannot be an intermediate in the major reaction pathway.[24]

Bach has found that the stereospecificity of oxymercuration of optically active 1,2-cyclononadiene (**34**) depends on the ligands on the mercury.[25] The racemization must occur in an intermediate which is formed reversibly, since partial racemization of the starting allene (**34**) was observed. For example, reaction of 1,2-cyclononadiene (**34**) with two-thirds of the theoretical amount of mercuric acetate caused 8% racemization of the recovered **34**, and reaction with mercuric chloride caused about 60% racemization. The optical purity of the final product is not known but the degree of racemization with mercuric chloride can be shown to be very high. The derived 3-ethoxycyclononene (**35**) showed a +15° rotation when the oxymercurating agent was ethylmercuric acetate, +12° when it was mercuric acetate, but only 0.1–0.2°

34 $\xrightarrow[EtOH]{Hg(OAc)_2}$ (HgOAc, OEt adduct) $\xrightarrow{NaCl}$ (HgCl, OEt adduct) $\xrightarrow{H^+}$ **35**

when it was mercuric chloride. These results were interpreted on the basis of a cyclic mercurinium ion (**36**) which isomerizes reversibly to a symmetrical allylic cation (**37**), which of course must revert to **36** and its enantiomer **38** with equal probability.[25] The rate of interconversion of the ions **36**, **37**, and **38** appears to be governed by the ligand X on the mercury. Ligands such as chloride, which would be expected to stabilize RHgX and destabilize the mercury(II)–olefin complex, increase the rate of formation of the allylic ion (**37**).

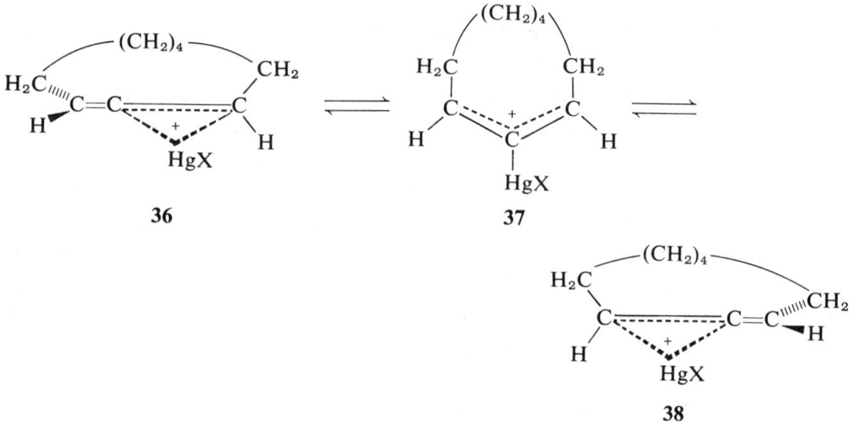

36 ⇌ **37** ⇌ **38**

Caserio *et al.* have also observed racemization in the deoxymercuration of 3-chloromercuri-4-methoxy-2-pentene (**39**) to 1,3-dimethylallene.[26] Their evidence indicates that under the reaction conditions used, an allylic cation

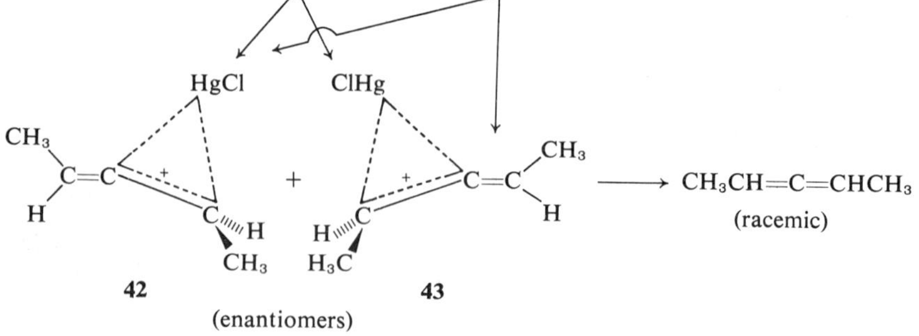

39 $\xrightarrow{HCl}$ **40** (symmetrical) or **41** (symmetrical)

42 + **43** (enantiomers) ⟶ $CH_3CH{=}C{=}CHCH_3$ (racemic)

(**40** or **41**) is the first intermediate. Recovered starting material (**39**) is partially racemized, and the product 1,3-dimethylallene is totally racemic. From other data on oxymercuration, it is unlikely that the allylic cation (**40** to **41**) can lose its mercury(II) cation directly to form the allene, but probably that it first rearranges to the mercurinium ion (enantiomers **42** and **43**), which has the right geometry to form the allene as the mercury(II) departs.

Further evidence for the direct formation of allylic cation intermediates (**40** and **41**) was provided by the conversion of optically active acetoxy compound (**44**) to the methoxy compound (**39**), which again resulted in complete racemization.[26] The rate is similar to that of a simple allyl acetate, indicating little neighboring-group assistance by the mercury.

$$\underset{\substack{\textbf{44}\\ \text{(optically active)}}}{CH_3(H)C{=}C(HgCl)CH(OAc)CH_3} \xrightarrow[HBF_4]{MeOH} \underset{\substack{\textbf{39}\\ \text{(racemic)}}}{CH_3(H)C{=}C(HgCl)CH(OMe)CH_3}$$

Any reader who is inclined to jump up with questions at the end of a seminar should by now be fidgeting with the question, "But what about the principle of microscopic reversibility? Is all this self-consistent?" The answer is, Yes. The conditions of oxymercuration, where the asymmetric mercurinium ion can go directly to asymmetric oxymercuration product, differ significantly from those of the deoxymercurations just described, where the formation of optically active mercurinium ion is not observed. The major differences are in the ligands on the mercury and in the activity of the nucleophile (methanol or acetic acid) in the reaction medium. As noted in Bach's study of the oxymercuration of the (somewhat strained) cyclic allene with mercuric chloride, racemization is almost complete.[25] The deoxymercuration conditions reported by Caserio and co-workers involved not only a chloride ligand on the mercury but concentrated aqueous hydrochloric acid in a two-phase system with benzene.[26] The effective nucleophilicity of the methanol would be reduced under these conditions or, conversely, its nucleofugicity would be improved, and the path requiring nucleophilic assistance by the mercury (mercurinium ion) to expel the methanol would be decreased in importance from slight to unobservable. The fact that the allenes used in the oxymercuration and deoxymercuration studies are different could also lead to different ratios of mercurinium and allylic ions.

A qualitative reaction coordinate profile, which represents a slight refinement of that suggested by Caserio and co-workers, is shown in Fig. 5-3.

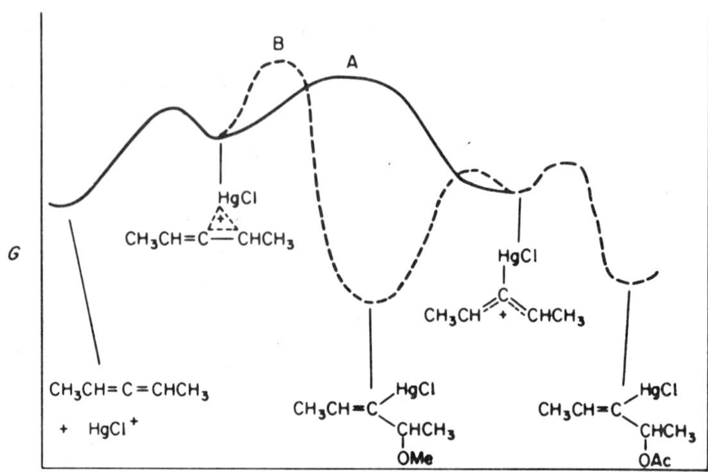

Fig. 5-3. Free-energy relationships between dimethylallene and oxymercuration products. Path A represents direct opening of the mercurinium cation to the allylic cation, and path B represents attack of methanol on the cyclic mercurinium ion. Where there is a chloride ligand on the mercury, as illustrated, path A by way of the allylic ion predominates, the allylic ion lies lower in energy than the mercurinium ion, and the conversion of the acetoxy to the methoxy compound therefore proceeds by way of the allylic ion. If the ligand on the mercury is changed to acetate, the direct path B to oxymercuration becomes lower in energy than path A, and the relative energies of the mercurinium and allylic ions are not known.

E. Oxymercuration of Sterically Rigid Cycloolefins

Pasto and Gontarz have reported evidence that the oxymercuration of 4-*t*-butylcyclohexene (**45**) proceeds by way of attack of water on an intermediate mercurinium ion.[27] The kinetically controlled oxymercuration of **45** yields exclusively diaxial oxymercuration products **46** and **47**, which were characterized by reduction with sodium borohydride to the alcohols **48** and **49**. On prolonged standing (10 days at 25°C) the diaxial oxymercuration products **46** and **47** rearrange to more stable equatorial products. The rapid formation of the diaxial products parallels the results found with the addition of bromine or bromine–methanol (BrOMe), but contrasts with the 11–16% equatorial bromide formed from HBr addition, where the intermediate is not a cyclic ion.

Oxymercuration of 3,5-di-*t*-butylcyclohexene (**50**) places the mercury mostly at the more-hindered 2-position, so that the nucleophilic attack by water on the mercurinium ion occurs at the less-hindered 1-position.[27] The 99% discrimination in favor of one isomer seems inconsistent with a transition

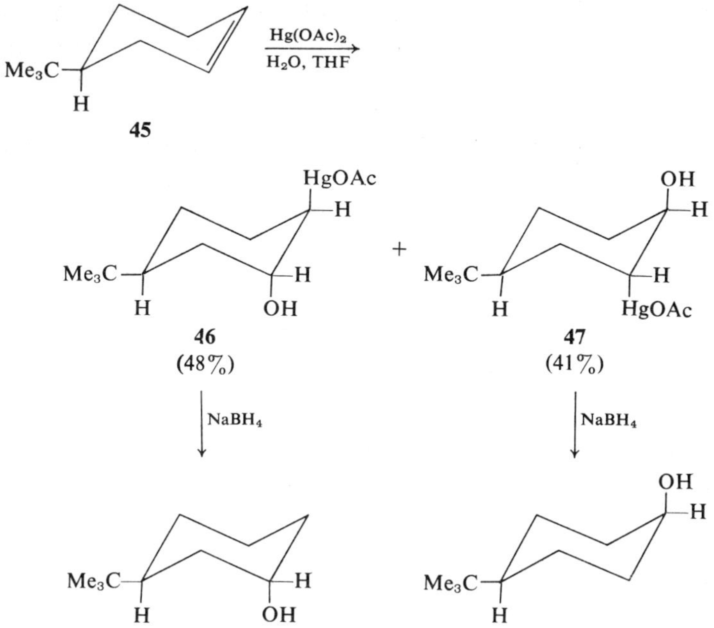

45

46 (48%) + 47 (41%)

48 49

state resembling an ordinary elimination–addition, but is consistent with opening a ring by nucleophilic displacement. Pasto and Gontarz suggested that eclipsing of the 3-*t*-butyl group and the 2-hydrogen would be required in the transition state leading to the less abundant product and would be avoided in the path to the favored isomer.

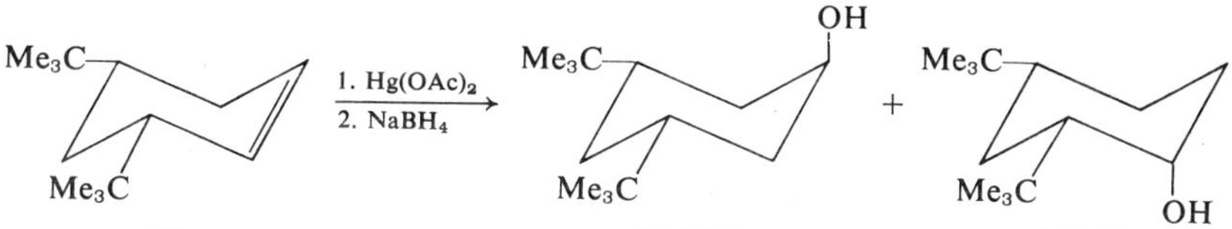

50 99.06% 0.94%

The oxymercuration of norbornenes shows some special features not seen elsewhere. Traylor and Baker found that norbornene yields the *cis,exo*-oxymercuration product **51**.[28] Oxythallation of norbornene also proceeds *cis,exo*,[29] as do various other electrophilic additions. Traylor has reviewed

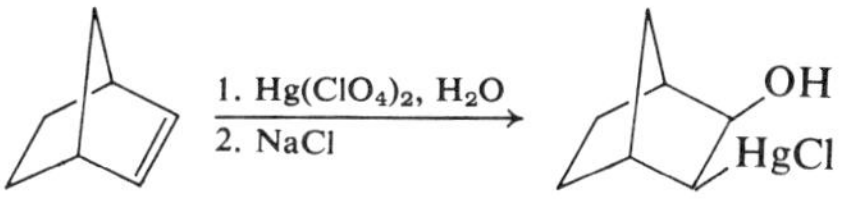

51

this topic.[30] The preference for *cis* addition evidently results from steric factors, since the norbornene would have to be twisted to allow *trans* addition with all groups coplanar. In spite of such steric strain, addition of arenesulfenyl chloride, ArSCl, to norbornene and related compounds goes *trans*.

The most peculiar aspect of the oxymercuration of norbornenes is that *syn* substituents in the 7-position do not alter the qualitative preference for *cis,exo* addition.[31] For example, 1,4,7,7-tetramethylnorbornene (**52**) yields only

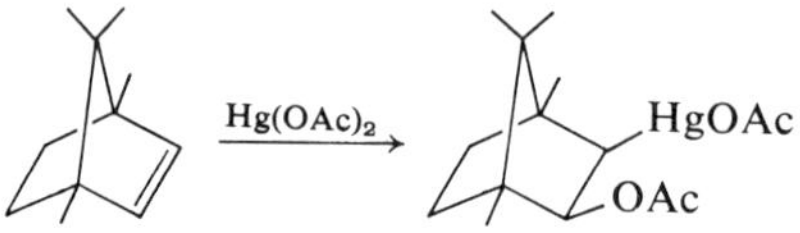

$Hg(OAc)_2$ → HgOAc, OAc

52

the *cis,exo* product. This preference for *cis,exo* addition can be reversed if there is an appropriate neighboring group present, as in the alcohols **53** and **54**.[32]

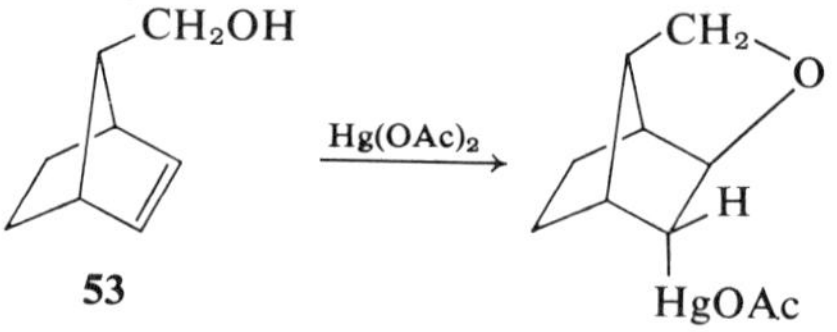

CH_2OH — $Hg(OAc)_2$ → CH_2, O, H, HgOAc

53

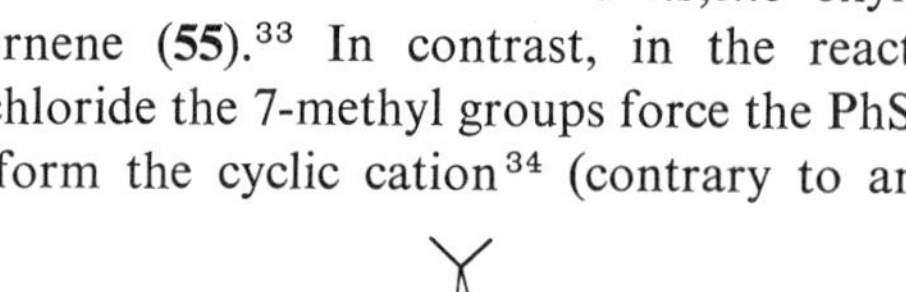

$Hg(OAc)_2$ → HgOAc

CH_2OH — CH_2—O

54

Brown and co-workers have also observed *cis,exo* oxymercuration with 7,7-dimethylnorbornene (**55**).[33] In contrast, in the reaction of **55** with benzene–sulfenyl chloride the 7-methyl groups force the PhS^+ to attack from the *endo* side to form the cyclic cation[34] (contrary to an earlier report).

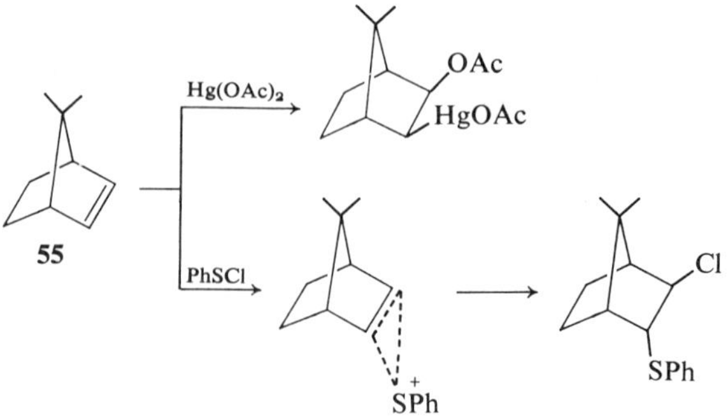

$Hg(OAc)_2$ → OAc, HgOAc

55

PhSCl → $\overset{+}{S}Ph$ → Cl, SPh

Clearly, the mechanisms of the two reactions differ significantly, and the oxymercuration pathway cannot be related to the relative energies of the *exo* and *endo* mercurinium ions, or in other words, there is probably no mercurinium ion intermediate in this case.

The evidence also shows that oxymercuration does not involve any sort of carbonium ion intermediate.[33] The reaction of 3-methylnorbornene (**56**) with mercuric acetate shows no significant influence by the methyl group and no rearrangement except for a small amount caused by the acidity of the reaction medium.

1. $Hg(OAc)_2$
2. $NaBH_4$

CH_3 **56** → OH, CH_3 48% + OH, CH_3 48% + OH, CH_3 4% but caused by H^+, not Hg^{2+}

Brown and Liu have found comparison of the norbornene and 7,7-dimethylnorbornene rates for *exo* attack to be a useful criterion for deciding whether the transition state is cyclic or acyclic.[35] The relative rates of addition to norbornene and 7,7-dimethylnorbornene where cyclic transition states are involved include PhSCl, 1820, and 9-borabicyclononane, 480. A typical noncyclic transition state is provided by HCl, which yields a ratio of 2.2. Oxymercuration with mercuric acetate in water comes out 58, which on a logarithmic scale is about half way in between. Allowing for the steric bulk of the mercury atom, the reasonable conclusion is that oxymercuration does not require a cyclic transition state or a mercurinium ion intermediate.

Do these results with norbornenes refute or at least undermine the supposition that cyclic mercurinium ions are intermediates in most oxymercuration reactions? Not at all. However, they do serve to underscore the difference between an intermediate and a transition state. In this case, the transition state requires a substantial distortion of the symmetrical mercurinium ion intermediate. A *syn*-7-methyl group (as in **55**) greatly raises the energy of the symmetrical *exo* mercurinium ion, perhaps to the point where it is no longer a stable species, but the *syn*-7-methyl group does not raise the energy of the unsymmetrical transition state nearly as much, and *cis,exo* addition proceeds in preference to *endo* attack in spite of the interference. It should be made clear that the *syn*-7-methyl group does interfere with oxymercuration to some degree. Equilibrium constants for adduct formation with mercuric trifluoroacetate in benzene are norbornene, 500,000; 7,7-dimethylnorbornene, 6,800; cyclohexene, 38.[36] The interference with complexing by silver ion appears to be greater. The equilibrium constants are norbornene, 62; 7,7-dimethylnorbornene, "0"; cyclohexene, 3.6. For the

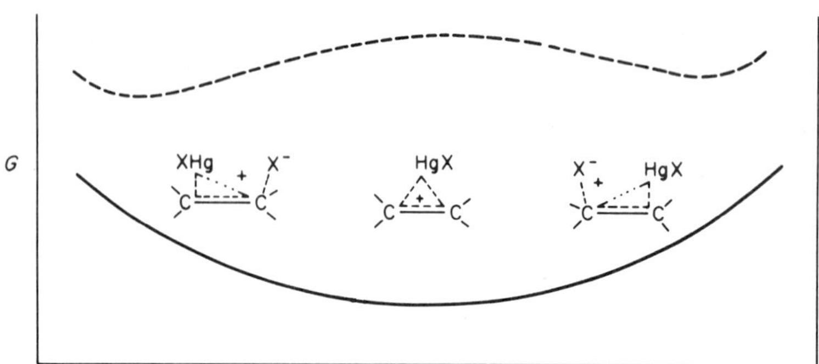

Fig. 5-4. Plot of energy vs. position of the mercury cation in mercurinium and distorted mercurinium ions. Bottom curve is for norbornene, top curve represents 7,7-dimethylnorbornene.

oxymercuration adducts, the difference in energy between the norbornene and 7,7-dimethylnorbornene compounds is 2.4 kcal/mole. For the silver ion complexes it is hard to tell how small "zero" really is, and about all that can be said about the energy difference is that it is >3 kcal/mole. The relative values for cyclohexene merely emphasize that the oxymercuration product does not itself resemble the silver ion complex.[19]

The energy relationships for oxymercuration of norbornenes are diagrammed qualitatively in Figs. 5-4 and 5-5. It can be seen from these figures that even if the *syn*-7-methyl group obliterates the energy valley corresponding

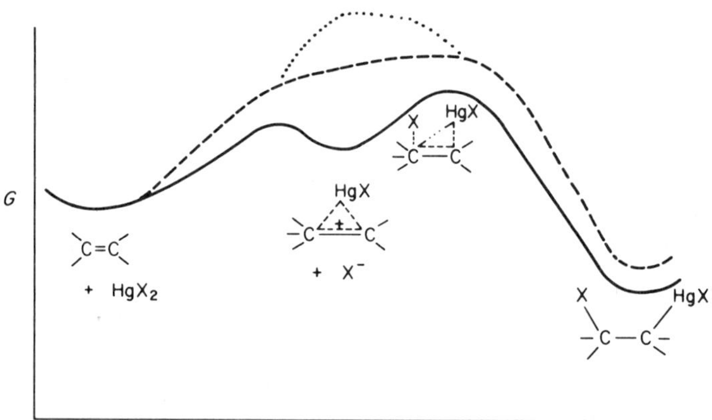

Fig. 5-5. Plot of G vs. reaction coordinate for oxymercuration. —, norbornene; ---, 7,7-dimethylnorbornene, allowing for path which avoids symmetrical cyclic cation; ····, 7,7-dimethylnorbornene if path required symmetrical mercurinium ion.

to the mercurinium ion, it need not necessarily do much to the energy or structure of the transition state for oxymercuration.

The transition state for the reaction of benzenesulfenyl chloride is greatly affected by the *syn*-7-methyl group, in contrast to that for oxymercuration. A possible explanation is that the cyclic sulfur cation complex is very near in energy and structure to the transition state for its opening by nucleophiles. The cyclic mercurinium ion is considerably "more stable" than the sulfur cation, that is, its energy and structure are farther from those of the transition state for oxymercuration. Consequently, the benzenesulfenyl chloride reaction behaves as expected for a cyclic transition state, and the oxymercuration behaves like an open transition state.

Considerable distortion of electron-deficient cyclic cations may be possible with little change in energy. Recent *ab initio* minimal basis set calculations on the 1-propyl cation and protonated cyclopropane suggest that the most stable configuration is neither the open-chain cation (**57**) nor the isosceles triangle (**59**) but an in-between structure (**58**) have an 83.4° C—C—C angle.[37]

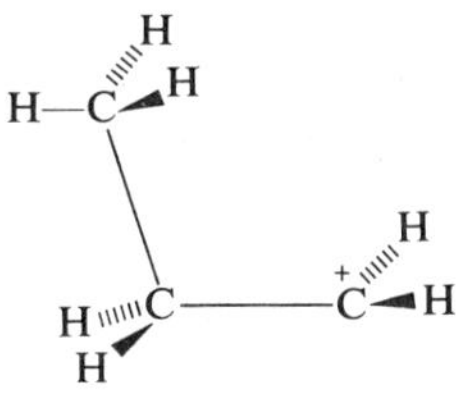

57 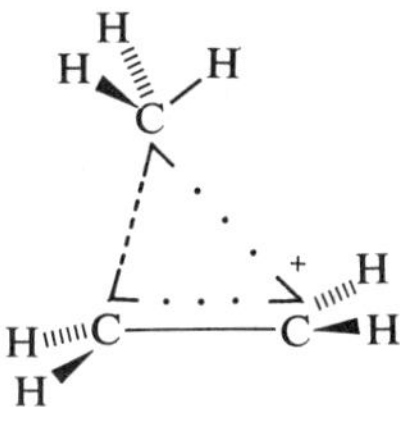 **58** 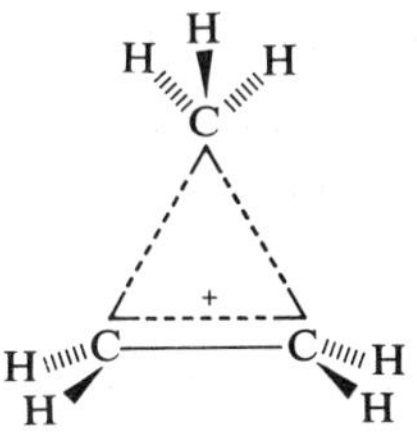 **59**

However, the energy difference between **58** and its isomers **57** and **59** was calculated to be only 0.5 kcal/mole. Within the accuracy of such calculations, this amounts to saying that all three species have about the same energy.

Jerkunica and Traylor have suggested that the interaction governing the stereochemistry of oxymercuration resembles hyperconjugation rather than cyclic three-center bonding.[10] However, these two categories do not differ in any sharply definable way. As the symmetrical structure **59** is distorted to **58** and then to **57**, there is a gradual shift from cyclic three-center bonding to the open-chain cation with its significant hyperconjugation. Replacing the methyl group of **57–58–59** with a mercury cation does not change the bonding. One small refinement is possible: the filled *d* orbitals of the mercury probably back-bond with the empty ethylenic antibonding orbital, stabilizing the symmetrical mercurinium ion to some extent. On the other hand, the mercury cation is relatively large and "soft," and easy distortion of its complex seems reasonable.

Much has been written about steric versus electronic influences in the oxymercuration of norbornenes.[30–35] Tidwell and Traylor suggested that the lack of reversal of stereochemistry by a *syn*-7-methyl or -bromo substituent pointed toward the cause of *cis,exo* oxymercuration being electronic rather than steric.[31] However, since the transition state involves considerable distortion of the cyclic mercurinium ion, as already discussed, it seems possible that the *syn*-7-methyl group may not be in as good a position for steric shielding as the *endo*-5- and -6-hydrogen atoms. Perhaps some norbornyl cation character is also involved, with the usual sort of stabilization by neighboring-group participation from the *endo* side requiring that the mercury cation and the other ligand remain *exo*. A true norbornyl cation is ruled out by the evidence,[30,33] but in view of the ability of the mercury cation to cause the transannular displacement of boron and closure of the cyclopropane ring in a norbornyl system (Chapter 4), some degree of interaction of this sort seems probable. In summary, there seem to be enough imaginable reasons to account for the *cis,exo*-oxymercuration of norbornenes but not clear enough evidence to prove which reason predominates.

F. Oxythallation

The mechanism of oxythallation resembles that of oxymercuration, except that the initially formed alkylthallium compounds generally decompose under the reaction conditions to form glycols, aldehydes, or ketones. Henry has reported that the oxidation of ethylene by thallium(III) in aqueous solution is first order in ethylene, first order in Tl^{3+}, zero order in H^+, and strongly accelerated by increased concentration of salts.[38] The acceleration by added salts resembles the mercuration of benzene (Chapter 3), in which dehydration of the metal cation occurs before reaction. The relative reactivities of Tl^{3+} salts are $ClO_4^- > SO_4^{2-} > NO_3^-$. The reaction probably proceeds by way of a three-membered cyclic thallinium ion intermediate (**60**), assuming that analogy to oxymercuration is appropriate. The further steps leading to the products, ethylene glycol (55%) and acetaldehyde (45%), are reasonable guesses.

$$Tl^{3+} + H_2C{=}CH_2 \longrightarrow \underset{\mathbf{60}}{H_2C\overset{Tl^{3+}}{\cdots\cdots}CH_2} \xrightarrow{H_2O} HOCH_2CH_2Tl^{2+} + H^+ \longrightarrow$$

$$\longrightarrow H_2C\overset{OH^+}{\text{———}}CH_2 + Tl^+ \xrightarrow{H_2O} HOCH_2CH_2OH + H^+$$

$$\longrightarrow H^+ + HOCH{=}CH_2 + Tl^+ \longrightarrow CH_3CHO$$

With higher alkenes, the major product of oxythallation is usually a ketone.[39] Propylene yields 75–85% acetone, 15–25% propylene glycol, and no propionaldehyde. The acetone cannot be formed from an epoxide intermediate, since such cleavage of an epoxide would have to be accompanied by propionaldehyde formation. Isobutylene cannot yield a ketone but yields isobutyraldehyde (40%) and the glycol $(CH_3)_2C(OH)CH_2OH$ (60%). These oxythallation studies were also extended to thallium(III) acetate in aqueous acid.[40] Two terms were found in the rate law, one first order in $Tl(OAc)_2^+$ and the other much slower term first order in $Tl(OAc)_3$.

The oxidation of acetylenes with thallium(III) nitrate appears to be another example of oxythallation followed by decomposition of the unstable thallium compound.[41] Diphenylacetylene yields benzyl in aqueous acidic glyme. Alkylarylacetylenes such as **61** undergo rearrangement in methanol.

$$\underset{\textbf{61}}{PhC{\equiv}CMe} + Tl(NO_3)_3 \longrightarrow Ph{-}\underset{MeO}{C}{=}\underset{Tl(NO_3)_2}{C}{-}Me \xrightarrow{MeOH} (MeO)_2C{=}C(Ph)(Me) \xrightarrow{H_2O} MeO_2C{-}CH(Ph)(Me)$$

IV. Group IV Halide and Related Eliminations

A. Silicon

The solvolysis of trimethyl(2-chloroethyl)silane (**62**) in aqueous ethanol yields ethylene.[42] With changing ethanol concentration, the Grunwald-Winstein *m* value is 1.02 (similar to *t*-butyl chloride solvolysis). Nucleophilicity of the solvent is unimportant in the reaction, the rate in formic acid being 2.7 times the value calculated from *m*. Sommer and Baughman concluded that the mechanism involved ionization to a trimethylsilyl cation.[42] The possibility that the transition state might resemble that for oxymercuration, with

$$\underset{\textbf{62}}{Me_3SiCH_2CH_2Cl} \longrightarrow \left[Me_3Si^{\delta+}\cdots H_2C{=}{=}CH_2\cdots Cl^{\delta-}\right] \longrightarrow Me_3Si^+ + H_2C{=}CH_2 + Cl^-$$

$$Me_3Si^+ \xrightarrow{H_2O} Me_3SiOH + H^+$$

formation of a cyclic siliconium ion intermediate, was also recognized but was not required by the data then available.

Jarvie, Holt, and Thompson have shown that the solvolysis of (*erythro*-1,2-dibromopropyl)trimethylsilane (**63**) in aqueous ethanol involves a *trans* elimination to form *cis*-1-bromopropene.[43] However, the reaction is not 100% stereospecific, and the amount of *trans*-1-bromopropene rises from

$$\underset{\mathbf{63}}{Me_3Si(H)(Br)C{-}C(H)(CH_3)(Br)} \xrightarrow{H_2O} (H)(Br)C{=}C(H)(CH_3) + Me_3SiOH + HBr$$

0.3% in a solvent mixture having a high ethanol content (Grunwald-Winstein polarity index $Y = -2.03$) to 15% in a highly aqueous medium ($Y = +2.05$). Jarvie and co-workers argued against a simple concerted elimination mechanism and in favor of a cyclic siliconium·ion intermediate. They also criticized Sommer and Baughman's representation of the transition state, "$R_3Si{\cdots}Y$," as seeming to imply an unlikely carbanion $Y^- = {}^-CH_2CH_2Cl$, even while noting that Baughman's thesis advocated concerted fragmentation into Me_3Si^+, C_2H_4, and Cl^-. It is inconceivable that Sommer ever meant to imply anything other than a concerted fragmentation, the "Y" notation arising from the pressure to save space in a communication to the editor. The stereospecificity data neatly refute the carbanion hypothesis, but it is a mistake to assume that anyone ever thought the idea likely in the first place.

The evidence for the cyclic siliconium ion is now very strong, based on scrambling of deuterium labels.[44,45] Eaborn *et al.* found that solvolysis of $Me_3SiCH_2CD_2Br$ in 30% aqueous methanol to 50% completion (half-life 300 seconds at 20°C) results in rearrangement of the recovered bromoalkylsilane to a 2:1 mixture of $Me_3SiCH_2CD_2Br$ and $Me_3SiCD_2CH_2Br$.[44] The cyclic siliconium ion **64** is evidently involved. Similarly, Bourne and Jarvie

$$Me_3SiCH_2CD_2Br \rightleftharpoons \underset{\mathbf{64}}{[H_2C{-}CD_2 \text{ bridged by } SiMe_3]^+} \rightleftharpoons BrCH_2CD_2SiMe_3$$

$$\mathbf{64} + Br^- \xrightarrow{H_2O} Me_3SiOH + H_2C{=}CD_2 + H^+Br^-$$

observed that reaction of $Me_3SiCH_2CD_2OH$ with PBr_3 or $SOCl_2$ results in almost complete scrambling of the label to form $Me_3SiCH_2CD_2X$ and $Me_3SiCD_2CH_2X$.[45] As is usual in such cases, it is not possible to decide whether **64** is a true intermediate or a transition state lying on a low-energy barrier between two hyperconjugatively stabilized primary carbonium ions, $Me_3SiCH_2CD_2^+$ and $^+CH_2CD_2SiMe_3$. In an operational sense, there must be an intermediate which on the time scale of its average lifetime has the symmetry properties of **64**. The fact that the stereospecificity of elimination is less than 100% in polar media[43] suggests that the transition state leading from the haloalkylsilicon compound to **64** resembles that for oxymercuration, with a compromise between hyperconjugation and cyclic three-center bonding allowing some *cis* departure of halide to compete with the dominant *trans* elimination.

An alternative elimination mechanism is available to β-haloalkylsilicon compounds under nonpolar conditions. Haszeldine and co-workers have found that the thermal decomposition of tributoxy(2,2-difluoroethyl)silane (**65**) is unimolecular in the gas phase at 240°–320°C,[46] and in silicone oil solution at 230°–266°C.[47] The gas-phase reaction is unaffected by the addition of nitric oxide or propene, and the solution reaction is unaffected by $FSi(OBu)_3$ or the radical inhibitor *trans*-stilbene. A simple four-center mechanism is

$F_2CHCH_2Si(OBu)_3$ (**65**) ⟶ [CHF---CH₂ / F------Si(OBu)₃] ⟶ $CHF{=}CH_2$ + $FSi(OBu)_3$

indicated. In the gas-phase reaction, $\Delta H^* = 34.1$ kcal/mole, $\Delta S^* = -14.8$ eu, and $k = 4.1 \times 10^{-5}$ sec^{-1} at 250°C. In the liquid, $\Delta H^* = 29.3$, $\Delta S^* = -22.6$, and $k = 7.4 \times 10^{-5}$ at 250°C. The lower ΔH^* in solution was interpreted as suggesting some polar character in the transition state, though the effect is not large and the polar character must be slight.

Musker and Larson found that elimination reactions of β-methoxyalkyltrimethylsilanes can be either *cis* or *trans*, depending on the reaction conditions.[48] The thermal decomposition of *cis*-2-trimethylsilylcyclohexyl methyl ether (**66**)

66 (OMe, H, SiMe₃, H) ⟶ cyclohexene + $MeOSiMe_3$

at 233°–276°C is first order and evidently proceeds through a four-center transition state. The *trans* isomer (**67**) decomposes slower than the *cis* by two competing paths, the major reaction being elimination of methanol (75%) and the minor reaction elimination of $MeOSiMe_3$ (25%). The rate

67 (H, OMe, SiMe₃, H) ⟶ 1-(trimethylsilyl)cyclohexene (SiMe₃) or 3-(trimethylsilyl)cyclohexene (SiMe₃) + MeOH

+ Me_3SiOMe

constant for elimination of the *cis* isomer (**66**) is 9.5×10^{-5} sec^{-1} at 276°C, that for $Me_3SiCH_2CH_2OMe$ is 1.6×10^{-5} sec^{-1} at 276°C. In contrast, the *trans*-cyclohexyl compound **67** does not give reproducible rates because the glass tube walls catalyze the reaction. However, at 25°C in benzene the toluenesulfonic acid-catalyzed elimination of the *trans* isomer (**67**) to form cyclohexene is very rapid, and the *cis* isomer (**66**) is inert.

B. Tin and Lead

Tin and lead compounds undergo eliminations analogous to those of the silicon compounds. $Ph_3SnCH_2CH_2OH$, $Ph_3PbCH_2CH_2OH$, and related compounds studied by Davis and Gray yield ethylene in 73% aqueous methanol with 0.006–0.09 *M* perchloric acid.[49] The rate is proportional to the Hammett acidity function h_0 and is first order in the organometallic compound. In acetic acid, the rate is proportional to the water concentration in the 0–5% H_2O range. The probable transition state **68** is a typical E2 type.

$$Ph_3SnCH_2CH_2OH + H_3O^+ \longrightarrow Ph_3SnCH_2CH_2OH_2^+ + H_2O \longrightarrow$$

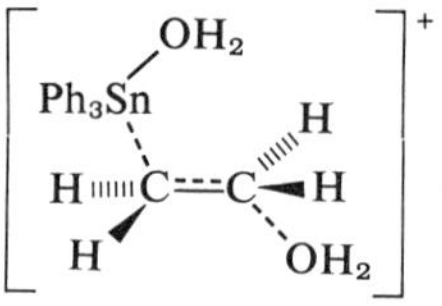

$$\left[\text{(}Ph_3Sn(OH_2)\text{)}\cdots CH_2 \cdots CH_2 \cdots OH_2\right]^+ \longrightarrow Ph_3SnOH_2^+ + H_2C{=}CH_2 + H_2O$$

68

The requirement for a molecule of water, presumably bonded to the metal atom, in the transition state differs from the reactions of the haloethylsilicon compounds. Possible influences leading to this difference may include the greater reluctance of water compared to halide to act as a leaving group from the β-carbon, the larger size of tin and lead compared to silicon allowing easier entry of water into the inner coordination sphere in the transition state, or the choice of different reaction conditions where the water was shown to participate (acetic acid rather than formic acid). The solvation energy of Me_3Si^+ should be greater than that of Ph_3Sn^+, but the transition states may be too distant in structure from these cations to allow this influence to predominate.

The elimination was shown to be stereospecifically *trans* by the use of 3-triphenylstannyl-2-butanol (**69**) made from *cis*-2,3-epoxy butane.[49]

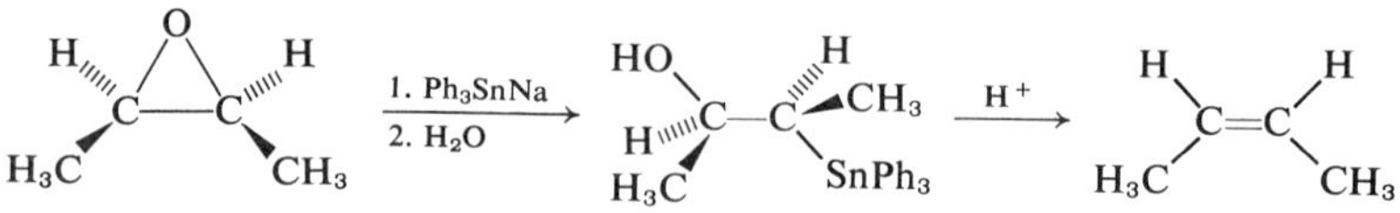

cis-2,3-epoxybutane $\xrightarrow[\text{2. } H_2O]{\text{1. } Ph_3SnNa}$ $CH_3CH(OH)CH(CH_3)SnPh_3$ (**69**) $\xrightarrow{H^+}$ $H_3C(H)C{=}C(H)CH_3$

C. δ Eliminations

Silicon and halogen more remote than β to each other can be eliminated if there is a carbon–carbon bond in between that can be broken, as in *cis*-1-

methyl-2-(trimethylsilylmethyl)-3,3-dichlorocyclopropane (**70**).[50] The ring opening was shown to be disrotatory from the geometry of the stereo-

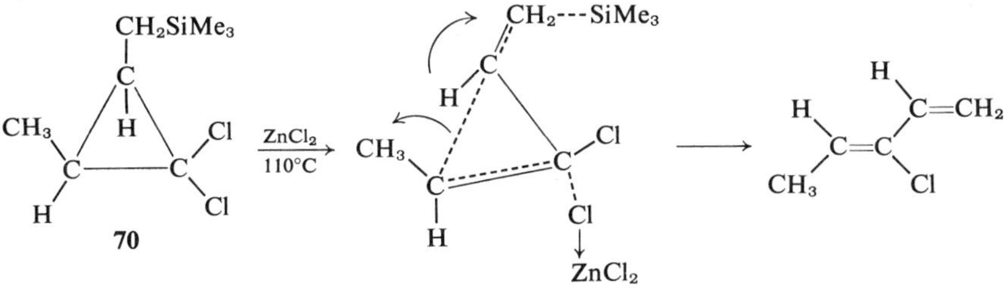

Disrotatory opening

specifically formed product. The isomer of **70** having the methyl group *trans* reacts much more slowly and gives a mixture of isomers in low yield because the disrotatory opening is hindered by the methyl group. A tin compound analogous to **70** undergoes elimination without the zinc chloride catalyst.

The ease of elimination of tin halides has been utilized in the generating of tetraphenylbutadiene (**71**), which is, of course, unstable but can be trapped by various other reagents.[51] The pyrolysis of the tin compound alone follows

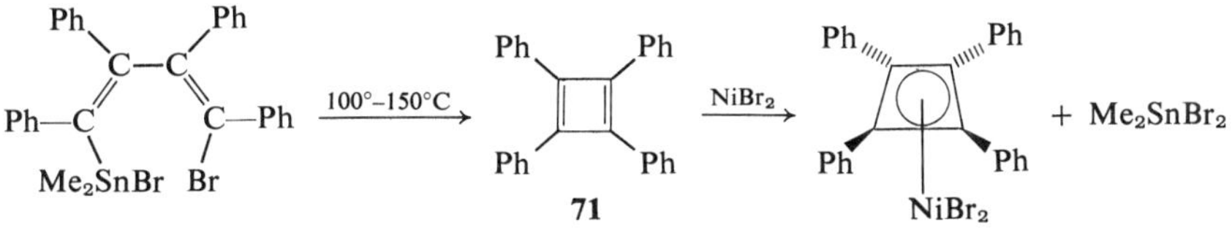

first-order kinetics, and addition of dimethyl maleate (which undergoes Diels-Alder condensation with tetraphenylcyclobutadiene) does not affect the rate. The Arrhenius activation energy is 24 kcal/mole and ΔS^* is -17 eu. The activation energy is much lower than the 50 kcal/mole required for breaking a carbon–tin bond and the entropy is consistent with a cyclic transition state.

V. Boron Halide Eliminations

A. Boronic Esters

The author and Liedtke found that β-haloethaneboronic esters are generally useless for synthetic purposes because they eliminate boron and halide extremely easily, precluding the possibility of displacing the halide by other nucleophiles.[52] We therefore decided to salvage what we could from this disappointment, and our subsequent study of mechanisms of deborono-bromination has become the classic paper in this (understandably) neglected field.[53]

Deboronobromination was shown to be stereospecifically *trans* by the fact that dibutyl *erythro*-2,3-dibromobutaneboronate (**72**) yields only 2-bromo-*cis*-2-butene (**73**) on treatment with water, sodium hydroxide, or pyridine.[53] The *threo* isomer yields 2-bromo-*trans*-2-butene. The proof of *trans* elimina-

$(BuO)_2B(CH_3)C{=}C(CH_3)H \xrightarrow{Br_2}$ **72** $\xrightarrow{\text{internal rotation}}$ **72** $\xrightarrow{\text{base}}$ **73** $(BrC(CH_3){=}C(H)CH_3)$

tion rests on the assumption that the addition of bromine to the carbon–carbon double bond is *trans*, there being no other way to determine which dibromo compound is *erythro* and which is *threo*, but in spite of the findings that hydrogen halides add *cis* to some double bonds there appears to be no reason to doubt that *trans* addition of bromine is the usual rule.[54]

Kinetic studies were carried out with dibutyl 2-bromoethaneboronate, $BrCH_2CH_2B(OBu)_2$, which evolves ethylene by a pseudo-first-order process in aqueous ethanol.[53] The Grunwald-Winstein slope m is 0.40 in 70–90% ethanol, which indicates a low dependence on solvent polarity comparable to the behavior of concerted nucleophilic displacements (for example, solvolysis of ethyl bromide, $m = 0.34$).[55] The rate is much slower in 100% ethanol than predicted by extrapolation from aqueous ethanol, and it is also much slower than predicted by the solvent polarity parameter Y in 80% formic acid. In the presence of sodium acetate the rate became too fast to measure. These observations all point toward direct involvement of one molecule of solvent as a nucleophile in the transition state, and are inconsistent with solvolysis to a boronium cation, $(RO)_2B^+$. Interpretation is only slightly complicated by the fact that the oxygen ligands on boron exchange rapidly with the solvent, so that the species being examined in pure ethanol is $BrCH_2CH_2B(OEt)_2$, in aqueous formic acid probably $BrCH_2CH_2B(OH)_2$, and in aqueous ethanol a mixture of both ligands.

Second-order kinetics were obtained in the reaction of $BrCH_2CH_2B(OR)_2$ with substituted anilines in 100% ethanol.[53] A Brønsted plot of log k_2 vs. pK_b for a series of $ArNH_2$ was linear, but dimethylaniline reacted at only 1/40 the rate expected from its basicity in relation to the $ArNH_2$ series. Thus, the function of the base is not removal of a proton in the transition state. This conclusion was confirmed by running the reaction with dimethylaniline in C_2H_5OD, which slowed the reaction by only 6%, not necessarily more than experimental error and certainly incompatible with rate-determining proton transfer. Likewise, in the solvolytic reaction in 79% ethanol, $k_H/k_D = 1.7$ (comparing ordinary with fully *O*-deuterated solvent), which is a secondary deuterium isotope effect and not compatible with rate-determining proton transfer.

From these results, the solvolysis of a β-bromoalkaneboronic ester (**74**) is a straightforward concerted elimination of the E2 type, with a molecule of solvent functioning as the base. The reversible first step to form **75** is postulated because it is known that the boron can exchange alkoxy ligands without significant elimination occurring.[52] The possibility that there might be a reversibly formed anion intermediate $BrCH_2CH_2B(OR)_3^-$ is ruled out by the failure of the kinetics to show inverse dependence on the H^+ concentration.

$$BrCH_2CH_2B(OR)_2 + ROH \rightleftharpoons BrCH_2CH_2B(OR)_2(\leftarrow O(H)R) \longrightarrow$$

74 (R = H or alkyl) **75**

$$[(RO)_2B\cdots O(H^{\delta+})R \cdots H_2C\cdots CH_2 \cdots Br^{\delta-}]^{\ddagger} \longrightarrow Br^- + H_2C{=}CH_2 + (RO)_2B{-}O(H^+)R \longrightarrow B(OR)_3 + H^+$$

A corollary to these conclusions is that intramolecular proton exchange between alkoxy groups in the solvated boronic ester **75** must be faster than proton transfer from **75** to the solvent. This is required by the fact that proton transfer is required in the exchange of ligands OR and OR′ on boron, and this exchange is faster than the elimination, which is in turn faster than formation of $BrCH_2CH_2B(OR)_3^-$ in neutral or acid solution. (Under basic conditions where formation of $BrCH_2CH_2B(OR)_3^-$ would be expected, elimination is of course extremely rapid.)

The amine-catalyzed elimination has the same fundamental mechanism as the solvolytic elimination, merely substituting $ArNH_2$ or $PhNMe_2$ for ROH in the transition state, and there seems no need to write it out in detail.

B. Trialkylboranes

Pasto and co-workers have studied the stereochemistry of β eliminations in various hydroboration products.[56] The major hydroboration products from 3-chlorocyclohexene, *trans*-2-chlorocyclohexylboranes (**76**), are stable in ether, in contrast to the usual rapid elimination observed in β-haloalkylboranes.[56a] This unusual stability can be attributed to the reluctance of the

$BH_3 \cdot Et_2O$; $ArCO_3H$

76 (R = H or chlorocyclohexyl)

compound to assume the necessary diaxial conformation necessary for β elimination. If there is a *t*-butyl group to prevent the boron and chlorine from assuming equatorial conformations, elimination is rapid, as illustrated with 1-chloro-4-*t*-butylcyclohexene (**77**).[56b]

77 R = H or cycloalkyl (hydroborates further)

Where the β substituent is alkoxy, elimination under neutral conditions is *cis* and presumably intramolecular.[56c] Acidic and basic catalysts cause *trans* elimination in the same system. The deuteroboration of *cis*-β-ethoxystyrene (**78**) with dipropyldeuteroborane is a representative example of the reactions studied. These results are analogous to those found with β-alkoxysilicon

78 + Pr_2BD ; neutral ; BuLi or BF_3 ; Pr_2BOEt

compounds, in which the uncatalyzed unimolecular reaction gives *cis* elimination and the acid-catalyzed reaction goes *trans* (Section IV,A).

C. The β-Transfer Question

In addition to β elimination, Pasto and co-workers have postulated another process which they call "β transfer", which supposedly involves rearrangement of an X—C—C—B—H system to H—C—C—B—X without any intermediate elimination to C=C and X—B—H. One of the more convincing examples is provided by the deuteroboration of 3-methoxycyclohexene (**79**).[56a] The stereochemistry of the dideuterocyclohexanol was established by

OMe → OMe, BD_2, D → D, BD_2, D → D, OH, D

79 **80**

nmr. If this conclusion that there is a nucleophilic displacement with inversion is correct, then the displacement must be greatly aided by the neighboring boron (since more distant alkoxy groups are not displaced by boron-bound hydride). Carbon–boron hyperconjugation might provide the necessary electronic interaction in the transition state.

It must be emphasized that all these so-called β-transfer reactions are found in systems which give rise to several products, and the reactions involve several steps between starting material and isolated product. The need to postulate β transfer in many cases rests on a discrepancy between the ratios of products predicted on the basis of elimination and rehydroboration and the ratios actually found. β Elimination and β transfer both occur in these systems, and slight errors at any stage of the way could supply enough discrepancy to account for all the β transfer. The one example which cannot be wished away on this basis is the conversion of **79** to **80** just cited, which could hardly be stereospecific enough to yield the results claimed if elimination to 3-deuterocyclohexene occurred to any great extent. On the other hand, one must consider it very strange that displacement should occur in preference to elimination in a cyclohexyl system. That leaves the possibility that the interpretation of the nmr spectrum (which was not itself published) could have been incorrect. In summary, the idea of β transfer is an interesting one, but the author feels that there is still considerable doubt whether this is a real phenomenon.

VI. Hydroboration

A. Introduction

The mechanism of hydroboration has been studied by a variety of methods, and a self-consistent general mechanism emerges. The stereochemistry of hydroboration has been well established as *cis* in a large number of cases, with the boron atom always preferring the least sterically hindered position.[57a] Brown and Zweifel postulated a four-center transition state (**81**).[58,59] It is

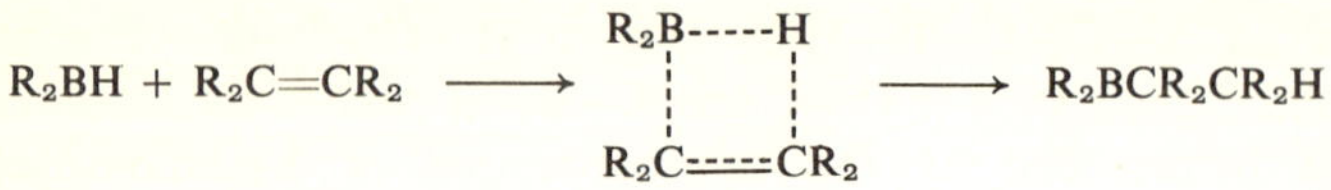

$$R_2BH + R_2C{=}CR_2 \longrightarrow \begin{array}{c} R_2B\text{-----}H \\ \vdots \qquad \vdots \\ R_2C\text{=====}CR_2 \end{array} \longrightarrow R_2BCR_2CR_2H$$

81

clear that this four-center transition state (**81**) is the essence of the hydroboration mechanism. However, it is equally clear that R_2BH monomer is not a free species if R is H or alkyl.[57b] Thus, the actual mechanism must involve transfer of an R_2BH group from its complex with a molecule of the ether solvent or with another molecule of R_2BH (the dimer $R_2BH_2BR_2$), either by direct displacement or dissociation.

B. Solution Kinetics

Brown and Moerikofer carried out the first kinetic study of hydroboration.[60] Diborane itself proved too rapid a hydroborating agent to permit convenient kinetic studies, but the hindered dialkylborane, "disiamylborane," yielded good results. It was first shown that disiamylborane is dimeric in tetrahydrofuran. It was then shown that the hydroboration is first order in cyclopentene (or various other olefins) and first order in the dimer of disiamylborane (**82**). (The reactions were followed by quenching aliquots with 1-decene and

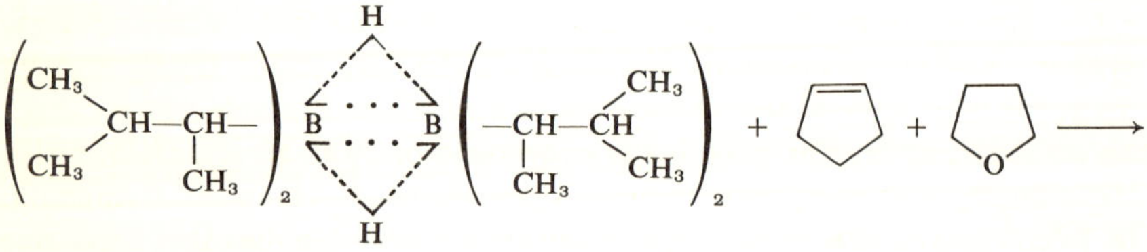

$$\left((CH_3)_2CH\text{—}CH(CH_3)\text{—}\right)_2 B \begin{array}{c} H \\ \cdots \\ H \end{array} B \left(\text{—}CH(CH_3)\text{—}CH(CH_3)_2\right)_2 + \text{cyclopentene} + \text{THF} \longrightarrow$$

82 (abbreviate as $R_2BH_2BR_2$)

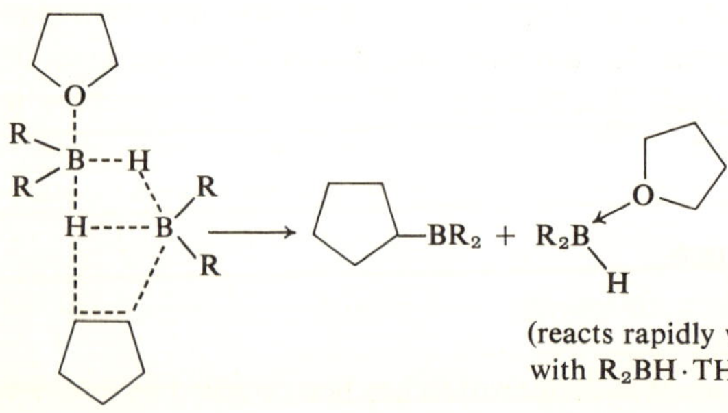

O
R
B---H
R
R
H-----B
R
—BR_2 + R_2B
O
H

83

(reacts rapidly with olefin or with $R_2BH \cdot THF$ to form dimer) analyzing for unreacted cyclopentene by gas chromatography.) The transition state (**83**) must involve liberation of one molecule of disiamylborane monomer from the dimer, and this is probably assisted by coordination of the monomer with tetrahydrofuran. Some rate constants found at 0°C include cyclopentene, 14.0×10^{-4}, cyclohexene, 0.134×10^{-4}; cycloheptene, 72×10^{-4}; *cis*-2-butene, 23×10^{-4}; *trans*-2-butene, 3.8×10^{-4} liter/mole second.

Competitive studies have indicated that diborane is not a very selective hydroborating agent.[61] In diglyme at 0°C, only about a factor of 20–30 separates the most reactive olefins (2-methyl-1-butene) from the least reactive studied (2,3-dimethyl-2-butene, 2,4,4-trimethyl-2-pentene). This lack of selectivity contrasts with the behavior of disiamylborane, which reacts 10,000 times faster with 1-octene than with cyclohexene. The differences in behavior are obviously steric in origin.

Pasto and Kang found that chloroborane in tetrahydrofuran yields readily measurable hydroboration rates, first order in olefin and first order in $ClBH_2 \cdot THF$.[62] Hydroboration of substituted styrenes correlates with σ (rather than σ^+) for a series of substituents H, *p*-Me, iso-Pr, F, Cl, Br, MeO, EtO, and NO_2, and *m*-NO_2. For attack of boron at the terminal position, $\rho = -1.43$, and at the internal position, $\rho = -0.65$, with some scatter. Correlation of the difference between terminal and internal addition yielded a better plot with $\rho = -0.65$. These results suggest a small degree of positive charge on the carbon in the transition state, in accord with the four-center mechanism.

The isotope effect k_H/k_D is 1.92 for internal, 1.90 for terminal boron addition to styrene in THF at 25°C.[62] This isotope effect is consistent with some degree of hydride transfer in the transition state and supports Brown's four-center model. Assuming the chloroborane is still complexed with the THF, transition state **84** may be written for addition of boron to the terminal carbon.

H(Ph)C=CH₂ + THF→BH₂Cl ⟶ [84] ⟶ $PhCH_2CH_2$—BHCl (THF complex)

84

The two hydroborations just cited were simplified by the existence of the hydroborating agent as essentially 100% dimer in the case of disiamylborane in ether, 100% monomer–solvent complex in the case of BH_2Cl in tetrahydrofuran. In other cases, mixtures may be expected. Pasto and co-workers have estimated monomer–dimer equilibria by ^{11}B nmr.[63] The K for association of BH_3 and n-$C_3H_7BH_2$ to $PrBH_2BH_3$ in THF appears to be ~0.6; for 2 n-Pr_2BH = n-Pr_2BH_2B-n-Pr_2, ~2.3; and for 2 iso-Pr_2BH = iso-Pr_2BH_2B-iso-Pr_2, ~20. These figures make sense only if the monomer is stabilized by strong complexing with the tetrahydrofuran. BH_3 is known to be monomeric in THF,[57b] undoubtedly as a result of complexing. The fact that BH_2Cl is a slower hydroborating agent than BH_3 in tetrahydrofuran is also consistent with the borane–solvent complex model. Extended Hückel calculations suggest that the positive charge on boron is greater in BH_2Cl than in BH_3,[64] but the BH_2Cl–THF complex could reasonably be more tightly bonded and less reactive than the BH_3–THF complex.

Klein, Dunkelblum, and Wolff have reported the kinetics of hydroboration of substituted styrenes with diborane itself in tetrahydrofuran.[65] The reaction was carried out in the presence of excess $BH_3 \cdot THF$, and the disappearance of the styrene was followed spectrophotometrically. The reaction is first order in borane and first order in styrene. Rate constants ranged from 0.186 liter/mole second with m-chlorostyrene through 0.225 with styrene to 0.367 liter/mole second with p-methoxystyrene at 25°C. The boron goes mostly to the β-carbon. The rate constants found for α and β attack do not yield a self-consistent Hammett correlation. The results seem generally consistent with the four-center transition state previously discussed. However, it should be pointed out that Klein and co-workers did "one experiment too many," which leaves the meaning of the entire paper slightly uncertain.

The embarrassing result is that aged samples of m- and p-methoxystyrene showed an induction period in the reaction, as long as six half-lives in the case of aged samples of m-methoxystyrene. Freshly distilled samples showed no induction period. Klein and co-workers tried to explain the induction period on the basis of inhibition of the hydroboration by some impurity in the styrene. They envisioned that free BH_3 might be an intermediate in the reaction, and that peroxides in the methoxystyrene might act as scavenger for the BH_3. However, this explanation cannot work. All other data are inconsistent with BH_3 being formed in some sort of chain reaction, which is the only kind of reaction that a small amount of impurity could inhibit. Remember that the $BH_3 \cdot THF$ is present in tenfold excess of the styrene, and whether BH_3 were generated reversibly or irreversibly there is little that trace impurities could do to dry up the supply of BH_3 for six half-lives. It is very difficult to imagine an explanation for data of this sort. Perhaps the aged samples contained an ultraviolet-absorbing impurity which disappeared by a radical mechanism. Perhaps students can do some very strange laboratory operations which can only be discovered and corrected if the research director gets puzzled enough soon enough. At any rate, the data with the freshly distilled samples and all the other data appear self-consistent and reasonable.

Pasto and co-workers have obtained self-consistent and satisfying data on hydroborations with borane and alkylboranes in tetrahydrofuran.[66] The reaction of $BH_3 \cdot THF$ with tetramethylethylene is first order in the $BH_3 \cdot THF$ complex and first order in the olefin, with $\Delta H^* = 9.2$ kcal/mole and $\Delta S^* = -27$ eu. The k_H/k_D value for BH_3 versus BD_3 was found to be 1.18. Tritium isotope effects k_H/k_T were measured for several olefins and found to be in the range 10–11 for monosubstituted alkenes, 9 for disubstituted alkenes, 4 for trisubstituted alkenes, and about 3 for tetramethylethylene. Hydroboration also shows boron isotope effects with $k_{^{10}B}/k_{^{11}B}$ in the range 1.03–1.05. Hydroboration by RBH_2 in tetrahydrofuran is faster than hydroboration by BH_3 itself, apparently because BH_3 forms a stronger complex with the solvent. These data are in agreement with transition state **84** proposed earlier in this section for the reaction of chloroborane in tetrahydrofuran. It may be concluded that the general mechanism for hydroboration is well understood.

C. Gas-Phase Kinetics

It is clear that hydroboration in solution does not involve free BH_3 or any other uncomplexed reactive monomeric borane as an intermediate. Fehlner has studied the rate of addition of free BH_3 (from pyrolysis of BH_3PF_3) to ethylene in the gas phase.[67] Diborane was shown not to react with ethylene under the conditions of these experiments. Rate constants for the BH_3–C_2H_4 reaction were found to be 0.9×10^9 liters/mole second at 445°K, 1.4×10^9 liters/mole second at 530°K, and 1.6×10^9 liters/mole second at 590°K, corresponding to an activation energy of about 2 kcal/mole but with an uncertainty of about 3 kcal/mole. The k for the dimerization of the product, $C_2H_5BH_2$, is about 4×10^9 and that for the reaction of $C_2H_5BH_2$ with C_2H_4 is 1×10^8, with considerable uncertainty in both. The entropy of activation for the addition of BH_3 to C_2H_4, assuming the measurements have yielded the high-pressure limiting rate constant and not mere diffusion control, is about -13 eu, and that for the reverse reaction, loss of BH_3 from $C_2H_5BH_2$, may be estimated to be $\sim +11$ eu. Thus, the transition state (**85**) is a rather loose one, not a four-center structure. The results of this study indicate that if free BH_3 were involved in hydroborations in solution, the formation of the

$BH_3 + H_2C{=}CH_2 \longrightarrow$ [**85**: H—BH_2 loosely bonded above H_3C—CH_2] $\longrightarrow$ [H----BH_2 / H_2C—CH_2 four-center] $\longrightarrow H_3CCH_2BH_2$

85
transition state

downhill from transition state

BH_3 would have to be rate-controlling and its reaction with olefins would be extremely rapid, contrary to the kinetic results actually observed and described in Section VI,B.

D. Isomerization of Boranes

The thermal isomerization of alkylboranes (**86**)[68] can be explained on the basis of reversible loss and readdition of R_2BH. The principle of microscopic

$$CH_3CH_2CH(CH_3)—BR_2 \xrightarrow{200°C} CH_3CH_2CH_2CH_2BR_2$$

86

reversibility would then require that boron hydrides or ethers catalyze the rearrangement, because the absence of free R_2BH as a hydroboration intermediate indicates that transition states leading to or from $(R_2BH)_2$ or R_2BH-etherate have lower energies. Catalysis by boron hydrides has in fact been observed, and in their presence rearrangements proceed readily at 100°C.[69]

However, there is good evidence that actual separation of R_2BH from the olefin does not occur under the usual isomerization conditions. Rickborn and Wood have shown that the thermal rearrangement of 1-(dialkylboryl)-*cis*-1,2-dimethylcyclohexane (**87**) proceeds to yield *cis*-1-(dialkylborylmethyl)-2-methylcyclohexane, that is, the boron and hydrogen exchange places with retention of configuration. The equilibrium mixture should be mostly *trans*, and it is clear that the rearrangement occurs without dissociation to free olefin. A borane–olefin π complex seems the reasonable explanation.[70]

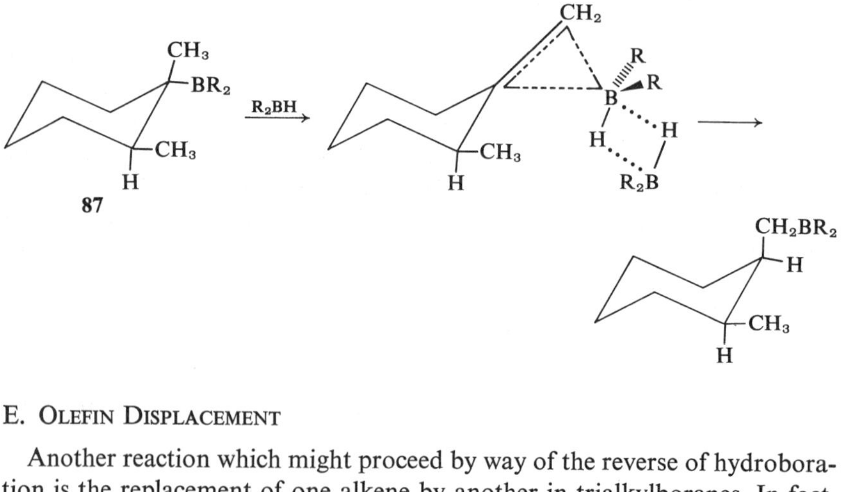

CH₃, BR₂, CH₃, H — **87** —(R_2BH)→ CH₂, R, R, B, H, H, CH₃, H, R₂B → CH₂BR₂, H, CH₃, H

E. Olefin Displacement

Another reaction which might proceed by way of the reverse of hydroboration is the replacement of one alkene by another in trialkylboranes. In fact, this mechanism does prevail in the gas phase at 407°–469°K.[71] Triisobutylborane decomposes by a first-order process, log $k = 12.3 - 30.4/2.3RT$ ($\Delta H^* = 29.5$ kcal/mole, $\Delta S^* = -5.0$ eu) for the loss of the first isobutylene. In the presence of ethylene, the iso-Bu_2BH reacts to form iso-Bu_2Et, which goes on to eliminate the second and third isobutyl groups at rates comparable to the first. The entropy of activation for the elimination is not very negative, suggesting that the transition state (**88**) is a loose π complex. These results

$$(CH_3)_2CHCH_2BR_2 \longrightarrow \begin{matrix} H\text{-----}BR_2 \\ (CH_3)_2C\text{---}CH_2 \end{matrix} \longrightarrow \begin{matrix} HBR_2 \\ (CH_3)_2C\text{---}CH_2 \end{matrix} \longrightarrow$$

Lower H than transiton state

88
Transition state

$$HBR_2 + (CH_3)_2C{=}CH_2$$

$$HBR_2 + C_2H_4 \xrightarrow{\text{fast}} C_2H_5BR_2$$

are in accord with those from gas-phase hydroboration studies and also with the existence of an intermediate borane–alkene π complex which can rearrange without dissociating.

However, in the liquid phase, which synthetic chemists find to be more compact and convenient, there appears to be another mechanism for the displacement of one alkene by another. Displacement occurs readily under conditions where isomerization of the trialkylborane does not occur.[72] Therefore, displacement does not require dissociation, since isomerization is faster than dissociation, and there must be a displacement mechanism entirely distinct from the reverse-hydroboration process. Mikhailov and co-workers have suggested a reasonable transition state model (**89**).[73] This transition state could be reached by way of a trialkylborane–alkene π complex.

$$(RCH_2CH_2)_3B + R'CH{=}CH_2 \longrightarrow \begin{matrix} CH_2 \\ (RCH_2CH_2)_2B \cdots CH{-}R \\ H_2C \cdots H \\ CH \\ R' \end{matrix} \longrightarrow$$

89

$$(RCH_2CH_2)_2BCH_2CH_2R' + CH_2{=}CH{-}R$$

Unfortunately, the kinetic plots presented by Mikhailov and co-workers in support of this mechanism are essentially zero order, the rate-limiting step probably being distillation of olefin from the mixture and not the exchange itself. In spite of the worthless kinetic data, the mechanism seems probable on other grounds.

F. Summary Description of Potential Surfaces

In summary, potential energy surfaces for hydroboration appear to involve the following features. For free R_2BH and alkene, the transition state is a loose π complex with little if any greater energy than the separated starting materials. There may be an energy minimum for a tighter π-complex intermediate, or there may merely be a forked energy trough falling off in both directions to the two possible directions of addition of B—H to the double bond. Even with the latter situation, the entropy loss required to organize the four-center species might raise its free energy to the status of a transition state. Thus, the tight π complex could have a finite lifetime sufficient to give selective isomer formation even without any enthalpy barriers to its collapse. The addition of a boron hydride or an ether molecule to this system will lower the energy of the separate R_2BH and alkene system the most (by complexing R_2BH), the loose π complex next (perhaps to the point where it is not the transition state in some systems, such as $BH_3 \cdot THF$), the tighter π complex somewhat less, and the four-center complex relatively little but not negligibly. The hydroboration product may also be lowered in energy by complexing with R_2BH or ethers, but its energy must not be lowered as much as those of the π complex and four-center species, since boranes catalyze isomerization. The π complex of a trialkylborane with an alkene cannot lead to hydroboration but apparently is a route for displacement of one alkene by another. Structures of these various species have been presented in the mechanisms outlined in the preceding paragraphs.

G. Selectivity of Hydroboration

Steric hindrance is the major factor influencing relative rates of hydroboration. To review this topic would require another book by itself, and anyway most readers will be familiar with the general results (boron goes to the least-hindered available position[59]). There are a few exceptions to the rule that electronic effects are small. The ethoxy group of 1-ethoxy-2-methylpropene (**90**) directs the boron to the tertiary carbon.[74] This directive effect can be attributed to the contribution of resonance structure **91** to the transition state, as in any polar addition to a double bond. An acetoxy group does not show this directive effect and the major product is $(CH_3)_2CHCH(BR_2)OC_2H_5$. A borane substituent also appears to have a considerable electronic influence on the course of hydroboration, $RCH{=}CHBR_2$ and HBR_2 yielding mostly $RCH_2CH(BR_2)_2$.[75] Here electron donation by boron to the α-carbon seems to be an important electronic influence.

Asymmetric induction is a steric effect of particular interest. Diisopinocampheylborane (**92**) from the hydroboration of pinene hydroborates *cis*-olefins stereoselectively to yield 50–90% optical purities.[76,77] From inspection

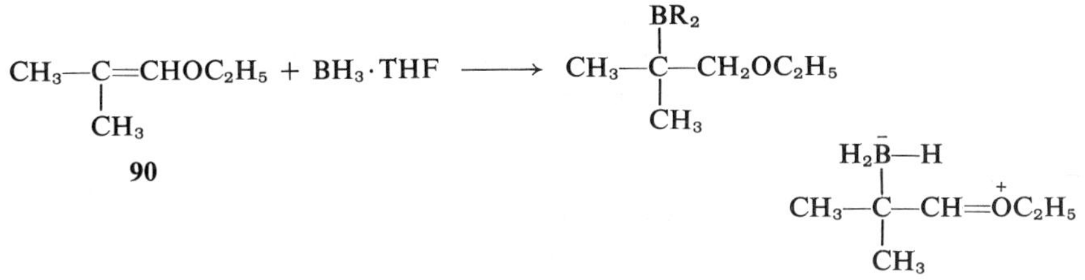

$$CH_3{-}\underset{\underset{CH_3}{|}}{C}{=}CHOC_2H_5 + BH_3\cdot THF \longrightarrow CH_3{-}\overset{\overset{BR_2}{|}}{\underset{\underset{CH_3}{|}}{C}}{-}CH_2OC_2H_5$$

90

$$CH_3{-}\overset{\overset{H_2\bar{B}{-}H}{|}}{\underset{\underset{CH_3}{|}}{C}}{-}CH{=}\overset{+}{O}C_2H_5$$

91

(a resonance structure)

of models, the conformation shown (**92**) is the only one which exposes the boron sufficiently to allow hydroboration to occur.[77,78] The reacting species is, of course, actually the borane dimer, but this does not change the essential features of the geometry and steric hindrance.

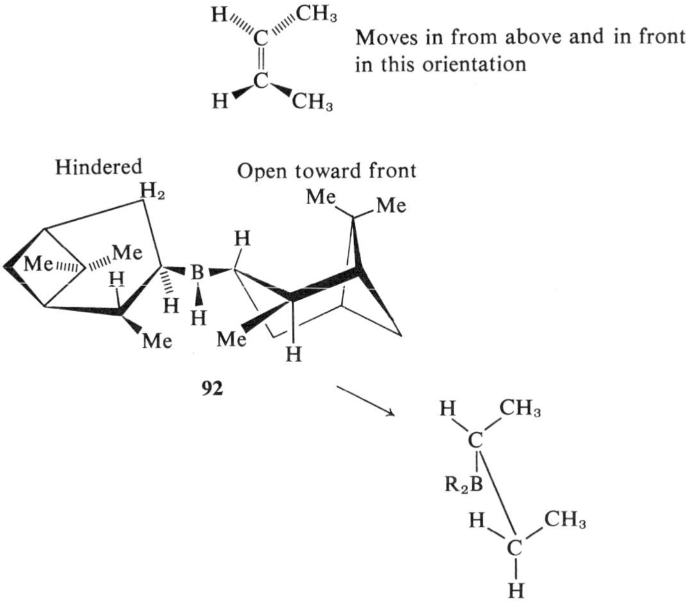

Moves in from above and in front in this orientation

Hindered

Open toward front

92

With *trans*-olefins, diisopinocampheylborane (**92**) is too hindered to enter into hydroboration.[77] Brown and co-workers found that the dialkylborane dimer $R_2BH_2BR_2$ slowly and reversibly eliminates pinene to form R_2BH_2BHR, which appears to be the active hydroborating agent in the very slow reaction with *trans*-olefins. The optical purities with *trans*-olefins are low (13–22%) and opposite in direction to those predicted on the basis of the model (**92**) which works with the *cis*-olefins.

Streitwieser and co-workers studied the hydroboration of *cis*-1-butene-1-*d* with diisopinocampheylborane.[79] From their inspection of models they concluded that Brown's transition state model would work only if the boron were centered over the double bond in a π complex. From other considerations this π-complex model seems likely, but the author in his own inspection of models was unable to see that the system would be particularly sensitive to this small a structural shift.[78] Others have challenged the Brown model for the transition state,[80,81] in part because *trans*-olefins do not fit the model, but it seems that results contrary to the model **92** can be accounted for by the known formation of monoisopinocampheylborane in these systems.[77] Moore and co-workers have recently presented a detailed critical review of this topic.[81a]

H. Oxidative Elimination of Boron Hydride

Davies and co-workers have found that 2-nitroso-2-methylpropane (**93**) or *cis*-azobenzene (but not *trans*) will abstract R_2BH very rapidly and stereospecifically from trialkylboranes.[82] A cyclic mechanism was postulated to account for the stereochemistry.

Et(H)(Me)C—C(BR₂)(Me)(H) + $(CH_3)_3C—N=O$ (**93**) ⟶ cyclic transition state [$(CH_3)_3C$—N—O, H, BR_2; Et, Me C---C Me, H]

↓ Et(Me)CH—C(=CH₂)H (67%)

↓ Et(Me)C=C(Me)H + $(CH_3)_3C—NH—OBR_2$

(33%) (100% stereospecific)

I. Diboron Tetrachloride

The addition of B_2Cl_4 to alkenes and acetylenes is evidently a four-center process mechanistically related to hydroboration. In accord with this mechanism, the addition of B_2Cl_4 to acetylene yields *cis*-1,2-bis(dichloroboryl)ethene (**94**),[83–85] and *cis* additions of B_2Cl_4 to *cis*- and *trans*-2-butene,[83,84] 2-butyne,[84] cyclohexene,[84] cyclopentene, cyclobutene, and cyclopropene[86] have been observed. This observation that the addition of cyclopentene is *cis* was checked by oxidation to *cis*-1,2-cyclopentanediol as well as by nmr evidence,[86] and directly contradicts an earlier report of *trans* addition to cyclopentene.[87]

$$Cl_2B{-}BCl_2 + HC{\equiv}CH \longrightarrow (Cl_2B)(H)C{=}C(BCl_2)(H)$$

94

VII. Hydralumination

The mechanism of hydralumination resembles that of hydroboration in being a four-center, *cis* addition to the double or triple bond, but there are some significant differences in the details. The *cis* hydralumination of alkynes was observed by Wilke and Müller[88] and has been shown to be synthetically useful by Zweifel and Steele.[89] 2-Butyne yields the *cis*-hydralumination product (**95**) with diisobutylaluminum hydride, but the *trans* product (**96**) results if a trialkylaluminum hydride anion is used.[89] It is also possible to

$$CH_3{-}C{\equiv}C{-}CH_3 \xrightarrow{HAl(iso\text{-}Bu)_2} (H)(CH_3)C{=}C(Al\text{-}iso\text{-}Bu_2)(CH_3) \xrightarrow[2.\ CO_2\ \ 3.\ H^+]{1.\ MeLi} (H)(CH_3)C{=}C(CO_2H)(CH_3)$$

95

$$CH_3{-}C{\equiv}C{-}CH_3 \xrightarrow{LiHAl(iso\text{-}Bu)_2Me} (H)(CH_3)C{=}C(CH_3)(Al(iso\text{-}Bu)_2^-{-}Me) \xrightarrow[2.\ H^+]{1.\ CO_2} (H)(CH_3)C{=}C(CH_3)(CO_2H)$$

96

condense 2 moles of 3-hexyne with one of diisobutylaluminum hydride to form a more exotic all-*cis* adduct (**97**), but this can be rationalized on the basis of a hydralumination followed by an addition of alkenylaluminum to a second molecule of alkyne[90] (see the following section for Al—R additions).

$$2C_2H_5C{\equiv}C{-}C_2H_5 + HAl(iso\text{-}Bu)_2 \longrightarrow (C_2H_5)(H)C{=}C(C_2H_5){-}C(C_2H_5){=}C(Al(iso\text{-}Bu)_2)(C_2H_5)$$

97

Eisch and Rhee have found that the hydralumination of 4-octyne with diisobutyaluminum hydride is 0.37 order in $[(iso\text{-}Bu)_2AlH]_3$ and first order in 4-octyne.[91] Since diisobutylaluminum hydride is trimeric,[92] these results correspond to direct reaction of monomeric $(iso\text{-}Bu)_2AlH$ with 4-octyne to form a four-center species (**98**). The rate law was determined from initial rates because the reaction product slows the reaction (see next paragraph)

$$[(\text{iso-Bu})_2\text{AlH}]_3 \rightleftharpoons 3(\text{iso-Bu})_2\text{AlH} \xrightarrow{C_3H_7C{\equiv}CC_3H_7} \begin{array}{c}(\text{iso-Bu})_2\text{Al}\text{----H} \\ \vdots \quad\quad \vdots \\ C_3H_7C{\equiv}CC_3H_7 \\ \mathbf{98}\end{array} \longrightarrow \begin{array}{c}(\text{iso-Bu})_2\text{Al} \quad\quad H \\ C{=}C \\ C_3H_7 \quad\quad C_3H_7\end{array}$$

and good long-term kinetic plots were not possible. The values $\Delta H^* = 20.2$ kcal/mole, $\Delta S^* = -11.8$ eu, and $k_{Al-H}/k_{Al-D} = 1.68$ were found. In a related study, Hay, Jones, and Robb found the reaction of diisobutylaluminum hydride with 1-butene to be half-order in the aluminum compound,[93] which is perhaps within experimental error of 0.33 order if the aluminum compound remains predominantly a trimer under the conditions used.

From nmr measurements, Clark and Zweifel have found that the initial products from the hydralumination of alkynes are dimers having one hydride bridge and one alkenyl bridge (**99**).[94] This, of course, does not conflict with

$$R{-}C{\equiv}CH + [(\text{iso-Bu})_2\text{AlH}]_3 \longrightarrow (\text{iso-Bu})_2\text{Al} \begin{array}{c} H \\ \cdots \\ C(H){=}CHR \end{array} \text{Al}(\text{iso-Bu})_2$$

99

the kinetic results, since the initially formed 1:1 adduct analogous to **98** could combine with a second molecule of (iso-Bu)$_2$AlH monomer in a subsequent fast step. The formation of adducts involving 2 moles of aluminum compound (**99**) would account for the inhibition of the reaction by-product mentioned previously.

The elimination of iscbutylene from triisobutylaluminum in the gas phase is unimolecular, much like the reaction of triisobutylborane described in the preceding section.[95] In order to study the kinetics, the diisobutylaluminum hydride formed in the reaction was captured with ethylene, which does not enter into the rate law, yielding EtAl(iso-Bu)$_2$ as the product.

The elimination of isobutylene from triisobutylgallium in the gas phase at 416°–500°K is similarly unimolecular, with $\Delta H^* = 29.5$ kcal/mole and $\Delta S^* = -8.4$ eu.[96]

VIII. Metal Alkyl Additions to Double Bonds

A. Introduction

The addition of a metal alkyl R—M to an alkene such as $CH_2{=}CH_2$ to form the homologous metal alkyl RCH_2CH_2M is the basis for the Ziegler-Natta type of polymerization. Really efficient polymerization of ethylene requires a transition metal catalyst and is outside the scope of this book. However, oligomerization can be carried out with aluminum alkyls and is an industrially important process for making higher straight-chain alkenes.[97] Alkyllithiums can also add to certain alkenes and can initiate polymerization of dienes. Finally, in view of the importance of the reactions and the secondary involvement of nontransition metal alkyls in the reactions, this section will conclude with a brief description of some of the types of alkene reactions with transition metal alkyls, though the coverage will be very selective and limited in scope.

B. Aluminum

Eisch and Hordis have studied the addition of triphenylaluminum dimer to substituted diphenylacetylenes.[98] Two isomeric products (**100** and **101**) are formed from $XC_6H_4C{\equiv}CPh$, where X is Me_2N, MeO, Me, MeS, H, or Cl. The addition was shown to be stereospecifically *cis* with the isomer in which *cis* addition differs from *trans* (**101**). A Hammett correlation of the ratio of isomers **101**/**100** yields $\rho = -0.713$, that is, electron-donating substituents

$$(Ph_3Al)_2 + XC_6H_4C{\equiv}CPh \longrightarrow \underset{\mathbf{100}}{(Ph_2Al)(XC_6H_4)C{=}C(Ph)(Ph)} + \underset{\mathbf{101}}{(Ph)(XC_6H_4)C{=}C(AlPh_2)(Ph)}$$

$$\mathbf{100} \xrightarrow{H_2O} (H)(XC_6H_4)C{=}C(Ph)(Ph) \qquad \mathbf{101} \xrightarrow{H_2O} (Ph)(XC_6H_4)C{=}C(H)(Ph)$$

direct the aluminum toward the β-carbon by stabilizing carbonium ion character at the α-carbon to a slight degree, as expected for an electrophilic addition of the four-center type.

The addition of triphenylaluminum to diphenylacetylene is half-order in triphenylaluminum dimer and first order in diphenylacetylene in mesitylene as solvent at 110°–140°C, and ΔH^* is about 18 kcal/mole.[99] The rate-limiting step must be attack of triphenylaluminum monomer on the triple bond to form a π complex or four-center transition state (**102**). Electron-donating

$$Ph_2Al(\mu\text{-}Ph)_2AlPh_2 \rightleftharpoons 2Ph_3Al \xrightarrow{PhC\equiv CPh}$$

$$\begin{array}{c} Ph\quad\quad Ph \\ C\equiv C \\ \vdots \quad\quad \vdots \\ Ph\text{----}AlPh_2 \end{array} \longrightarrow (Ph)_2C{=}C(Ph)AlPh_2$$

102

substituents on the diphenylacetylene interfered with getting good kinetic plots, evidently because they complex with triphenylaluminum. However, CH_3 accelerates the reaction and Cl retards it, and the effects are comparable in magnitude with the degree of regioselectivity shown in the reaction (preceding paragraph). Thus, the overall rate and the selectivity of direction of addition are governed by similar electronic factors, suggesting that the transition state may be four-center in character (**102**) rather than a loose π complex.

Ziegler and Hoberg reported in 1960 that the rate of absorption of ethylene by triethylaluminum in decalin at 110°C is half-order in triethylaluminum dimer, and absorption by triisobutylaluminum is first order in monomeric (iso-Bu)$_3$Al.[100] More recently, Hay and co-workers have found that the reaction of trimethylaluminum dimer with 1-octene in decalin at 400°–440°K is half-order in Al_2Me_6, following the reaction by dilatometry.[101] Triethylaluminum dimer also follows half-order kinetics with 1-alkenes.[102] Thus, the requirement of a monomeric trialkylaluminum in additions to unsaturated compounds is a general one.

Egger and Cocks have found that the addition of ethylene to monomeric trimethylaluminum in the gas phase initially yields propyldimethylaluminum, but this rapidly eliminates propylene and the resulting dimethylaluminum hydride adds rapidly to ethylene, yielding ethyldimethylaluminum as the major product.[103] For the first step, $\Delta H^* = 21.5$ kcal/mole and $\Delta S^* = -26.6$ eu or, if corrected for the number of methyl groups in $AlMe_3$, $\Delta S^* = -28.3$ eu, in the temperature range 455°–549°K. Although the reaction obviously must involve a four-center species at some stage, Egger and Cocks advocated a looser, π-complex type of transition state (**103**). The gas-phase reaction of triethylaluminum is similar. The rate-determining step is addition of C_2H_4 to $AlEt_3$, 1-butene is eliminated rapidly, and the resulting Et_2AlH

$$Me_3Al + H_2C{=}CH_2 \longrightarrow \underset{\substack{\textbf{103}\\ \text{Transition}\\ \text{state (?)}}}{[Me_3Al \cdots H_2C{-}CH_2]} \longrightarrow \underset{\substack{\text{Lower energy}\\ \text{than transition}\\ \text{state (?)}}}{[Me\text{----}AlMe_2 \,/\, H_2C{=}CH_2]} \longrightarrow$$

$$MeCH_2CH_2AlMe_2 \longrightarrow MeCH{=}CH_2 + HAlMe_2 \xrightarrow{C_2H_4} C_2H_5AlMe_2$$

reacts rapidly with another molecule of C_2H_4 to regenerate $AlEt_3$. The reaction is faster than with $AlMe_3$ because the ΔH^* is lower, 16.7 kcal/mole, even though the statistically corrected ΔS^* is more negative, -32.9 eu.[104] The ΔS^* values are in accord with the greater steric shielding of the aluminum in the ethyl compound, but the reason for the ethyl having a lower ΔH^* than the methyl is not obvious. Some sort of hyperconjugation is a possibility.

In favor of the existence of aluminum–olefin π complexes, compounds of the type $R_2Al(CH_2)_3CH{=}CHR'$ show infrared shifts of about $-22\ cm^{-1}$ in the C=C stretching band.[105]

The very negative ΔS^* values favor a four-center transition state in which rotational entropy has been lost. Egger and Cocks concluded that other arguments, such as the fact that ethylene reacts much faster than propylene, rule out a tight four-center transition state and favor the loose π-complex model.[103] However, this may be an inherently undecidable choice because the transition state could have some characteristics of both. Loss of rotational entropy might occur right at the beginning of organization of the four-center species, while the new carbon–carbon bond is still long and very weak, and the strongest bonding is a distorted aluminum–olefin π complex. Trying to fix the position of a transition state along a continuum of possibilities is not an either–or proposition.

C. Lithium

There appear to be two types of addition of alkyllithiums to double bonds, one involving undissociated alkyllithium tetramers and the other involving alkyllithium monomers or dimers and probably carbanion intermediates.

Bartlett and co-workers have found that ethylene adds quantitatively and cleanly to isopropyl-, *t*-butyl-, *sec*-butyl, and cyclohexyllithium, but not to cyclobutyl-, phenyl-, benzhydryl-, or triphenylmethyllithium.[106] Primary alkyllithiums are also unreactive. An ether or amine catalyst is required. No other acyclic, unconjugated olefin approaches ethylene in reactivity.

The fact that the alkyllithiums most able to dissociate into carbanions are least able to react with ethylene is evidence against a carbanion mechanism.

(Kinetic acidities of some of the parent hydrocarbons are Ph_3C, 1100; Ph_2CH, 270; Ph, 1; cyclobutyl, 10^{-6}; and cyclohexyl, 1.1×10^{-8}.[107]) This contrasts with the relative rates of addition of RLi to Michler's ketone, Ph > Et > iso-Pr.[108]

Bartlett *et al.* have studied the kinetics of the insertion of ethylene into the carbon–lithium bond of several alkyllithium reagents.[109] The reaction is strongly catalyzed by ethers. It was first shown that up to 4 moles of dimethyl ether can coordinate with *n*-butyllithium tetramer in *n*-heptane. Kinetic measurements were carried out on the more reactive isopropyl-, *sec*-butyl-, and *t*-butyllithiums. With these compounds, the reaction is first order in alkyllithium tetramer, first order in ethylene, and (for at least the first two) second order in diethyl ether at −25°C. With *t*-butyllithium as the reactant and tetrahydrofuran as the catalyst in pentane at −41°C, the reaction is roughly second order in THF at lower THF concentrations but levels off to a maximum at 2 *M* THF and becomes inversely first order in THF at higher concentrations, as if the THF were using up the last coordination site on the alkyllithium tetramer.

Kinetics on systems of this sort are always plagued by the foreseeable difficulties as well as a few extra that nature usually tosses in. Careful exclusion of air and moisture is necessary to avoid degradation of the alkyllithiums to lithium alkoxides. Another diversion was provided by a nichrome wire stirrer, which evidently incorporated traces of nickel into the alkyllithium. The nickel compound is a much more efficient, though ill-defined, catalyst for the ethylenation reaction, as might be expected by analogy to the Ziegler polymerization catalysts, and it was necessary to use a Teflon stirrer instead during the preparation of the alkyllithium reagents in order to get reproducible (and much slower) rates.

1,4-Dimethoxybutane, triethylamine, and *N*-methylpyrrolidine also proved to be effective catalysts, entering the rate expression in a first-order manner. The order in nucleophile merely shows that more molecules of nucleophile than the prevailing equilibrium level of coordination are required for optimum catalysis. Thus, second order in ether may mean that the transition state contains two ether molecules if the alkyllithium tetramer has none for the most part, but from the equilibrium studies it seems more likely that the alkyllithium tetramer coordinates strongly with one molecule of ether in the prevailing species, and there are three molecules of ether in the transition state.

Activation parameters for the ethylenation reaction are listed in Table 5-3.

In spite of the gradual change of the original secondary alkyllithium tetramer to an equilibrating mixture including both primary and secondary alkyl groups as the reaction proceeds, the reaction remains remarkably first order in alkyllithium tetramer out to 90% or more consumption of the

TABLE 5-3

ACTIVATION PARAMETERS FOR ETHYLENATION OF ALKYLLITHIUMS[a]

Alkyllithium	Ligand	ΔH^* (kcal/mole)	ΔS^* (cal/deg mole)
Isopropyl	Et_2O	9.4	−41.5
2-Butyl	Et_2O	9.3	−39.5
Isopropyl	THF	9.0	−30.6
Isopropyl	$MeO(CH_2)_4OMe$	10.0	−27.5
2-Butyl	Et_3N	11.5	−25.4

[a] From P. D. Bartlett, C. V. Goebel, and W. P. Weber, *J. Amer. Chem. Soc.* **91**, 7425 (1969).

sec-alkyl groups. This is rather surprising, but several factors which might tend to modify the rate may cancel accidentally.

The mechanism cannot involve any sort of dissociation of the alkyllithium compound. Carbanions cannot be involved, because excess tetrahydrofuran could not inhibit dissociation of RLi to R^- and solvated Li^+. The very negative entropies of activation are consistent with a concerted process. The most likely mechanism involves some sort of coordination of the ethylene with a vacant site on the alkyllithium tetramer (**104**) followed by migration of the alkyl group from lithium to one of the ethylene carbons, with subsequent migration of the other ethylene carbon to the face of the tetrahedron of lithium atoms which has been vacated by the migrating alkyl group.

Carbanionic polymerization of ethylene is also possible. Butyllithium in the presence of the chelating agent $Me_2N(CH_2)_4NMe_2$ has been reported to be an effective catalyst.[110]

Anionic polymerization has been reviewed in detail elsewhere.[111]

Welch and Magid have shown that the addition of phenyllithium to cyclopropene (**105**) is >99% *cis*.[112] This result, of course, rules out any free

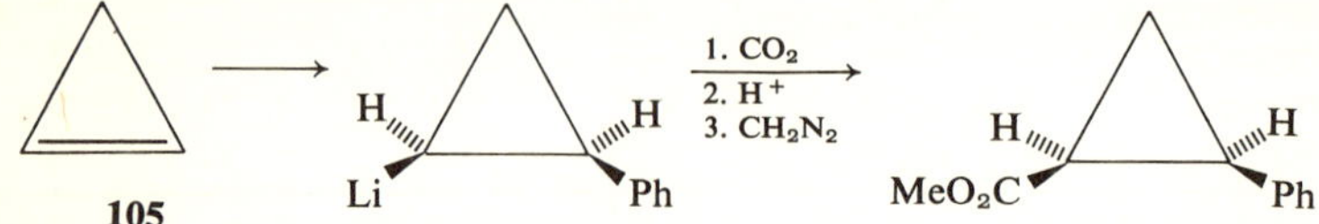

carbanion intermediate and supports some sort of four-center mechanism.

Waack and Doran have studied the addition of various alkyllithiums to 1,1-diphenylethylene (**106**).[113,114] In tetrahydrofuran or ether the reaction

$$(RLi)_n \rightleftharpoons nRLi; \quad \underset{\mathbf{106}}{Ph_2C{=}CH_2} + RLi \longrightarrow Ph_2\underset{\displaystyle Li}{\underset{|}{C}}{-}CH_2{-}R$$

orders suggest dissociation of the organolithium compound to monomer followed by reaction of the monomer with the 1,1-diphenylethylene. A summary of reaction orders and rate constants (22°C) for the various lithium compounds tested is shown in Table 5-4.

TABLE 5-4

REACTION ORDERS IN RLi AND RATE CONSTANTS FOR ADDITIONS OF ORGANOLITHIUM COMPOUNDS TO 1,1-DIPHENYLETHYLENE[a]

RLi	Reaction order		*k* in Appropriate units	
	Et_2O	THF	Et_2O	THF
MeLi	0.21	0.27	0.00017	0.12
PhLi	0.51	0.66	0.0028	0.25
$CH_2{=}CHLi$	—	0.34	—	0.11
$CH_2{=}CHCH_2Li$	1.3	1	1.1	110
BuLi	0.3	0.4	1.0	500
$PhCH_2Li$	1.2	1.1	1.8	3000

[a] From R. Waack and M. A. Doran, *J. Organometal. Chem.* **29**, 329 (1971).

The difference in behavior between 1,1-diphenylethylene and ethylene is understandable when the different natures of the products are considered. The products from 1,1-diphenylethylene have considerable carbanion character and are probably themselves monomeric in ether or tetrahydrofuran, but the products from ethylene cannot have delocalized carbanion character and remain tetrameric.

The very rapid reaction of menthyllithium with 1,1-diphenylethylene has been cited as evidence for the existence of this sterically hindered cycloalkyllithium as a dimer which probably dissociates readily to a monomer.[115]

Grovenstein and Wentworth have shown that the rearrangement of 2,2,3-triphenyl-1-propyllithium (**107**) involves dissociation and readdition, since isopropyllithium or ^{14}C-labeled benzyllithium in the reaction mixture become incorporated into the products.[116] Aryl groups can migrate by an intramolecular mechanism (Chapter 4, Section VI,D).

$$\underset{\mathbf{107}}{PhCH_2C(Ph)_2CH_2Li} \longrightarrow PhCH_2Li + Ph_2C{=}CH_2 \begin{cases} \xrightarrow{PhCH_2Li} Ph_2C(Li)CH_2CH_2Ph \\ \xrightarrow{Ph^{14}CH_2Li} Ph_2C(Li)CH_2{}^{14}CH_2Ph \\ \xrightarrow{(CH_3)_2CHLi} Ph_2C(Li)CH_2CH(CH_3)_2 \end{cases}$$

Grignard reagents are not reactive enough to add to ordinary carbon–carbon double bonds, but hindered Grignard reagents such as isopropylmagnesium bromide (**108**) add to the double bond of vinylsilicon compounds.[117] Less-hindered compounds, such as *n*-propylmagnesium bromide, tend to attack at silicon almost exclusively. The ratio of products also depends on the leaving nucleophile available on the silicon.

$$\underset{\mathbf{108}}{(CH_3)_2CHMgCl} + CH_2{=}CH{-}Si(OEt)_3 \longrightarrow \underset{(71\%)}{(CH_3)_2CHCH_2CH(MgBr)Si(OEt)_3} + \underset{(29\%)}{CH_2{=}CH{-}Si(OEt)_2CH(CH_3)_2}$$

An attempt to add a hindered Grignard reagent, mesitylmagnesium bromide, to the double bond of a hindered vinylboron compound, di-*t*-butyl vinylboronate, yielded only the product of attack of the Grignard reagent on the boron atom.[118]

D. Transition Metals

Though outside the intended scope of this book, some mention of transition metal-catalyzed additions to double bonds is necessary to round out the preceding discussion. Most of the cases considered here involve a nontransition metal as well, though where both are present the transition metal tends to become the center of catalytic activity.

The Ziegler-Natta polymerization of ethylene is generally carried out with solid catalysts prepared from trialkylaluminums and titanium(III) or (IV) chloride. It is difficult to prove the structure of such a catalyst, let alone its mechanism of action. However, insight into the nature of action of these catalysts is provided by studies of catalytically active soluble titanium–aluminum complexes reported by Henrici-Olivé and Olivé.[119] The active catalyst formed from dicyclopentadienyltitanium dichloride (Cp_2TiCl_2) and ethylaluminum dichloride ($EtAlCl_2$) in toluene was shown by epr studies to contain titanium(IV), and the titanium(III) complexes formed by reductive decomposition were found to be inactive in the polymerization of ethylene.

Bis(cyclopentadienyl)ethyltitanium chloride ($Cp_2TiEtCl$) can be prepared in crystalline form and is stable in solution. It is not catalytically active. Addition of $EtAlCl_2$ yields a highly active 1:1 adduct (**109**), which is much more active than the catalyst prepared from Cp_2TiCl_2 and $EtAlCl_2$.[119] A

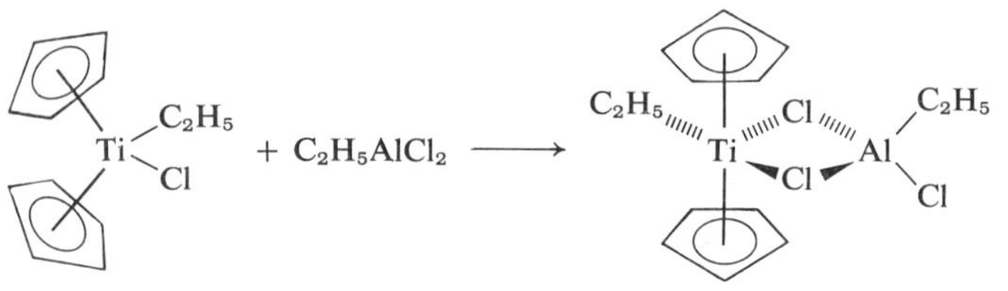

$Cp_2Ti(C_2H_5)Cl + C_2H_5AlCl_2 \longrightarrow$ **109**

complex prepared from $Cp_2TiEtCl$ and $AlCl_3$ proved to be inactive. It was concluded that both the titanium and the aluminum atoms must bear an alkyl group in order for there to be catalytic activity, and the active catalyst formed from Cp_2TiCl_2 and $EtAlCl_2$ is probably **109**, which can result from exchange of chloride and ethyl groups. This is in accord with the observed induction period for buildup of catalytic activity in the Cp_2TiCl_2–$EtAlCl_2$ system.

The active catalyst **109** shows no induction period and the polymerization is immediately first order in ethylene.[119] The probable mechanism involves a π complex of ethylene at the vacant site of octahedral titanium(IV) (**110**), which can rearrange to a new alkyltitanium complex (**111**) with restoration of a vacant site to the titanium atom. Thus, ethylene can continue to insert

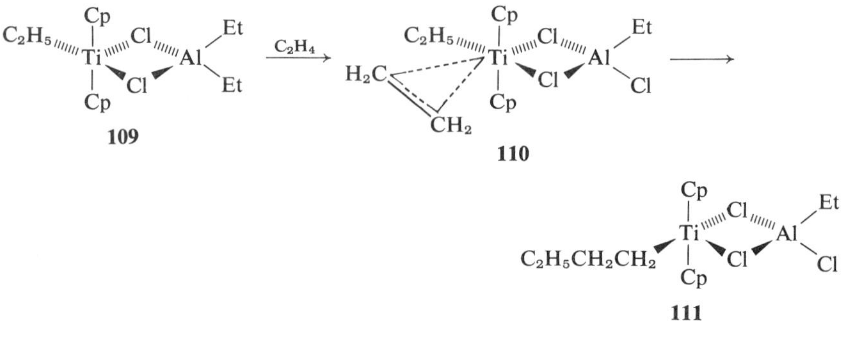

109 $\xrightarrow{C_2H_4}$ **110** $\longrightarrow$ **111**

repeatedly into the titanium–alkyl bond. In principle, this mechanism is the same as that for the insertion of ethylene into alkylaluminum or alkyllithium compounds, but the titanium allows unrestricted chain growth.

The active catalyst (**109**) undergoes bimolecular self-reduction to the inactive titanium(III) complex $Cp_2TiCl_2AlEtCl$, $k = 5 \times 10^{-3}$ liter/mole second at 20°C, and there is also a termination step from reaction of two growing-chain catalyst complexes, $k \simeq 0.6$ liter/mole second. It appears that the failure of soluble titanium(III) complexes to show any catalytic activity results from the titanium atom becoming tetrahedral. Heterogeneous catalysts based on titanium(III) are active, but it appears that the active sites are octahedral titanium atoms in the crystal lattice.[119]

Dimethylcopper lithium adds *cis* to the triple bond of $RC{\equiv}CCO_2Me$ at −78°C in THF[120] or −100°C in ether–hexane.[121] Some *trans* addition occurs at higher temperatures. Some sort of π complex between the copper atom and the triple bond (**112**) seems likely, and the mechanism thus resembles that of ethylene polymerization and related reactions.

$$R{-}C{\equiv}C{-}CO_2Me + Me_2CuLi \longrightarrow R{-}C{\equiv}C{-}CO_2Me \cdots Me_2CuLi \ (\mathbf{112}) \longrightarrow$$

$$(R)(Me)C{=}C(CO_2Me)(CuMeLi) \xrightarrow{H_2O} (R)(Me)C{=}C(CO_2Me)(H)$$

Nickel complexes catalyze the displacement of one olefin from an aluminum alkyl by another olefin.[122] Nickel acetylacetonate and nickel *N*-alkylsalicylaldimino complexes show different catalytic activities, and if the *N*-alkyl group of the latter is optically active 2-butyl, slight optical activity can be induced in a properly chosen product. These results indicate that at least one of the original chelating groups remains in the catalytically active nickel species.

Other evidence indicates that alkylnickel and nickel hydride complexes are not involved. Treatment of a dilute hydrocarbon solution of the aldimino nickel complex with triisobutylaluminum changes the color from green to brown, but addition of methanol converts the brown species quantitatively back to the original aldimino complex. More importantly, optically active tris(2-methylbutyl)aluminum is not racemized by the nickel complexes in the absence of free olefins at 25°C. If nickel alkyl and β elimination of nickel hydride to form olefin were involved, racemization would be expected. Addition of 2-methyl-1-butene to the system leads to alkene exchange and, necessarily, to racemization. Zinc alkyls alkylate the nickel complex, but do not undergo exchange with alkenes, the nickel alkyls merely decomposing to metallic nickel. The alkene exchange occurs with beryllium as well as aluminum alkyls, but not with zinc or boron alkyls.

From these results, it appears that the transition state must contain the nickel complex, the aluminum alkyl, and the olefin together. A π complex with the nickel seems the most likely explanation.

The order of reactivity of the metal alkyls, Al > Be > B or Zn, is the same as in the growth reaction between metal alkyls and ethylene.[123]

Heck has shown that olefins can be alkylated or arylated by organomercury, tin, or lead compounds in the presence of Li_2PdCl_4 and related transition metal catalysts.[124] The evidence indicates that an intermediate organopalladium compound is involved, as illustrated with the phenylation of 1,1-dimethylallyl alcohol (**113**). Evidence that the organopalladium complex reacts directly with the olefin, without involving radicals or carbonium ions,

$$PhHgCl + LiPdCl_3 \longrightarrow [PhPdCl] + HgCl_2 + LiCl$$

$$\underset{\mathbf{113}}{CH_2{=}CH{-}C(CH_3)_2{-}OH} + PhPdCl \longrightarrow Ph{-}CH{=}CH{-}C(CH_3)_2{-}OH + \underset{\text{(reoxidized by } CuCl_2)}{[HPdCl]}$$

is provided by the relative reactivities C_2H_4, 14,000; $MeO_2CCH{=}CH_2$, 970; $CH_3CH{=}CH_2$, 220; $PhCH{=}CH_2$, 42; $PhC(Me){=}CH_2$, 1. Otherwise, details of the mechanism are not known, and this very interesting synthetic reaction is an obvious candidate for mechanistic studies.

Hydrosilylation of olefins is somewhat related to additions of metal alkyls and is catalyzed by transition metals and their compounds. The platinum(II)-catalyzed reaction appears to involve conversion of a square-planar platinum complex to an octahedral complex (**114**).[125] When the catalyst is platinum

$$(H_2C{=}CH_2)PtL_3 + R_3SiH \longrightarrow \underset{\mathbf{114}}{(H_2C{=}CH_2)Pt(H)(SiR_3)L_3} \longrightarrow CH_3CH_2{-}Pt(SiR_3)L_3 \xrightarrow{C_2H_4} CH_3CH_2SiR_3 + (H_2C{=}CH_2)PtL_3$$

metal itself, the silicon atom retains its configuration,[126] and the platinum-catalyzed addition of R_3SiH to diphenylacetylene is stereospecifically *cis*.[127] However, Pd and Ni catalysts lead to 63–99% inversion at silicon in hydrosilylations.[128] The addition of Et_3SiH to olefins such as 1-pentene is catalyzed by dicobalt octacarbonyl. It has been shown that the active catalyst is

$HCo(CO)_4$ and that isomerization of the olefin is faster than the hydrosilylation.[129]

References

1. W. Hanstein, H. J. Berwin, and T. G. Traylor, *J. Amer. Chem. Soc.* **92**, 829 (1970).
2. W. Hanstein, H. J. Berwin, and T. G. Traylor, *J. Amer. Chem. Soc.* **92**, 7476 (1970).
3. W. L. Waters and E. F. Kiefer, *J. Amer. Chem. Soc.* **89**, 6261 (1967).
4. A. R. Bassindale, C. Eaborn, D. R. M. Walton, and D. J. Young, *J. Organometal. Chem.* **20**, 49 (1969).
5. M. A. Cook, C. Eaborn, and D. R. M. Walton, *J. Organometal. Chem.* **24**, 293 (1970).
6. A. J. Smith, W. Adcock, and W. Kitching, *J. Amer. Chem. Soc.* **92**, 6140 (1970).
7. H. Bock and H. Alt, *J. Amer. Chem. Soc.* **92**, 1569 (1970).
8. B. G. Ramsey, "Electronic Transitions in Organometalloids," pp. 88–94. Academic Press, New York, 1969.
9. W. M. Schubert and D. F. Gurka, *J. Amer. Chem. Soc.* **91**, 1443 (1969); W. M. Schubert and R. G. Minton, *ibid.* **82**, 6188 (1960).
10. J. M. Jerkunica and T. G. Traylor, *J. Amer. Chem. Soc.* **93**, 6278 (1971).
11. D. Seyferth and S. S. Washburne, *J. Organometal. Chem.* **5**, 389 (1966).
12. J. A. Landgrebe and D. A. Thurman, *J. Amer. Chem. Soc.* **91**, 1759 (1969).
13. (a) J. A. Landgrebe and D. A. Thurman, *J. Amer. Chem. Soc.* **89**, 3542 (1967); (b) **90**, 6256 (1968).
14. M. M. Kreevoy and F. R. Kowitt, *J. Amer. Chem. Soc.* **82**, 739 (1960).
15. M. M. Kreevoy and L. T. Ditsch, *J. Amer. Chem. Soc.* **82**, 6127 (1960).
16. L. L. Schaleger, M. A. Turner, T. C. Chamberlain, and M. M. Kreevoy, *J. Org. Chem.* **27**, 3421 (1962).
17. G. A. Olah and P. R. Clifford, *J. Amer Chem. Soc.* **93**, 1261 (1971); **95, 6067 (1973).**
18. G. A. Olah and P. R. Clifford, *J. Amer. Chem. Soc.* **93**, 2320 (1971).
19. H. C. Brown, M.-H. Rei, and K.-T. Liu, *J. Amer. Chem. Soc.* **92**, 1760 (1970).
20. R. G. Parker and J. D. Roberts, *J. Amer. Chem. Soc.* **92**, 743 (1970).
21. R. D. Bach and H. F. Henneike, *J. Amer. Chem. Soc.* **92**, 5589 (1970).
22. J. Halpern and H. B. Tinker, *J. Amer. Chem. Soc.* **89**, 6427 (1967).
23. J. E. Byrd and J. Halpern, *J. Amer. Chem. Soc.* **92**, 6967 (1970).
24. W. L. Waters, W. S. Linn, and M. C. Caserio, *J. Amer. Chem. Soc.* **90**, 6741 (1968).
25. R. D. Bach, *J. Amer. Chem. Soc.* **91**, 1771 (1969).
26. W. S. Linn, W. L. Waters, and M. C. Caserio, *J. Amer. Chem. Soc.* **92**, 4018 (1970).
27. D. J. Pasto and J. A. Gontarz, *J. Amer. Chem. Soc.* **92**, 7480 (1970); **93**, 6902, 6909 (1971).
28. T. G. Traylor and A. W. Baker, *Tetrahedron Lett. No. 19*, 14 (1959).
29. K. C. Pande and S. Winstein, *Tetrahedron Lett.* 3393 (1964).
30. T. G. Traylor, *Accounts Chem. Res.* **2**, 152 (1969).
31. T. T. Tidwell and T. G. Traylor, *J. Org. Chem.* **33**, 2614 (1968).
32. A. Factor and T. G. Traylor, *J. Org. Chem.* **33**, 2607 (1968).
33. H. C. Brown, J. H. Kawakami, and S. Ikegami, *J. Amer. Chem. Soc.* **89**, 1525 (1967).
34. H. C. Brown and K.-T. Liu, *J. Amer. Chem. Soc.* **92**, 3502 (1970).
35. H. C. Brown and K.-T. Liu, *J. Amer. Chem. Soc.* **93**, 7335 (1971).
36. H. C. Brown and M.-H. Rei, *J. Chem. Soc. D* 1296 (1969).

37. L. Radom, J. A. Pople, V. Buss, and P. v. R. Schleyer, *J. Amer. Chem. Soc.* **94**, 311 (1972).
38. P. M. Henry, *J. Amer. Chem. Soc.* **87**, 990 (1965).
39. P. M. Henry, *J. Amer. Chem. Soc.* **87**, 4423 (1965).
40. P. M. Henry, *J. Amer. Chem. Soc.*, **88**, 1597 (1966).
41. A. McKillop, O. H. Oldenziel, B. P. Swann, E. C. Taylor, and R. L. Robey, *J. Amer. Chem. Soc.* **93**, 7331 (1971).
42. L. H. Sommer and G. A. Baughman, *J. Amer. Chem. Soc.* **83**, 3346 (1961).
43. A. W. P. Jarvie, A. Holt, and J. Thompson, *J. Chem. Soc. B* 852 (1969).
44. M. A. Cook, C. Eaborn, and D. R. M. Walton, *J. Organometal. Chem.* **24**, 301 (1970).
45. A. J. Bourne and A. W. P. Jarvie, *J. Organometal. Chem.* **24**, 335 (1970).
46. R. N. Haszeldine, P. J. Robinson, and J. A. Walsh, *J. Chem. Soc. B* 578 (1970).
47. G. Fishwick, R. N. Haszeldine, P. J. Robinson, and R. F. Simmons, *J. Chem. Soc. B* 1236 (1970).
48. W. K. Musker and G. L. Larson, *J. Amer. Chem. Soc.* **91**, 514 (1969).
49. D. D. Davis and C. E. Gray, *J. Organometal. Chem.* **18**, P1 (1969).
50. D. Seyferth and T. F. Jula, *J. Amer. Chem. Soc.* **90**, 2938 (1968).
51. V. R. Sandel and H. H. Freedman, *J. Amer. Chem. Soc.* **90**, 2059 (1968).
52. D. S. Matteson and J. D. Liedtke, *J. Org. Chem.* **28**, 1924 (1963).
53. D. S. Matteson and J. D. Liedtke, *J. Amer. Chem. Soc.* **87**, 1526 (1965).
54. M. J. S. Dewar and R. C. Fahey, *J. Amer. Chem. Soc.* **85**, 2248 (1963).
55. S. Winstein, E. Grunwald, and H. W. Jones, *J. Amer. Chem. Soc.* **73**, 1120 (1951).
56. (a) D. J. Pasto and J. Hickman, *J. Amer. Chem. Soc.* **90**, 4445 (1968); (b) D. J. Pasto and R. Snyder, *J. Org. Chem.* **31**, 2773 (1966); (c) **31**, 2777 (1966).
57. H. C. Brown, "Hydroboration." Benjamin, New York, 1962. (a) pp. 123–135. (b) pp. 173–190.
58. H. C. Brown and G. Zweifel, *J. Amer. Chem. Soc.* **82**, 4708 (1960).
59. H. C. Brown and G. Zweifel, *J. Amer. Chem. Soc.* **83**, 2544 (1961).
60. H. C. Brown and A. W. Moerikofer, *J. Amer. Chem. Soc.* **83**, 3418 (1961).
61. H. C. Brown and A. W. Moerikofer, *J. Amer. Chem. Soc.* **85**, 2063 (1963).
62. D. J. Pasto and S.-Z. Kang, *J. Amer. Chem. Soc.* **90**, 3797 (1968).
63. D. J. Pasto, V. Balasubramaniyan, and P. W. Wojtkowski, *Inorg. Chem.* **8**, 594 (1969).
64. D. J. Pasto and P. Balasubramaniyan, *J. Amer. Chem. Soc.* **89**, 295 (1967).
65. J. Klein, E. Dunkelblum, and M. A. Wolff, *J. Organometal. Chem.* **7**, 377 (1967).
66. (a) D. J. Pasto, B. Lepeska, and T.-C. Cheng, *J. Amer. Chem. Soc.* **94**, 6083 (1972); (b) D. J. Pasto, B. Lepeska, and V. Balasubramaniyan, *ibid.* **94**, 6090(1972).
67. T. P. Fehlner, *J. Amer. Chem. Soc.* **93**, 6366 (1971).
68. G. F. Hennion, P. A. McKusker, E. C. Ashby, and A. J. Rutkowski, *J. Amer. Chem. Soc.* **79**, 5194 (1957).
69. H. C. Brown and G. Zweifel, *J. Amer. Chem. Soc.* **88**, 1433 (1966).
70. B. Rickborn and S. E. Wood, *J. Amer. Chem. Soc.* **93**, 3940 (1971).
71. A. T. Cocks and K. W. Egger, *J. Chem. Soc. A* 3606 (1971).
72. H. C. Brown, M. V. Bhatt, T. Munekata, and G. Zweifel, *J. Amer. Chem. Soc.* **89**, 567 (1967).
73. B. M. Mikhailov, M. E. Kuimova, and E. A. Shagova, *Dokl. Akad. Nauk SSSR* **179**, 1344 (1968).
74. H. C. Brown and R. L. Sharp, *J. Amer. Chem. Soc.* **90**, 2915 (1968).
75. G. Zweifel and H. Arzoumanian, *J. Amer. Chem. Soc.* **89**, 291 (1967).

76. G. Zweifel, N. R. Ayyangar, and H. C. Brown, *J. Amer. Chem. Soc.* **84**, 4342 (1962).
77. H. C. Brown, N. R. Ayyangar, and G. Zweifel, *J. Amer. Chem. Soc.* **86**, 1071 (1964).
78. D. S. Matteson, *Organometal. Chem. Rev. B* **6**, 323 (1970).
79. A. Streitwieser, L. Verbit, and R. Bittman, *J. Org. Chem.* **32**, 1530 (1967).
80. D. R. Brown, S. F. A. Kettle, J. McKenna, and J. M. McKenna, *Chem. Commun.* 667 (1967).
81. K. R. Varma and E. Caspi, *Tetrahedron* **24**, 6365 (1968).
81a. W. R. Moore, H. W. Anderson, and S. D. Clark, *J. Amer. Chem. Soc.* **95**, 835 (1973).
82. A. G. Davies, K. G. Foot, B. P. Roberts, and J. C. Scaiano, *J. Organometal. Chem.* **31**, C1 (1971).
83. R. W. Rudolph, *J. Amer. Chem. Soc.* **89**, 4216 (1967).
84. M. Zeldin, A. R. Gatti, and T. Wartik, *J. Amer. Chem. Soc.* **89**, 4217 (1967).
85. T. D. Coyle and J. J. Ritter, *J. Amer. Chem. Soc.* **89**, 5739 (1967).
86. A. Rosen and M. Zeldin, *J. Organometal. Chem.* **31**, 319 (1971).
87. H. K. Saha, L. T. Glickenstein, and G. Urry, *J. Organometal. Chem.* **8**, 37 (1967).
88. G. Wilke and H. Müller, *Justus Liebigs Ann. Chem.* **629**, 222 (1960).
89. G. Zweifel and R. B. Steele, *J. Amer. Chem. Soc.* **89**, 5085 (1967).
90. G. Zweifel, J. T. Snow, and C. C. Whitney, *J. Amer. Chem. Soc.* **91**, 7139 (1969).
91. J. J. Eisch and S. G. Rhee, *J. Organometal. Chem.* **31**, C49 (1971).
92. E. G. Hoffmann, *Justus Liebigs Ann. Chem.* **629**, 104 (1960).
93. J. N. Hay, G. R. Jones, and J. C. Robb, *J. Organometal. Chem.* **15**, 295 (1968).
94. G. M. Clark and G. Zweifel, *J. Amer. Chem. Soc.* **93**, 527 (1971).
95. K. W. Egger, *J. Amer. Chem. Soc.* **91**, 2867 (1969).
96. K. W. Egger, *J. Chem. Soc. A* 3603 (1971).
97. Anonymous, *Chem. Eng. News*, April 10, 1972, p. 10.
98. J. J. Eisch and C. K. Hordis, *J. Amer. Chem. Soc.* **93**, 2974 (1971).
99. J. J. Eisch and C. K. Hordis, *J. Amer. Chem. Soc.* **93**, 4496 (1971).
100. K. Ziegler and H. Hoberg, *Chem. Ber.* **93**, 2938 (1960).
101. J. N. Hay, P. G. Hooper, and J. C. Robb, *Trans. Faraday Soc.* **66**, 2800 (1970).
102. J. N. Hay, P. G. Hooper, and J. C. Robb, *Trans. Faraday Soc.* **66**, 2045 (1970).
103. K. W. Egger and A. T. Cocks, *J. Amer. Chem. Soc.* **94**, 1810 (1972).
104. K. W. Egger, *Trans. Faraday Soc.* **67**, 2638 (1971).
105. G. Hata, *Chem. Commun.* 7 (1968).
106. P. D. Bartlett, S. J. Tauber, and W. P. Weber, *J. Amer. Chem. Soc.* **91**, 6362 (1969).
107. A. Streitwieser, Jr., R. A. Caldwell, and M. R. Granger, *J. Amer. Chem. Soc.* **86**, 3578 (1964).
108. C. G. Swain and L. Kent, *J. Amer. Chem. Soc.* **72**, 518 (1950).
109. P. D. Bartlett, C. V. Goebel, and W. P. Weber, *J. Amer. Chem. Soc.* **91**, 7425 (1969).
110. E. G. Eberhardt and W. R. Davis, *J. Polym. Sci. Part A* **3**, 3753 (1965).
111. M. Swarc, "Carbanions, Living Polymers, and Electron-Transfer Processes." Wiley, New York, 1968.
112. J. G. Welch and R. M. Magid, *J. Amer. Chem. Soc.* **89**, 5300 (1967).
113. R. Waack and M. A. Doran, *J. Amer. Chem. Soc.* **91**, 2456 (1969).
114. R. Waack and M. A. Doran, *J. Organometal. Chem.* **29**, 329 (1971).
115. W. H. Glaze and C. H. Freeman, *J. Amer. Chem. Soc.* **91**, 7198 (1969).

116. E. Grovenstein, Jr., and G. Wentworth, *J. Amer. Chem. Soc.* **89**, 1852 (1967).
117. G. R. Buell, R. Corriu, C. Guerin, and L. Spialter, *J. Amer. Chem. Soc.* **92**, 7424 (1970).
118. D. S. Matteson and R. W. H. Mah, *J. Org. Chem.* **28**, 2171 (1963).
119. G. Henrici-Olivé and S. Olivé, *Angew. Chem. Int. Ed.* **6**, 790 (1967).
120. E. J. Corey and J. A. Katzenellenbogen, *J. Amer. Chem. Soc.* **91**, 1851 (1969).
121. J. B. Siddall, M. Bishop, and J. H. Fried, *J. Amer. Chem. Soc.* **91**, 1853 (1969).
122. L. Lardicci, G. P. Giacomelli, P. Salvadori, and P. Pino, *J. Amer. Chem. Soc.* **93**, 5794 (1971).
123. K. Ziegler, *Angew. Chem.* **64**, 323 (1952).
124. R. F. Heck, *J. Amer. Chem. Soc.* **90**, 5518, 5526, 5531, 5535, 5538, 5546 (1968).
125. A. J. Chalk and J. F. Harrod, *J. Amer. Chem. Soc.* **87**, 16 (1965).
126. L. H. Sommer, K. W. Michael, and H. Fujimoto, *J. Amer. Chem. Soc.* **89**, 1519 (1967).
127. A. G. Brook, K. H. Pannell, and D. G. Anderson, *J. Amer. Chem. Soc.* **90**, 4374 (1968).
128. L. H. Sommer and J. E. Lyons, *J. Amer. Chem. Soc.* **89**, 1521 (1967).
129. A. J. Chalk and J. F. Harrod, *J. Amer. Chem. Soc.* **89**, 1640 (1967).

CHAPTER 6

Carbene Transfer Agents, Carbenes, and Related Topics

I. Introduction

The famous Simmons-Smith reagent, ICH_2ZnI, readily transfers its CH_2 group to olefins to form cyclopropanes,[1] and may be classed as a carbene transfer reagent even though free carbene, CH_2, is not involved. Seyferth and co-workers have shown that $PhHgCCl_2Br$ is an efficient source of dichlorocarbene, CCl_2, which adds to a wide variety of olefins to form 1,1-dichlorocyclopropanes.[2] These two reagents have tended to dominate the thinking of chemists on the subject of α-haloalkylmetal compounds, which are often thought of as potential carbene sources regardless of their actual properties. It is therefore appropriate to put in a reminder that some kinds of α-haloalkylmetal compounds, of which boron compounds are the best known example,[3] show no tendency to undergo α elimination or carbene transfer, but instead react with nucleophiles to yield displacement products, with the neighboring metal atom greatly assisting the nucleophilic displacement. Reactions of this type have been covered in Chapter 4.

II. Methylene Transfer Reagents

A. Zinc

The reaction of zinc–copper couple with methylene iodide in ether yields a solution of iodomethylzinc iodide, ICH_2ZnI, which can be filtered and stored several hours at 0°C with retention of its ability to react with olefins to form cyclopropanes.[1] The formation of cyclopropanes is rigorously stereospecific.

The ICH_2ZnI reacts somewhat selectively with different olefins, favoring less-hindered olefins in competition experiments[4] and adding exclusively *exo* to norbornene.[5] Some insertion of a CH_2 group into the α-C—H bond of the ether occurs, but insertion into other C—H bonds was not observed. These results are incompatible with free carbene (CH_2) as an intermediate and also incompatible with multiple-step carbanion or radical processes. The most likely mechanism appears to be a direct transfer of the CH_2 group from ICH_2ZnI to the olefin, as illustrated in transition state **1**.

ICH_2ZnI + (H)(R)C=C(H)(R) ⟶ [I-----ZnI / CH_2 / H(R)C-----C(H)(R)] **1** ⟶ IZnI + cyclopropane (CH_2, H, R, H, R)

Several features of the mechanism remain undefined. The geometries of the CH_2 group and the departing ZnI_2 in the transition state are unknown. The possibility that the zinc is involved in π-complexing with the double bond in or just prior to the transition state has not been fully explored, though Simmons, Blanchard, and Smith did show that there is no measurable concentration of any intermediate $ClCH_2CHEtCHEtZnI$ in the reaction of $ClCH_2I$ and zinc with *cis*-3-hexene.[5] (See aluminum for an example of this type of mechanism, Section II,C.)

Because of the confusion existing over the structure of the Grignard reagent at that time, Blanchard and Simmons discussed at some length the possibility that the structure of iodomethylzinc iodide might be $(ICH_2)_2Zn \cdot ZnI_2$.[4] Now that Grignard reagents are known to be RMgX and not $R_2Mg \cdot MgX_2$ there is no precedent for assuming any structure other than ICH_2ZnI. Japanese workers have made iodoethylzinc compounds from CH_3CHI_2 and Et_2Zn or EtZnI and added them to cyclohexene to form the *endo*- and *exo*-methylnorcaranes **2** and **3**.[6] The *endo*/*exo* ratio was 1.5 from diethylzinc and 1.0 from

$CH_3CHI_2 + Et_2Zn \longrightarrow CH_3CH(I)ZnEt + EtI$

$CH_3CHI_2 + EtZnI \longrightarrow CH_3CH(I)ZnI + EtI$

(both + cyclohexene ⟶ **2** (7-CH_3, 7-H) + **3** (7-H, 7-CH_3))

ethylzinc iodide, the difference suggesting that $CH_3CHIZnI$ differs from $(CH_3CHI)_2Zn$ or from $CH_3CHIZnEt$. Such a difference does not really prove that $(ICH_2)_2Zn \cdot ZnI_2$ or its homolog does not exist, but does show that no such secondary structural feature is necessary for cyclopropane formation. This work also nicely demonstrates that the cyclopropane-forming reaction does not require traces of copper as a catalyst, though Simmons and Smith did previously present evidence that the only role of the copper was to promote the reaction of methylene iodide with the zinc.[1]

Hydroxyl groups normally react faster than olefins with ICH_2ZnI, but if the hydroxyl and olefin groups are properly situated in the same molecule, the hydroxyl group can assist the cyclopropane formation to be faster than protolysis and to be sterically controlled by the complexing of the oxygen with the zinc.[4,7,8] For example, cyclopenten-4-ol yields only the *cis* isomer of bicyclo[3.1.0]hexan-3-ol (**4**)[4,7] and cyclopenten-3-ol likewise yields *cis* product (**5**).[8] The steric contortions required of larger cycloalkene-3-ols lead to 10% *trans* product with cyclohepten-3-ol and 99.5% *trans* product (**6**) with cycloocten-3-ol.[8]

OH $\xrightarrow{ICH_2ZnI}$ $\xrightarrow{H_2O}$ OH

4

OH ⟶ ⟶ OH

(cyclohexen-3-ol similar)

5

CH_2

OH ⟶ ⟶ OH

6

In other work with iodomethylzinc iodide, Zweifel and co-workers have found that the reagent reacts unusually rapidly with vinylalanes to form cyclopropylaluminum compounds, hexenyldiisobutylalane reacting much faster than 1-octene in a competition experiment.[9] The stereospecificity of the hydralumination of alkynes (Chapter 5) combined with the stereospecificity of the ICH_2ZnI–olefin reactions makes this a very useful synthetic procedure. Seyferth and Andrews have used iodomethylzinc iodide to prepare other iodomethylmetal compounds such as $Me_2Sn(CH_2I)_2$,[10] though these have turned out to be much less useful than the zinc compound for making cyclopropanes.

B. Mercury

Mercury compounds are much less reactive than the zinc reagents, but bis(bromomethyl)mercury (7) does react slowly with olefins above 80°C.[11]

$$\underset{}{RCH{=}CHR} + \underset{\mathbf{7}}{BrCH_2HgCH_2Br} \longrightarrow \underset{RCH\text{—}CHR}{\overset{CH_2}{\triangle}} + BrCH_2HgBr$$

The reaction stops at the $BrCH_2HgBr$ stage unless diphenylmercury, Ph_2Hg, is added to convert $BrCH_2HgBr$ to the more reactive $BrCH_2HgPh$. Better reactivity is obtained with ICH_2HgI, especially in the presence of Ph_2Hg, but several days are often required for completion. The reaction is stereospecific and qualitative data indicate that the rate is dependent on the olefin. Electronegative substituents such as acetoxy or chloro retard CH_2 transfer, and steric effects are of lesser importance. Steric effects have a greater influence on the ICH_2ZnI reaction, perhaps because the zinc is strongly solvated by ether. The halomethylmercury reactions are carried out in benzene or other inert solvents. The evidence indicates that free CH_2 cannot be involved in the reaction, and Seyferth and co-workers therefore postulated a concerted extrusion of CH_2 from the mercury reagent to the olefin, completely analogous to the transition state written for the zinc reagent (**1**).[11]

C. Aluminum

Halomethylaluminum compounds react with olefins to form cyclopropanes by a different mechanism from that of the zinc and mercury compounds.

$$\underset{\mathbf{8}}{Et_2AlCH_2Cl} \xrightarrow{\text{cyclohexene}} \text{bicyclo[4.1.0]heptane}$$

$$\underset{\mathbf{8}}{Et_2AlCH_2Cl} \xrightarrow{CH_2{=}CHCH_3} Et_2AlCH_2CH(CH_2Cl)(CH_3) \xrightarrow{ROH} (CH_3)_2CH\text{—}CH_2Cl$$

$$\underset{\mathbf{8}}{Et_2AlCH_2Cl} \xrightarrow{CH_3C{\equiv}CCH_3} (Et_2Al)(CH_3)C{=}C(CH_2Cl)(CH_3) \longrightarrow \text{1,2-dimethylcyclopropene } (CH_2 \text{ over } CH_3C{=}CCH_3)$$

$$(Et_2Al)(CH_3)C{=}C(CH_2Cl)(CH_3) \xrightarrow{ROH} CH_3CH{=}C(CH_3)CH_2Cl$$

Hoberg prepared diethyl(chloromethyl)alane (**8**) from diethylaluminum chloride and diazomethane and studied its reactions with various unsaturated compounds.[12,13] The first step was shown to be addition of $Et_2Al–CH_2Cl$ across the carbon–carbon double bond (see Chapter 5) by protolysis of representative $Et_2Al—C—C—CH_2Cl$ intermediates at $-50°C$. The cyclopropane ring closure must resemble that of boron compounds of the type R_2BCH_2-CH_2CH_2Cl (Chapter 4), except that the aluminum compounds are so much more reactive that the reaction proceeds without base catalysis at low temperatures.

D. Other Methylene Transfer Reagents

Several other organometallic methylene transfer reagents have been reported, including $LiCH_2Br$ from MeLi and CH_2Br_2,[14] ICH_2MgI,[15] $Zn(CH_2I)_2$, $Cd(CH_2I)_2$, and $In(CH_2I)_3$,[16,17] $(C_5H_5)Fe(CO)_2CH_2^+$ from $(C_5H_5)Fe(CO)_2CH_2OCH_3$ and HBF_4,[18] and $(Ph_3P)_2Ir(CO)CH_2Cl$, which was postulated to react by way of $(Ph_3P)_2Ir(CO)CH_2^+$.[19]

All of these methylene transfer reagents probably transfer the CH_2 group to the olefin by one or the other of the mechanisms outlined for the zinc and aluminum compounds. There is no reason to believe that free methylene, CH_2, is formed in any case. If a free carbanion $ClCH_2^-$ were formed, it is hard to see any reason why the energetically favorable C—Cl bond should break to yield Cl^- and unstabilized CH_2. There has been speculation that the intramolecular insertion reactions of α-sodioalkyl halides are carbene rearrangements,[20] but the evidence presented does not rule out intramolecular proton or hydrogen atom transfers in carbanion or radical species instead, and there is no evidence that the chloride departs prior to initiation of the rearrangements.

The reaction of ethyl diazoacetate, EtO_2CCHN_2, with olefins is catalyzed by trialkyl or triaryl phosphite copper(I) chloride, $(RO)_3PCuCl$.[21] Changing the phosphite ligand changes the *endo/exo* ratio in carbethoxynorcarane formation with cyclohexene, and therefore the copper atom is directly involved in the transition state. There is no very detailed mechanistic evidence, but it seems possible that this reaction has something in common with halomethylmetal methylene transfers. There is also a possibility of involvement of free carbethoxycarbene, but the involvement of the copper atom in the transition state suggests that a copper–carbon bond is formed from the diazo compound, perhaps in an intermediate such as $EtO_2CCH(N_2^+)$-$CuClP(OR)_3^-$ or $EtO_2CCHClCuP(OR)_3$.

III. Dihalocarbenes

A. Sources

In contrast to CH_2 itself, which probably cannot be generated in the free state from organometallic compounds, halogenated carbenes are stabilized by interaction of the unshared electrons of the halogens with the electron deficiency on the carbon, and such species as CCl_2 and CF_2 are easily generated from a variety of trihalomethylmetal compounds. The first source of CCl_2 to be discovered was the simple sodium salt, $Na^+CCl_3^-$, formed as a nonisolable intermediate in the hydrolysis of chloroform, and in a very lucky guess, CCl_2 was proposed as an intermediate in this reaction as early as 1862.[22] Hine provided definitive evidence that CCl_2 is indeed an intermediate in this reaction,[23] and Doering,[24,25] Skell,[26] and their co-workers showed that CCl_2 adds stereospecifically to olefins to form dichlorocyclopropanes.

The most interesting organometallic source of dichlorocarbene is the series of phenyl(trihalomethyl)mercury compounds, $PhHgCCl_2X$, developed by Seyferth and co-workers. The relative reactivities of a series of olefins toward dichlorocarbene generated from $PhHgCCl_3$ and sodium iodide were found to be the same as toward CCl_2 generated from NaO_2CCCl_3,[27,28] showing that the CCl_2 is formed as a free species. The anion CCl_3^- appears to be a precursor to CCl_2 in this system, and the CCl_3^- was trapped by vinyl acetate, which yielded some of the product of addition of H—CCl_3 across the double bond.

The reaction of $PhHgCCl_2Br$ with olefins is first order in $PhHgCCl_2Br$ and only slightly dependent on the relative reactivity and quantity of the olefin.[29] The rate-determining step is formation of CCl_2, except that the CCl_2 can recombine with PhHgBr to form $PhHgCCl_2Br$ if the olefin is fairly inert. PhHgBr suppresses the rate considerably with 1-heptene, less with cyclooctene, and not at all with a more reactive olefin, 2,3-dimethyl-2-pentene. ΔS^* is -2.8 eu. A unimolecular transition state (**9**) was postulated for the extrusion of CCl_2 from $PhHgCCl_2Br$.

$$PhHgCCl_2Br \underset{}{\overset{\text{slow}}{\rightleftharpoons}} \left[\begin{array}{c} Br \\ \diagup \quad \diagdown \\ Ph\text{—}Hg \cdots\cdots CCl_2 \end{array} \right] \underset{\text{fast}}{\rightleftharpoons} PhHgBr + CCl_2$$

9

$$CCl_2 + RCH{=}CH_2 \xrightarrow{\text{fast}} \begin{array}{c} CCl_2 \\ \diagup \quad \diagdown \\ RCH\text{——}CH_2 \end{array}$$

The concerted transition state (**9**) is further supported by a study of a series of reagents p-$ZC_6H_4HgCCl_2Br$, where Z is H, Cl, F, CH_3, and CH_3O.[30]

The rate constants vary randomly in the narrow range $1.1\text{–}1.5 \times 10^{-5}\ \text{sec}^{-1}$ at 39°C in benzene, corresponding to $\rho = 0$. The ρ^+ value for the addition of CCl_2 to $ZC_6H_4CH{=}CH_2$ was found to be -0.619, where Z is p-CH_3, H, p-F, p-Cl, and m-F.

$PhHgCF_3$ reacts with sodium iodide to form CF_2, which yields difluorocyclopropanes with olefins.[31] The reagent is prepared from $PhHgCBr_3$ and PhHgF. Since PhHgF does not exchange fluorine for bromine with $CHBr_3$, $BrCCl_2H$, CBr_4, or 7,7-dibromonorcarane, the $PhHgCBr_3$ must be activated by a neighboring-group effect involving interaction of mercury and bromine.

Trimethyl(trichloromethyl)tin, Me_3SnCCl_3, adds CCl_2 to cyclooctene at 140°C.[32] With vinyl acetate, only the cyclopropane product was found, which indicates that CCl_3^- is not an intermediate and the extrusion of CCl_2 is unimolecular with a transition state analogous to **9**. Another tin reagent tested was trimethyltin trichloroacetate, which was found to yield CCl_2 by way of Me_3SnCCl_3.[33] For synthetic purposes, these organotin reagents are much less reactive and less useful than the mercury reagents.

A number of monohalocarbenes have also been generated from organometallic reagents. Free chlorocarbene is evidently generated from alkyllithiums and methylene chloride by way of $LiCHCl_2$.[34,35] Other halocarbene transfer reagents include $Zn(CHCl_2)_2$,[36] PhHgCHClBr, and $PhHgCHBr_2$.[37,38] $PhHgCCl_2CO_2Me$ decomposes thermally to $ClCCO_2Me$, which adds to olefins in the usual way.[39] The ethylene ketal $PhHgCCl_2CH(O_2C_2H_4)$ likewise decomposes to $ClCCH(O_2C_2H_4)$, but this carbene is unstable and rearranges by hydrogen migration to form $ClCH{=}C(O_2C_2H_4)$.[40] Of more interest to organometallic chemists is $Hg(CCl_2SiMe_3)_2$, which decomposes thermally to the silylchlorocarbene $ClCSiMe_3$.[41] The addition of Ph_2Hg permits utilization of both Me_3SiCCl_2 groups, and the $ClCSiMe_3$ adds to cyclohexene in the usual manner to form the cyclopropane or inserts into Si—H bonds to form such derivatives as $Et_3SiCHClSiMe_3$.

B. Reactions of CCl_2

Although the most familiar reaction of dichlorocarbene is the addition to double bonds, insertions of CCl_2 into metal–metal, carbon–metal, and metal–hydrogen bonds are also known. The facile insertion of CCl_2 into certain carbon–hydrogen bonds β to a metal atom is evidently promoted by hyperconjugation involving the carbon–metal bond and has been discussed in part in Chapter 5, Section II,B.

Dichlorocarbene from $PhHgCCl_2Br$ in benzene inserts into the Sn—Sn bond of hexamethylditin, $Me_3Sn{-}SnMe_3$, to form $Me_3SnCCl_2SnMe_3$.[42] The rate is similar to that for reaction of $PhHgCCl_2Br$ under similar conditions, and it appears that the mechanism involves a direct insertion of CCl_2 into the Sn—Sn bond. There is no evidence of any products that would be expected if a radical intermediate such as $Me_3SnCCl_2\cdot$ were involved.

Insertion of CCl_2 into the Si—Hg bond of bis(trimethylsilyl)mercury (**10**) leads to a complex reaction sequence.[43]

$$\underset{\mathbf{10}}{Me_3Si{-}Hg{-}SiMe_3} \xrightarrow[\text{from } PhHgCCl_2Br]{CCl_2} Me_3Si{-}Hg{-}CCl_2{-}SiMe_3 \xrightarrow[\text{into C—Hg bond}]{CCl_2}$$

$$Me_3Si{-}Hg{-}CCl_2CCl_2SiMe_3 \longrightarrow Me_3SiCl + Hg^0 + Cl_2C{=}CClSiMe_3$$

Dichlorocarbene also initiates a complex reaction sequence on insertion into the boron–carbon bond of tributylborane, ultimately yielding 4-nonene (**11**).[44] Although the 4-nonene (**11**) is what would be expected if the precursor

$$(CH_3CH_2CH_2CH_2)_3B \xrightarrow[\text{from } PhHgCCl_2Br]{CCl_2}$$

$$(CH_3CH_2CH_2CH_2)_2B{-}CCl_2{-}CH_2CH_2CH_2CH_3 \xrightarrow{\text{interchange of alkyl and Cl}}$$

$$\begin{matrix} CH_3CH_2CH_2CH_2B\diagdown^{Cl} \\ \quad C{-}CH_2CH_2CH_2CH_3 \\ CH_3CH_2CH_2CH_2\diagup \quad \diagdown Cl \end{matrix} \xrightarrow[\text{chloride migration}]{\text{hydride shift}} CH_3CH_2CH_2CH_2BCl_2 +$$

$$\underset{\substack{\mathbf{11} \\ (58\%\ cis,\ 42\%\ trans)}}{CH_3CH_2CH_2CH_2CH{=}CHCH_2CH_2CH_3}$$

α-chloroalkylborane underwent α elimination to a carbene, which would rapidly rearrange,[44] the general mechanism already discussed for rearrangement of α-haloalkylboron compounds (Chapter 4, Section III, B) is adequate to explain the results. Indeed, Seyferth and Prokai showed that the stabilized butylchlorocarbene, BuCCl, is not an intermediate in the first rearrangement step by reacting a mixture of tributylborane and tripropylborane with CCl_2, which yielded only 3-heptene and 4-nonene and none of the octenes which would result from scrambling of the butyl and propyl groups if there were a free carbene intermediate.[44] It seems unlikely that an α-chloroalkylborane system would fail to generate the carbene BuCCl in one step and then go on to generate the less-stabilized carbene Bu_2C in another. Reinvestigation (G. L. Larson, personal communication, 1973) has indicated that **11** is not formed unless moisture is present.

A simpler set of rearrangements follows the insertion of difluorocarbene

$$HCF_2Cl + LiOCEt_3 \longrightarrow CF_2 \xrightarrow{Bu_3B} Bu_2B{-}CF_2{-}Bu \longrightarrow \underset{\mathbf{12}}{F_2B{-}CBu_3}$$

into the carbon–boron bond of tributylborane, and the product (**12**) is that of migration of all three butyl groups from boron to carbon.[45]

An unresolved question is whether these dihalocarbene insertions into carbon–metal bonds are a two-step process, involving initial attack of the carbene on the metal atom to form structure **13** followed by alkyl migration to the electron-deficient carbon by way of structure **14**, or whether the insertion is a one-step process by way of transition state **14**. From the

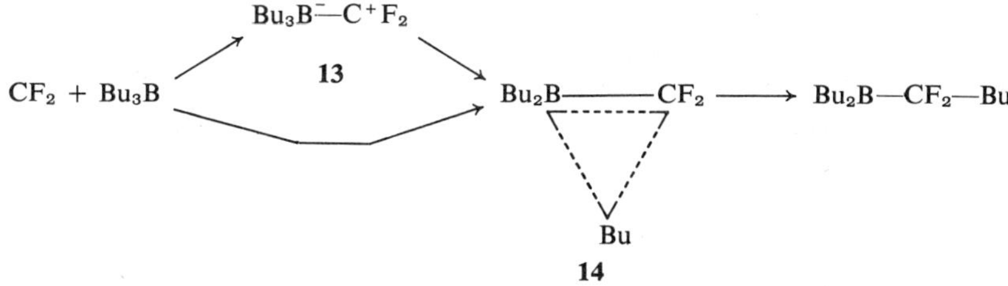

thermodynamic arguments presented in Chapter 4, **13** is probably higher in energy than **14**, and it is possible to see how **13** might be formed from collision of the starting materials, but not possible to decide unequivocally whether **13** is in the potential energy trough or somewhere up the side of it. In view of the thermodynamic trend with boron, it does seem unlikely that the new carbon–boron bond in **13** could be fully formed before some carbon–carbon bond formation leading toward **14** would take place.

Dichlorocarbene inserts into the C—Hg bond of diethylmercury (**15**).[46]

$$\underset{\mathbf{15}}{(C_2H_5)_2Hg} + CCl_2 \longrightarrow (C_2H_5CCl_2)_2Hg$$

With di-*n*-propylmercury and higher alkylmercuries, insertion into the β-C—H bond also occurs, evidently promoted by carbon–mercury bond hyperconjugation, as discussed in Chapter 5, Section II,B.

With silicon and tin compounds such as **16** and **17**, CCl_2 insertion into the β-C—H bond becomes the main reaction, and insertion into the carbon–metal bond is not usually observed.[47] The β insertion is inhibited by electro-

$$\underset{\mathbf{16}}{Me_2Si\langle\text{ring}\rangle} \xrightarrow{PhHgCCl_2Br} Me_2Si\langle\text{ring}\rangle\text{—}CCl_2H$$

$$\underset{\mathbf{17}}{Me_3SnCH_2CH_2CH_3} \xrightarrow{PhHgCCl_2Br} Me_3SnCH_2\underset{\displaystyle CCl_2H}{\underset{|}{C}}HCH_3$$

negative groups on silicon, and $Me_2SiClCH_2CH(CH_3)_2$ is inert. Primary β-hydrogens are inert, tertiary are the most reactive. Tin is a better activator than silicon. All these facts fit the hyperconjugation hypothesis described in Chapter 5.

Insertion of dichlorocarbene into a C—Si bond does occur if a strained four-membered ring (**18**) can be expanded to a five-membered ring in the process.[48] However, β-C—H insertion still competes, and Si—H insertion takes precedence over all if there is an Si—H bond as in **19**.

Si—Me, Me (**18**) $\xrightarrow[\text{from PhHgCCl}_2\text{Br}]{CCl_2}$ CCl_2, Si, Me Me (58%) + HCl_2C, Si—Me, Me (12%)

Si—Me, H (**19**) $\xrightarrow{CCl_2}$ Si—Me, CCl_2H (68% yield)

Seyferth and co-workers have shown that the rate-determining step in these insertions into Si—H bonds is the generation of CCl_2.[49] The reaction of $PhHgCCl_2Br$ with Et_3SiH in benzene at 30°C is first order in $PhHgCCl_2Br$ and zero order in Et_3SiH. It was also found that the Hammett ρ is only -0.63 for competition between a series of arylsilanes $ZC_6H_4SiMe_2H$, where Z is H, *m*-CF_3, *p*-CH_3, *p*-F, *p*-CH_2SiMe_3, and *p*-Cl. The slight acceleration by electron-donating groups was considered to be evidence that the mechanism is a concerted insertion of CCl_2 into the Si—H bond, not addition of CCl_2 to the silicon (for which the electronic effect is in the wrong direction, as in hypothetical structure **13**) and not formation of an $R_3Si^+HCCl_2^-$ ion pair. However, it should be noted that the activation energy for the insertion is probably very low, and therefore the differences between activation energies for differently substituted silanes must also be small. Thus, it is not really possible to rule out some or all of the insertion proceeding by way of an ion-pair intermediate. The fact that the stereospecificity is less than 100% (next paragraph) supports some sort of partial dissociation.

Sommer and co-workers have found 93% retention in the insertion of CCl_2 from $PhHgCCl_3$ into an optically active silane, R_3SiH, to form R_3SiCCl_2H.[50] Retention has also been observed in the insertion of CBr_2 into Si—H and Ge—H bonds.[51]

Methylene (CH_2) generated by photolysis of diazomethane inserts into the Si—H bond of CH_3SiH_3 8.9 times faster than into the C—H bond in the gas phase.[52] This is a rather large rate difference for a reagent as indiscriminate as

CH_2. The isotope effect k_H/k_D for SiH insertion is 1.15, somewhat smaller than the effect for C—H insertion. Some radical attack of CH_2 on CH_3SiH_3 to form $CH_3\cdot$ and $CH_3SiH_2\cdot$ was also evident from the mixture of products obtained, and 25–30% of the total reaction appeared to be free radical in nature.

IV. Silenes and Fluoroborene

Divalent silicon intermediates or silenes have been generated by several routes. Skell and Goldstein found that Me_2SiCl_2 reacts with sodium–potassium vapor at 260°–280°C to form Me_2Si, which inserts into the C—H bond of ethylene to yield $CH_2{=}CH{-}SiMe_2H$.[53] Atwell and Weyenberg have investigated the thermal decomposition of 1,2-dimethoxytetramethyldisilane (**20**) and found it to be unimolecular at 225°C.[54] The product must result from insertion of Me_2Si into the Si—O bond. Insertion into the Si—O rather than the Si—Si bond was postulated because the relative rates based on product analysis are all about the same up to formation of the Si_5 compound, which would not be expected if Si—Si bonds differing in whether they were adjacent to oxygen or to silicon were involved.[55] In the presence of diphenylacetylene a dimeric adduct (**21**) having a six-membered heterocyclic ring results.[54] With a mixture of diphenylacetylene and 2-butyne the three products are **21**, **22**, and **23**, showing that if there is a silacyclopropene intermediate (**24**) it opens at the Si—C bond to dimerize. Failure to form a silacyclopentadiene derivative (**25**) with excess diphenylacetylene is evidence against a diradical intermediate. Me_2Si also reacted with 2,3-dimethylbutadiene to form a five-membered heterocycle (**26**).

$$MeO{-}SiMe_2{-}SiMe_2{-}OMe \xrightarrow{225°C} MeO{-}SiMe_2{-}OMe + Me_2Si:$$

20

$$Me_2Si + MeO{-}SiMe_2{-}SiMe_2{-}OMe \longrightarrow MeO{-}SiMe_2{-}SiMe_2{-}SiMe_2{-}OMe \longrightarrow Si_4 \text{ compound, } etc.$$

$$Me_2Si + PhC{\equiv}CPh \longrightarrow \text{1,1,4,4-tetramethyl-2,3,5,6-tetraphenyl-1,4-disilacyclohexa-2,5-diene}$$

21

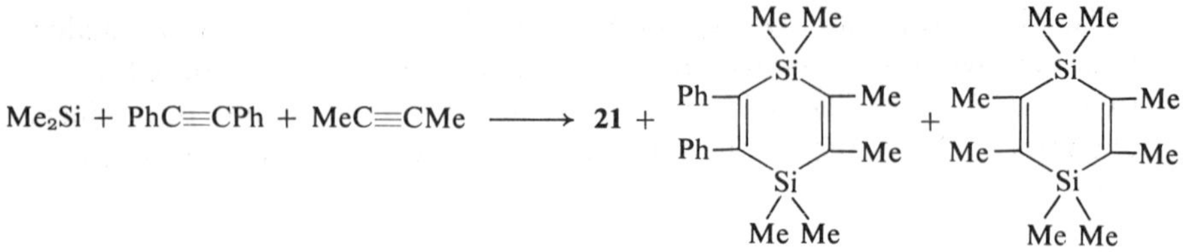

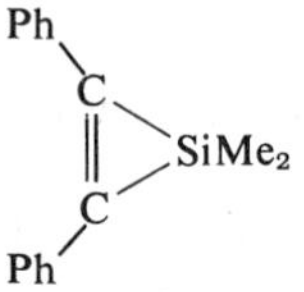

$$Me_2Si + PhC{\equiv}CPh + MeC{\equiv}CMe \longrightarrow \mathbf{21} + \ldots$$

Me Me Si Ph Me Ph Me Si Me Me + Me Me Si Me Me Me Me Si Me Me

Ph C C SiMe$_2$ Ph

24

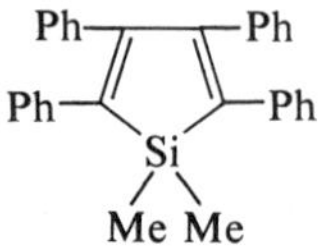

Ph Ph Ph Ph Si Me Me

25
(not formed)

Me_2Si + Me Me C—C H_2C CH_2 ⟶ Me Me Si Me Me

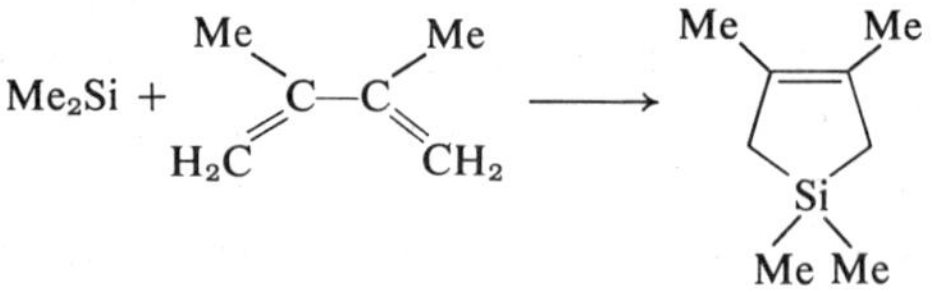

26

Photolysis of methylsilane, CH_3SiH_3, in the gas phase at 124 or 147 nm appears to generate the silenes SiH_2 and CH_3SiH.[56,57] With CH_3SiD_3 a major product is $CH_3SiD_2SiD_2CH_3$ from insertion of CH_3SiD into the Si—D bond of CH_3SiD_3, and there was little scrambling of hydrogen and deuterium between carbon and silicon. Final products in addition to $CH_3SiH_2SiH_2CH_3$ included H_2, $CH_3SiH_2SiH_3$, CH_4, C_2H_6, $CH_3SiH_2CH_3$, SiH_4, and polymer. From the effects of pressure, time, use of scavengers such as NO and C_2D_4, and deuterium labeling on the product yields, a number of mechanistic details were deduced, and it was concluded that the CH_3SiH is initially generated in an excited state.

Silicon difluoride, SiF_2, has been generated from Si and SiF_4 at 1100°C.[58] Boron monofluoride has similarly been generated from B and BF_3 at 1800°–2000°C,[59] which can be cocondensed with acetylene at −190°C to form several products, including $(F_2BCH{=}CH)_2BF$, which cyclizes with loss of BF_3 in the gas phase to form **27**.[60]

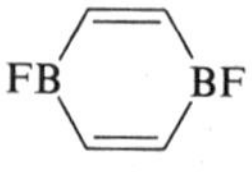

FB BF

27

Zavistoski and Zuckerman (to finish off this chapter alphabetically) have prepared the dimethylgermene, Me_2Ge, transfer reagent **28**.[61] The products of its reaction are those expected from Me_2Ge, but there is no evidence that this is actually liberated as a free species.

$$\text{(Ph)}_4\text{C}_4\text{GeMe}_2 + MeO_2C{-}C{\equiv}C{-}CO_2Me \longrightarrow \mathbf{28}$$

28

$$\mathbf{28} \xrightarrow{CH_2Cl_2} (Me_2Ge)_n + \text{(Ph)}_4\text{C}_6(CO_2Me)_2$$

$$\mathbf{28} \xrightarrow{Ph_3P} (\text{—}Ge(Me)_2\text{—}PPh_3\text{—})_n$$

References

1. H. E. Simmons and R. D. Smith, *J. Amer. Chem. Soc.* **81**, 4256 (1959).
2. D. Seyferth, J. M. Burlitch, R. J. Minasz, J. Y.-P. Mui, H. D. Simmons, Jr., A. J.-H. Treiber, and S. R. Dowd, *J. Amer. Chem. Soc.* **87**, 4259 (1965).
3. D. S. Matteson and R. W. H. Mah, *J. Amer. Chem. Soc.* **85**, 2599 (1963).
4. E. P. Blanchard and H. E. Simmons, *J. Amer. Chem. Soc.* **86**, 1337 (1964).
5. H. E. Simmons, E. P. Blanchard, and R. D. Smith, *J. Amer. Chem. Soc.* **86**, 1347 (1964).
6. J. Furukawa, N. Kawabata, S. Taniguchi, and J. Nishirura, *Kogyo Kagaku Zasshi* **72**, 1673 (1969); *Chem. Abstr.* **72**, 11851r (1970).
7. S. Winstein and J. Sonnenberg, *J. Amer. Chem. Soc.* **83**, 3235 (1961).
8. C. D. Poulter, E. C. Friedrich, and S. Winstein, *J. Amer. Chem. Soc.* **91**, 6892 (1969).
9. G. Zweifel, G. M. Clark, and C. C. Whitney, *J. Amer. Chem. Soc.* **91**, 1305 (1971).
10. D. Seyferth and S. B. Andrews, *J. Organometal Chem.* **18**, P21 (1969).
11. D. Seyferth, R. M. Turkel, M. A. Eisert, and L. J. Todd, *J. Amer. Chem. Soc.* **91**, 5027 (1969).
12. H. Hoberg, *Angew. Chem.* **73**, 114 (1961).
13. H. Hoberg, *Justus Liebigs Ann. Chem.* **656**, 1 (1962).
14. W. T. Miller, Jr., and C. S. Y. Kim, *J. Amer. Chem. Soc.* **81**, 5009 (1959).
15. G. Wittig and F. Wingler, *Chem. Ber.* **97**, 2139 (1964).
16. G. Wittig and K. Schwarzenbach, *Justus Liebigs Ann. Chem.* **650**, 1 (1961).
17. G. Wittig and F. Wingler, *Chem. Ber.* **97**, 2146 (1964).
18. P. W. Jolly and R. Pettit, *J. Amer. Chem. Soc.* **88**, 5044 (1966).
19. F. D. Mango and I. Dvoretzky, *J. Amer. Chem. Soc.* **88**, 1654 (1966).
20. L. Friedman and J. G. Berger, *J. Amer. Chem. Soc.* **83**, 492 (1961).
21. W. R. Moser, *J. Amer. Chem. Soc.* **91**, 1135, 1141 (1969).
22. A. Geuther, *Justus Liebigs Ann. Chem.* **123**, 121 (1862).
23. J. Hine, *J. Amer. Chem. Soc.* **72**, 2438 (1950).

24. W. v. E. Doering and A. K. Hoffmann, *J. Amer. Chem. Soc.* **76**, 6162 (1954).
25. W. v. E. Doering and P. M. LaFlamme, *J. Amer. Chem. Soc.* **78**, 5447 (1956).
26. P. S. Skell and A. Y. Garner, *J. Amer. Chem. Soc.* **78**, 3409, 5430 (1956).
27. D. Seyferth and J. M. Burlitch, *J. Amer. Chem. Soc.* **86**, 2730 (1964).
28. D. Seyferth, M. E. Gordon, J. Y.-P. Mui, and J. M. Burlitch, *J. Amer. Chem. Soc.* **89**, 959 (1967).
29. D. Seyferth, J. Y.-P. Mui, and J. M. Burlitch, *J. Amer. Chem. Soc.* **89**, 4953 (1967).
30. D. Seyferth, J. Y.-P. Mui, and R. Damrauer, *J. Amer. Chem. Soc.* **90**, 6182 (1968).
31. D. Seyferth, S. P. Hopper, and K. V. Darragh, *J. Amer. Chem. Soc.* **91**, 6536 (1969).
32. D. Seyferth and F. M. Armbrecht, Jr., *J. Amer. Chem. Soc.* **91**, 2616 (1969).
33. F. M. Armbrecht, Jr., W. Tronich, and D. Seyferth, *J. Amer. Chem. Soc.* **91**, 3218 (1969).
34. G. L. Closs and L. E. Closs, *J. Amer. Chem. Soc.* **82**, 5723, 5729 (1960).
35. G. L. Closs and L. E. Closs, *J. Amer. Chem. Soc.* **85**, 99 (1963).
36. G. Köbrich and H. R. Merkle, *Chem. Ber.* **99**, 1782 (1966).
37. D. Seyferth, H. D. Simmons, Jr., and L. J. Todd, *J. Organometal. Chem.* **2**, 282 (1964).
38. D. Seyferth, H. D. Simmons, Jr., and G. Singh, *J. Organometal. Chem.* **3**, 337 (1965).
39. D. Seyferth, D. C. Mueller, and R. L. Lambert, Jr., *J. Amer. Chem. Soc.* **91**, 1562 (1969).
40. D. C. Mueller and D. Seyferth, *J. Amer. Chem. Soc.* **91**, 1754 (1969).
41. D. Seyferth and E. M. Hanson, *J. Amer. Chem. Soc.* **90**, 2438 (1968).
42. D. Seyferth, F. M. Armbrecht, Jr., and B. Schneider, *J. Amer. Chem. Soc.* **91**, 1954 (1969).
43. D. Seyferth, E. M. Hanson, B. Prokai, and R. J. Cross, *J. Organometal. Chem.* **24**, 33 (1970).
44. D. Seyferth and B. Prokai, *J. Amer. Chem. Soc.* **88**, 1834 (1966).
45. H. C. Brown, B. A. Carlson, and R. H. Prager, *J. Amer. Chem. Soc.* **93**, 2071 (1971).
46. J. A. Landgrebe and D. A. Thurman, *J. Amer. Chem. Soc.* **91**, 1759 (1969).
47. D. Seyferth, S. S. Washburne, C. J. Attridge, and K. Yamamoto, *J. Amer. Chem. Soc.* **92**, 4405 (1970).
48. D. Seyferth, R. Damrauer, and S. S. Washburne, *J. Amer. Chem. Soc.* **89**, 1538 (1967).
49. D. Seyferth, R. Damrauer, J. Y.-P. Mui, and T. F. Jula, *J. Amer. Chem. Soc.* **90**, 2944 (1968).
50. L. H. Sommer, L. A. Ullard, and A. Ritter, *J. Amer. Chem. Soc.* **90**, 4486 (1968).
51. A. G. Brook, J. M. Duff, and D. G. Anderson, *J. Amer. Chem. Soc.* **92**, 7567 (1970).
52. C. J. Mazoc and J. W. Simons, *J. Amer. Chem. Soc.* **90**, 2484 (1968).
53. P. S. Skell and E. J. Goldstein, *J. Amer. Chem. Soc.* **86**, 1442 (1964).
54. W. H. Atwell and D. R. Weyenberg, *J. Amer. Chem. Soc.* **90**, 3438 (1968).
55. W. H. Atwell, L. G. Mahone, S. F. Hayes, and J. G. Uhlmann, *J. Organometal. Chem.* **18**, 69 (1969).
56. O. P. Strausz, K. Obi, and W. K. Duholke, *J. Amer. Chem. Soc.* **90**, 1359 (1968).
57. K. Obi, A. Clement, H. E. Gunning, and O. P. Strausz, *J. Amer. Chem. Soc.* **91**, 1622 (1969).
58. H. P. Hopkins, J. C. Thompson, and J. L. Margrave, *J. Amer. Chem. Soc.* **90**, 901 (1968).
59. P. L. Timms, *J. Amer. Chem. Soc.* **89**, 1629 (1967).
60. P. L. Timms, *J. Amer. Chem. Soc.* **90**, 4585 (1968).
61. J. G. Zavistoski and J. J. Zuckerman, *J. Amer. Chem. Soc.* **90**, 6612 (1968).

CHAPTER 7

Free-Radical and Photochemical Reactions

I. Introduction

Many carbon–metal bonds can be broken homolytically, either by attack of a free radical on the metal atom or by thermal or photochemical cleavage. The classical work of Paneth and Hofeditz on the thermolysis of tetramethyllead to form free methyl radicals[1] is probably familiar to most readers. Aryls of mercury,[2,3] bismuth,[4] and lead[3,5] can be photolyzed to the free metals and aryl radicals. There are a few known photochemical reactions of organometallic compounds which involve intramolecular rearrangements rather than radical cleavages, but these are likely to involve triplet states and may be grouped with radical reactions for convenience. There are also radical reactions at centers other than the carbon–metal bond which are influenced by the neighboring metal atom, and these are included in this chapter.

II. Principles of Radical Chemistry

Most radical reactions are chain reactions. They follow a reasonably simple set of mechanistic rules, but these are quite different from the rules for typical polar reactions. Once the rules are learned, it is often very easy to apply them to any new radical reaction and be reasonably sure of its gross mechanism without doing any kinetic or other mechanistic studies beyond establishing that it is a radical chain reaction. No attempt will be made to cover all organometallic examples for which such mechanistic insight has been claimed, but a number of illustrative cases will be cited. Extra attention will be given to those cases where actual detection of intermediate radicals by electron spin resonance (esr) or other serious mechanistic investigations have been made.

The brief summary of the principles of radical chain reactions which follows is based on more extended treatments in books such as those by Walling[6] and Pryor.[7]

The initial bond breaking which forms a pair of free radicals is called "initiation." The reaction of these free radicals with nonradicals is called "propagation." The combination of two radicals is called "termination."

Most of the product is formed by way of the propagation steps. Termination yields only minor by-products, often undetectable without special effort. It is always wrong to postulate formation of a major product by way of combination of two radicals in a radical chain reaction, but this mistake is often made. It is a quick way to put a referee in a foul temper.

Combination of two radicals is rare because the concentration of radicals is always very low. (This rule does not apply to highly stabilized radicals such as triphenylmethyl, but such stable radicals have nothing whatever to do with typical radical chain reactions, unless they are put in as inhibitors.) The rate constant for combination of two radicals is very high, but collisions are infrequent because of the low concentrations. Instead, the radical reacts with nonradical compounds present in much higher concentration, the concentration effect overruling the difference in rate constants. Rate constants for reactions of free radicals with appropriately chosen nonradical reagents can be very high, though usually several orders of magnitude below diffusion control.

Combination of a radical (odd number of electrons) with a nonradical (even number of electrons) always produces a new radical, since the total number of electrons is odd. To obtain a nonradical product, at least two propagation steps are required, the second of which regenerates the original radical to start the cycle over again.

The two principal types of radical chain reactions are substitutions and additions to double bonds. Initiation is the same in either case and involves the generation of some initiating radical Z· as in Eqs. (7-1) through (7-3).

$$\text{Z—Z} + \text{heat or light} \longrightarrow 2\text{Z}\cdot \tag{7-1}$$

$$\text{Z—N=N—Z} + \text{heat or light} \longrightarrow 2\text{Z}\cdot + \text{N}_2 \tag{7-2}$$

$$\text{Z—C(=O)—O—O—C(=O)—Z} + \text{heat or light} \longrightarrow 2\text{ZCO}_2\cdot \longrightarrow 2\text{Z}\cdot + 2\text{CO}_2 \tag{7-3}$$

These merely illustrate some of the possibilities, and of course the proper choice of the group Z is essential.

In a typical substitution, R—M and X—Y react to form R—X and M—Y, where R is an organic group, M may be a metal-containing group but in the classical case is hydrogen, X is usually a halogen atom, and Y is a nonmetal. Initiation must be followed by one of two processes leading toward propagation, Eqs. (7-4) and (7-5). Propagation then consists of an indefinite repetition

$$Z\cdot + R{-}M \longrightarrow Z{-}M + R\cdot \tag{7-4}$$
$$Z\cdot + X{-}Y \longrightarrow Z{-}X + Y\cdot \tag{7-5}$$

of Eqs. (7-6) and (7-7), each of which recycles the radical product from the

$$Y\cdot + R{-}M \longrightarrow M{-}Y + R\cdot \tag{7-6}$$
$$R\cdot + X{-}Y \longrightarrow R{-}X + Y\cdot \tag{7-7}$$

other. Termination may involve Eqs. (7-8) through (7-10). Termination often

$$R\cdot + R\cdot \longrightarrow R{-}R \text{ or (alkene + alkane)} \tag{7-8}$$
$$R\cdot + Y\cdot \longrightarrow R{-}Y \tag{7-9}$$
$$Y\cdot + Y\cdot \longrightarrow Y{-}Y \tag{7-10}$$

involves disproportionation of two radicals to saturated and unsaturated product rather than mere combination.

For addition to double bonds, the repeating chain-propagation steps are represented by Eqs. (7-11) and (7-12). With unsaturated compounds, forma-

$$Y\cdot + R{-}CH{=}CH_2 \longrightarrow R{-}\dot{C}H{-}CH_2Y \tag{7-11}$$
$$R{-}\dot{C}H{-}CH_2Y + X{-}Y \longrightarrow R{-}\underset{\substack{|\\ X}}{CH}{-}CH_2Y + Y\cdot \tag{7-12}$$

tion of telomer (n small) or polymer (n large) is an alternative possibility, depending on relative concentrations and rate constants, as illustrated in Eqs. (7-13) and (7-14). There are, of course, steps leading from initiation to

$$R{-}\dot{C}H{-}CH_2Y + R{-}CH{=}CH_2 \longrightarrow \begin{array}{l} R{-}CH{-}CH_2Y \\ \quad | \\ \quad CH_2{-}\dot{C}H{-}R \end{array} \longrightarrow \longrightarrow \begin{array}{l} R{-}CH{-}CH_2Y \\ \quad | \\ \quad (CH_2CHR)_n \\ \qquad | \\ \qquad CH_2{-}\dot{C}H{-}R \end{array} \tag{7-13}$$

$$\begin{array}{l} R{-}CH{-}CH_2Y \\ \quad | \\ \quad (CH_2CHR)_n \\ \qquad | \\ \qquad CH_2{-}\dot{C}H{-}R \end{array} + X{-}Y \longrightarrow \begin{array}{l} R{-}CH{-}CH_2Y \\ \quad | \\ \quad (CH_2CHR)_n \\ \qquad | \\ \qquad CH_2{-}CH{-}R \\ \qquad\qquad | \\ \qquad\qquad X \end{array} + Y\cdot \tag{7-14}$$

propagation and a number of possible termination steps, but these add no new principles.

Kinetic analysis of these chain reactions requires application of the steady-state approximation that the concentration of chain-carrying radicals is small and therefore does not change much, compared to the concentrations of the reactants in macroscopic amounts, during the course of the reaction. The details will be postponed until an actual example supported by experimental data is encountered, Section III,B. It may be noted here that the rate is always half-order in radical concentration (or half-order in initiator, which corresponds). It is also usually first order in one or the other of two reactants or a fractional order in both, and information on this point will indicate which radicals are involved in chain termination.

It should be emphasized that one radical never displaces another radical from a tetrahedral carbon atom. (An exception can, of course, be found in the strange world of strained ring compounds,[8] but this has nothing to do with organometallic compounds. Radical displacements or addition–eliminations are allowed on aromatic ring carbon.) In fact, radical displacements normally take place on monovalent atoms such as hydrogen or halogen. These may alternatively be viewed as transfers of hydrogen or halogen from one radical to another. The reason for this tendency is that all radical reactions must take place very quickly or else lose out in the competition with other possible rapid reactions. Approach to a tetrahedral carbon atom is sterically blocked, but the radical collides easily with the peripheral hydrogen atoms, and if any of the C—H bonds is relatively weak for some reason, that hydrogen is likely to be pulled off.

Radical displacements do take place readily on metal atoms, as implied in Eq. (7-6), even if there is some degree of hindrance to attack on the metal. Oxidative cleavage of the carbon–metal bond is generally very favorable thermodynamically (Chapter 1, Section II,B), and the metal atom often has a vacant orbital to facilitate approach of the radical. The question of an intermediate $R—\dot{M}—Y$ must be raised in such cases, but such structures appear to have little if any finite lifetimes and are probably transition states rather than intermediates in the cases that have been studied, except in the case of Group V elements.

Another point which should be emphasized is that ordinary radical reactions in solution never generate free hydrogen atoms. There is always a lower-energy alternative. Any other free atom or radical delocalizes the odd electron better than does the hydrogen atom. Hydrogen atoms are often transferred but are never set free.

Finally, recombination of solvent-caged radical pairs must be considered. For example, in the generation of 2 Z· from Z—N=N—Z (where Z is alkyl) the 2 Z· and N_2 are initially formed together inside a cage of solvent molecules. The cage is a flimsy one and the Z· radicals can break loose within a few collisions, but in the meantime they may run into each other to form the coupling product, Z—Z. It is common to find that half the radicals formed from an azo initiator are lost by this route. However, radical coupling within a solvent cage should not normally be considered a likely mechanism for any reaction which goes in high yield. In one of the most favorable cases, the photolysis of $CH_3—N{=}N—CH_3$ in carbon tetrachloride at 0°C, recombination of the methyl radicals to form ethane is favored over diffusion apart by a factor of 5.60 (85% cage recombination).[9] Less viscous solvents give more diffusion apart. In 2,3-dimethylbutane the ratio of recombination to diffusion is 2.13 at 0°C and 1.04 at 90°C.[9] It should be emphasized that the combination of two methyl radicals is among the most efficient of radical combinations, with nearly every collision yielding coupling product. The coupling of $CF_3\cdot$ radicals from $CF_3—N{=}N—CF_3$ is slightly less efficient, evidently requiring proper orientation of the $CF_3\cdot$ radicals before they can combine. In this case, the ratio of recombination to diffusion apart in carbon tetrachloride at 0°C is 1.74, in 2,3-dimethylbutane at 0°C it is 0.74, and in the more viscous decalin at 0° it is 3.34.[10]

The foregoing discussion perhaps does not rule out all possibility of there being a radical mechanism which involves coupling of solvent-caged radicals and gives high yields of products, provided the radicals are extremely reactive. However, such mechanisms are likely to be extremely rare, and the possibility of a chain reaction should always be considered. There is clearly a simple experimental test for the cage recombination mechanism: The yields of cage recombination products will be significantly higher in more viscous solvents. If the proposer of such a mechanism has not provided experimental evidence of this sort, the referee should send the paper back where it came from, and should feel free to compose a few grouchy insults to send with it.

III. Replacements of Metal Atoms

A. Mercury Compounds

Jensen and co-workers observed radical cleavage of organomercury compounds by bromine in nonpolar solvents in their attempts to establish the stereochemistry of electrophilic displacement reactions.[11,12] In a study of the radical reaction itself, bromine cleavage of *cis*- and *trans*-4-*t*-butylcyclohexylmercuric bromide (**1** and **2**) in carbon tetrachloride was found to yield nearly a statistical distribution of bromides from either starting material.[13] The propagation steps of the probable radical chain mechanism are outlined.

Chlorination of *cis*- and *trans*-4-*t*-butylcyclohexylmercuric chlorides (analogous to **1** and **2**) with sulfuryl chloride, SO_2Cl_2, yields 70% *cis*-4-*t*-butylcyclohexyl chloride from either isomer of the mercury compound.[13] Again, the loss of stereochemistry is evidence for a free-radical intermediate. It appears that the different ratio of products arises in the reaction of the 4-*t*-butylcyclohexyl radical with SO_2Cl_2, which has a higher heat of activation than the reaction with Br_2 and thus, in accord with Hammond's postulate,[14] has greater bond formation in the transition state, which would result in greater sensitivity to torsional strain in the cyclohexyl system.

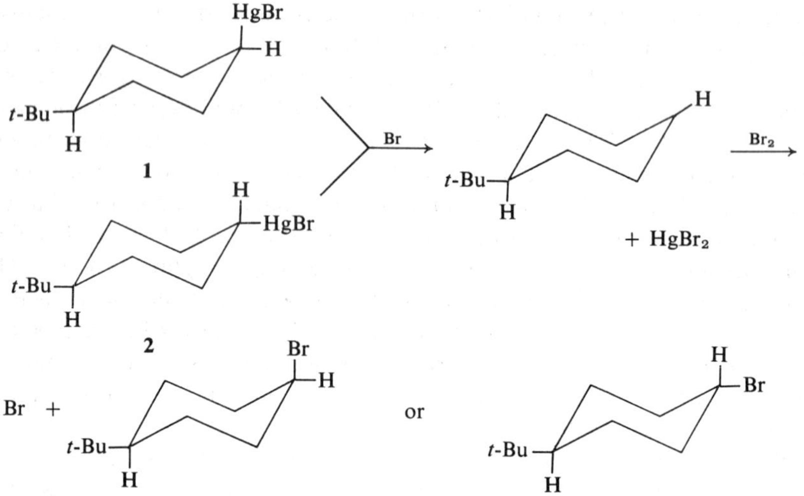

The reduction of alkylmercuric halides with sodium borohydride appears to proceed by way of an organomercuric hydride intermediate, RHgH, which decomposes to RH and Hg by a free-radical mechanism.[15-17] An earlier suggestion that the decomposition of RHgH was four-center in nature[18] was based on inadequate information, a substituted *exo*-norbornylmercuric chloride yielding *exo*-deuterated product from $NaBD_4$ because of steric factors in the intermediate radical. This once more points out the need for using both diastereoisomers before drawing any mechanistic conclusions from formation of a product that should be favored anyway, no matter what the mechanism.

In support of the radical mechanism, Pasto and Gontarz found that the diastereoisomeric 3-acetoxymercuri-2-butanols (**3** and **4**) (from oxymercuration of *cis*- and *trans*-2-butene) both yield 50:50 mixtures of *erythro*- and *threo*-3-deutero-2-butanols with $NaBD_4$.[15] However, *trans*-2-

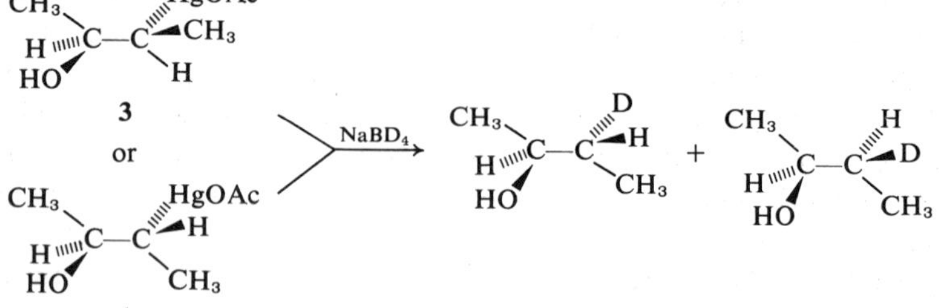

acetoxymercuricyclopentanol gave 95% *trans*-deuterated cyclopentanol, indicating either some sort of inherent stereospecificity in the reaction or a considerable steric effect in the 2-hydroxycyclopentyl radical. Sodium borohydride converted $PhCH{=}CHCH_2HgBr$ mostly to the allylic rearrangement product $PhCH_2CH{=}CH_2$ (96%) with a little of the unrearranged isomer $PhCH{=}CHCH_3$ (4%).[15]

Both the norbornenylmercuric chloride **5** and its nortricyclyl isomer **6** yielded the same mixture of reduction products with sodium borohydride. An initial discrepancy in the results between Pasto and Gontarz[15] and Gray and Jackson[16] has been resolved in favor of the latter.[16] The products are accounted for on the basis of the rapidly equilibrating radicals, **7**, **8**, and **9**.

OAc, HgCl **5** or ClHg, OAc **6** $\xrightarrow{NaBH_4}$ **7** (OAc) + **8** (OAc) + **9** (AcO)

↓ (60%) ↓ (6%) ↓ (34%)

Whitesides and San Filippo have carried out an extensive investigation of all stages of the reduction.[17] Neophylmercuric bromide (**10**) yields deutero-*t*-butylbenzene and not rearrangement product from reduction with $NaBD_4$, Et_2AlD, Bu_3PCuD, or Bu_3SnD, indicating that the neophyl radical is too short-lived to rearrange in all cases. Both *exo*- and *endo*-2-norbornylmercuric

$$Ph{-}C(CH_3)_2{-}CH_2HgBr \longrightarrow Ph{-}C(CH_3)_2{-}\dot{C}H_2 \longrightarrow Ph{-}C(CH_3)_2{-}CH_2D \quad \text{not} \quad D{-}C(CH_3)_2{-}CH_2{-}Ph$$

10 $<0.1\%$

bromide (**11** and **12**) react with the same reducing agents to yield a mixture of 90% *exo*- and 10% *endo*-2-deuteronorbornane in all cases. This differs from the composition of the mixture formed from *exo*- or *endo*-2-norbornylchloride and Bu_3SnD, which yields 84% *exo*- and 16% *endo*-2-deuteronorbornane. Thus, the reductions of RHgBr proceed not only through a common radical intermediate R· but must also share a common type of hydrogen atom donor, presumably RHgH. Confirmation of the presence of norbornyl radicals was provided by trapping with di-*t*-butylnitroxyl radical, $t\text{-}Bu_2NO\cdot$

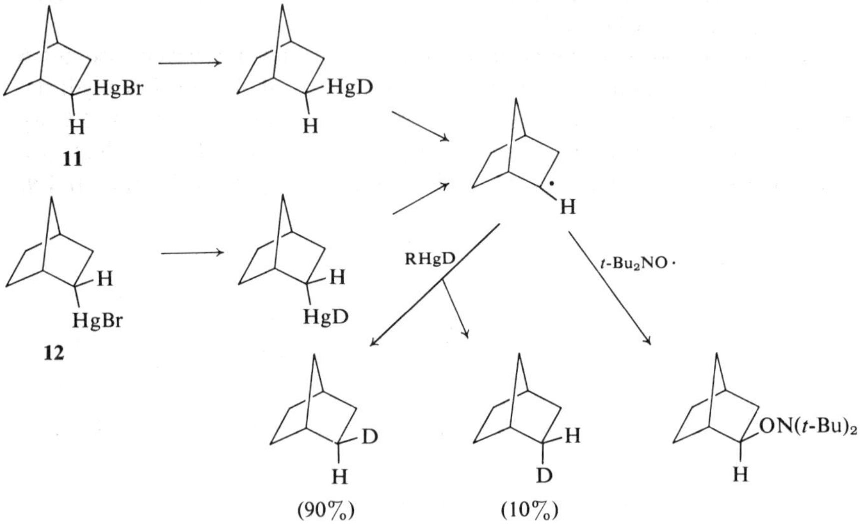

HgBr H **11** → HgD H ↘ · H

12 H HgBr → H HgD ↗

RHgD → D H (90%) ; H D (10%)

$t\text{-}Bu_2NO\cdot$ → $ON(t\text{-}Bu)_2$ H

A single radical chain propagation step

$$RHgH + R\cdot \longrightarrow R\cdot + Hg + HR$$

can account for all these decompositions of RHgH. It is also possible that there is a short-lived alkylmercurous radical, RHg·, which would make the sequence

$$RHgH + R\cdot \longrightarrow RHg\cdot + HR \longrightarrow R\cdot + Hg + HR$$

but there is no way of distinguishing this possibility on the basis of the available information.

The yield of capture product with di-*t*-butylnitroxyl was only 20% in the norbornyl system (**11** and **12**), which implies that RHgH is an extremely active hydrogen atom source.[17] A mechanism involving recombination of R· and HHg· inside a solvent cage has been proposed to account for the stereoselectivity found in some cases,[15] and extensive cage recombination might provide an alternative explanation of the low yield of nitroxyl radical coupling product.[17] However, this suggestion suffers from all the usual difficulties associated with cage recombination mechanisms. The high yields of hydrocarbons are incompatible with the amount of escape from the solvent cage that would be expected of HHg·, which would probably be reactive enough to attack the solvent or cause other side reactions. If HHg· were not quite as reactive as expected, it might survive to attack RHgH, but this would result in the chain sequence

$$RHgH + HHg\cdot \longrightarrow HgH_2 + RHg\cdot \longrightarrow HgH_2 + Hg + R\cdot$$

followed by

$$HgH_2 + R\cdot \longrightarrow HHg\cdot + HR$$

which is different in detail but not in principle from the postulated chain mechanism already described. Combination of HHg· and R· outside a solvent cage is, of course, a very minor pathway at most. The fact that *exo*- and *endo*-norbornylmercury compounds yield identical product mixtures suggests that the stereoselectivity is a property of the free norbornyl radical, not a solvent-caged pair of radicals. This becomes especially apparent when it is considered that the reaction within a solvent cage would have to be extremely rapid, recombination occurring with nearly every collision,[9] if the bulk of the product is to be formed by this route rather than some pathway involving diffusion apart. There would be no time for the norbornyl radical to tumble around to reach a 90% *exo* and 10% *endo* mixture of radical pairs regardless of whether the source was *exo* or *endo* mercury compound, as observed experimentally.[17] The solvent mixture used, tetrahydrofuran and water, is not particularly viscous and would not promote cage recombination at the expense of diffusion, and of course the effect of solvent viscosity on the reaction was not tested. Finally, the 20% yield of coupling product from norbornyl radicals and di-*t*-butylnitroxyl radical implies that at least 20% of the norbornyl radicals escape from the solvent cage, and there is no reason to believe that significantly less than 100% of them do. In summary, only a chain mechanism can be considered reasonable.

B. Oxidative Boron–Carbon Bond Cleavage and Related Reactions

Cleavage of carbon–boron bonds by oxidizing radicals occurs readily and has been studied in a number of cases. Cleavages of more reactive Group II and III alkylmetallic compounds appear to resemble those of alkylboron compounds but are generally more difficult to study quantitatively.

Krusic and Kochi have reported electron spin resonance (esr) studies which provide fundamental information about the carbon–metal bond-breaking process.[19] Photolytically generated *t*-butoxy radicals attack trimethylborane, $(CH_3)_3B$, at $-140°$ to $-110°C$ to produce an easily detectable steady-state concentration of methyl radicals, $CH_3\cdot$. The other product is presumably $(CH_3)_2B$-O-*t*-Bu. The esr spectrum of the methyl radicals is shown in Fig. 7-1. Ethyl radicals were similarly produced from triethylborane, triethylaluminum (Fig. 7-2), or triethylgallium. Butyl radicals were obtained from tributylaluminum, butyllithium, or dibutylmercury. Attempts were

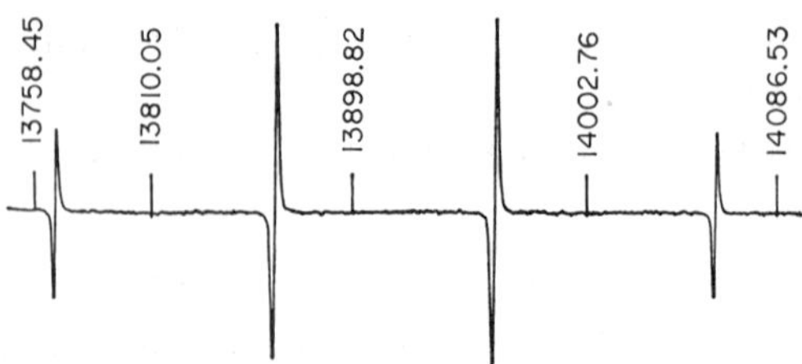

Fig. 7-1. The esr spectrum of methyl radicals from Me_3B and *t*-BuO· at −138°C. The field is marked in kilohertz. [From P. J. Krusic and J. K. Kochi, *J. Amer. Chem. Soc.* **91**, 3942 (1969).]

made to detect possible intermediate radicals $R_3\dot{B}$-O-*t*-Bu or $R_3\dot{A}l$-O-*t*-Bu at the lower end of the temperature range, but no splitting patterns of the type expected of ^{11}B or ^{27}Al could be detected, and such radicals must therefore have an exceedingly short lifetime or be transition states, not intermediates. The characteristic splitting pattern caused by ^{11}B may be seen in the spectrum of the radical $(CH_3CH_2O)_2B—O—\dot{C}HCH_3$ formed from *t*-BuO· and $(CH_3CH_2O)_3B$ (Fig. 7-3).

Since many readers will probably be less familiar with esr than with nmr spectra, a brief interpretation is presented here. The CH_3· spectrum (Fig. 7-1) is clearly the quartet with intensity ratios 1:3:3:1 that would be expected from splitting of the electron signal by the three hydrogen nuclei. The splitting constant is 22.89 gauss at −110°C. The CH_3CH_2· spectrum (Fig. 7-2) looks complex at first glance, but is easily interpreted when it is remembered that only one electron gives rise to the absorption signal, which is split simultaneously by two nonequivalent groups of protons. In nmr

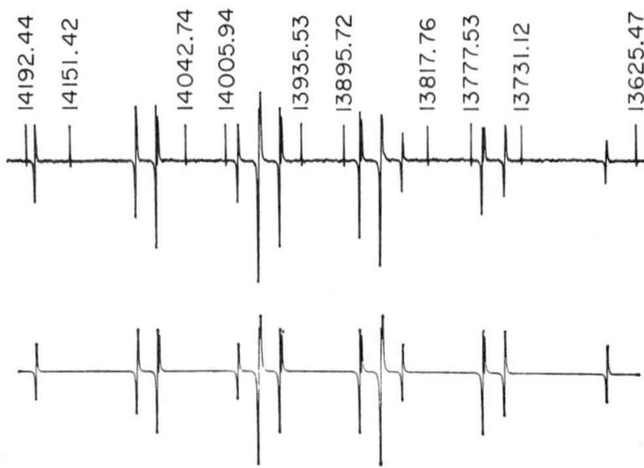

Fig. 7-2. Experimental and calculated esr spectra of ethyl radicals from Et_3Al and *t*-BuO· at −114°C with partially resolved second-order splittings. [From P. J. Krusic and J. K. Kochi, *J. Amer. Chem. Soc.* **91**, 3942 (1969).]

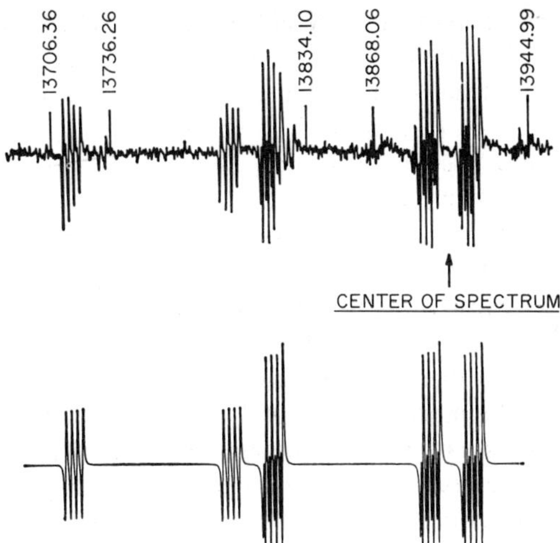

Fig. 7-3. Experimental and calculated esr spectra of $(CH_3CH_2O)_2BOCHCH_3$ from $(EtO)_3B$ at $-54°C$. Part of the fine structure is due to second-order effects. [From P. J. Krusic and J. K. Kochi, *J. Amer. Chem. Soc.* **91**, 3942 (1969).]

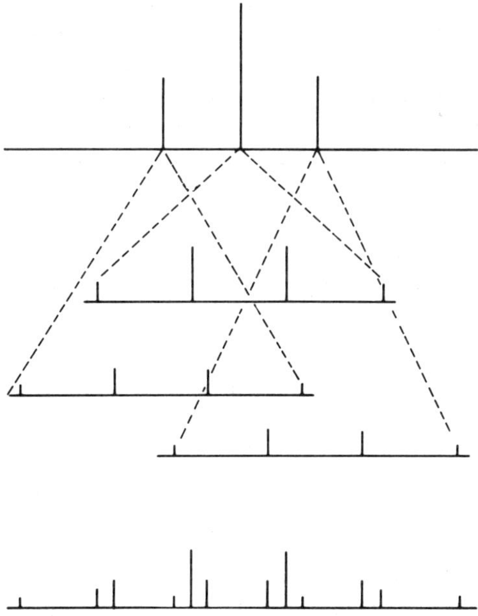

Fig. 7-4. Graphic analysis of the splittings in the esr spectrum of the ethyl radical.

language, this is the X part of an A_2B_3X spectrum. The two α-protons split the signal into a 1:2:1 triplet (splitting constant 22.37 gauss at $-90°C$) and the three β-protons further split each line of the triplet into a 1:3:3:1 quartet, as illustrated in Fig. 7-4. (The order in which the splittings are considered is, of course, immaterial.) Second-order splittings resulting from proton–proton spin couplings are relatively small and barely visible in Fig. 7-2, and are analogous to the second-order splittings in complex nmr spectra. The primary feature of interest in the spectrum of $(CH_3CH_2O)_2BO\dot{C}HCH_3$ (Fig. 7-3) is the 1:1:1:1 quartet caused by the 3/2 spin of the boron atom (splitting constant 0.66 gauss). Additional splittings due to the α-hydrogen (doublet, 18.55 gauss) and the three β-hydrogens (1:3:3:1 quartet, 23.52 gauss) are superimposed, and the more remote ethyl hydrogens yield no splittings.

Davies and Roberts have carried out similar esr studies on a wider variety of alkylmetal compounds.[20] The *t*-butoxy radicals were generated either by photolysis of di-*t*-butyl peroxide or, if the alkylmetal compound was light-sensitive, thermolysis of di-*t*-butyl hyponitrite, *t*-BuO—N═N—O—*t*-Bu, at 80°C. Cleavage of diethylzinc was carried out in tetrahydrofuran, and hydrocarbon solvents were used with dimethylcadmium, di-*n*-butyl- and di-*sec*-butylborane, tri-*n*-, tri-*sec*-, and tri-*t*-butylboroxine, trimethyl-aluminum, diethylaluminum butoxide, triethylantinomy, and triethylbismuth. The corresponding alkyl radical was observed by esr spectroscopy in each case.

Radical reactions carried out to yield macroscopic amounts of products are, of course, usually chain reactions. Kinetics have been measured on only a few of the many examples of such processes that have been found. The kinetics of chain reactions differ in some respects from those of familiar nonchain processes, and in view of the general utility and power of the kinetic method it seems appropriate to begin with a deboronation that has been studied in detail, the reaction of BEt_3 with I_2 to form EtI and $IBEt_2$.

The thermal iodination of triethylborane in cyclohexane solution at 100°–140°C was found by Lissi and Sanhueza to follow the rate law shown in Eq. (7-15).[21] The activation energy is 9.8 kcal/mole. The mechanism is outlined in Eqs. (7-16) through (7-18).

$$\frac{-d[I_2]}{dt} = k[I_2]^{1/2}[BEt_3] \tag{7-15}$$

$$I_2 \underset{k_{-1}}{\overset{k_1}{\rightleftharpoons}} 2I\cdot \quad \text{(initiation and termination)} \tag{7-16}$$

$$I\cdot + BEt_3 \xrightarrow{k_2} IBEt_2 + Et\cdot \quad \text{(propagation)} \tag{7-17}$$

$$Et\cdot + I_2 \xrightarrow{k_3} EtI + I\cdot \quad \text{(propagation)} \tag{7-18}$$

This iodination shows the typical kinetic behavior of a radical chain reaction in particularly simple fashion. Either chain-propagation step may be chosen as rate-determining, since they must proceed alternately and therefore at equal overall rates. The overall rate is thus given by Eq. (7-19). To solve

$$\frac{-d[\mathrm{BEt_3}]}{dt} = \frac{-d[\mathrm{I_2}]}{dt} = k_2[\mathrm{I}\cdot][\mathrm{BEt_3}] = k_3[\mathrm{Et}\cdot][\mathrm{I_2}] \qquad (7\text{-}19)$$

Eq. (7-19) in terms of measurable quantities, an expression for either $\mathrm{I}\cdot$ or $\mathrm{Et}\cdot$ must be found from the steady-state relationships. If all chain termination is by recombination of 2 $\mathrm{I}\cdot$ [Eq. (7-16)], then $k_1[\mathrm{I_2}] = k_{-1}[\mathrm{I}\cdot]^2$ because initiation and termination must have equal rates to maintain a low steady-state concentration of radicals ($d[\mathrm{I}\cdot]/dt \simeq 0$). Solving for $[\mathrm{I}\cdot]$ yields $[\mathrm{I}\cdot] = (k_1/k_{-1})^{1/2}[\mathrm{I_2}]^{1/2}$, which may be substituted into Eq. (7-19) to yield (7-20).

$$\frac{-d[\mathrm{I_2}]}{dt} = k_2\left(\frac{k_1}{k_{-1}}\right)^{1/2}[\mathrm{I_2}]^{1/2}[\mathrm{BEt_3}] \qquad (7\text{-}20)$$

Equation (7-20) is clearly equivalent in form to the experimental rate law, Eq. (7-15).

In general, all radical chain reactions will turn out to be half-order in initiator concentration and to have some dependence on one or both major reactants. In the example just cited, if the major chain-termination step were combination of 2 $\mathrm{Et}\cdot$, then by the steady-state condition $k_1[\mathrm{I_2}] = k_4[\mathrm{Et}\cdot]^2$ and $[\mathrm{Et}\cdot] = (k_1/k_4]^{1/2}[\mathrm{I_2}]^{1/2}$. Substituting for $[\mathrm{Et}\cdot]$ in Equation (7-19) yields the rate $= k_3(k_1/k_4)^{1/2}[\mathrm{I_2}]^{1/2}[\mathrm{I_2}]$, the three-halves-power dependence on $\mathrm{I_2}$ being the result of half-order in $\mathrm{I_2}$ as initiator and first-order in $\mathrm{I_2}$ as major reactant. This calculated result is, of course, contrary to the experimental observations, and it may be concluded that chain termination is not by recombination of 2 $\mathrm{Et}\cdot$ but by recombination of 2 $\mathrm{I}\cdot$. Other calculations may be made assuming two or more termination steps operate simultaneously. For example, the steady-state condition might be $k_1[\mathrm{I_2}] = k_{-1}[\mathrm{I}\cdot]^2 + k_4[\mathrm{Et}\cdot]^2$, which leads to a complex solution half-order in the initiator and something less than first order in each of the two major reactants, again contrary to experimental observation in the present case.

Bromine behaves differently from iodine and abstracts an α-hydrogen from an alkyl group of a trialkylborane, to be discussed in Section VI,A.

Alkoxy radicals are particularly efficient at displacing carbon from boron, as noted in the discussion of esr studies. One of the earlier examples of a chain reaction evidently involving attack of an alkoxy radical on boron was found by the author.[22] Naturally, the reaction desired was entirely different from what was found. Ultraviolet irradiation of a mixture of *t*-butyl hypochlorite and trimethylboroxine (**13**) was carried out in the hope of making a chloromethylboron compound, but instead yielded methyl chloride and tris(*t*-butoxy)boroxine. Probable chain-propagation steps are illustrated.

Alkylperoxy radicals formed in air oxidations of alkylboron compounds behave much like alkoxy radicals. Davies and Roberts found that the

$$\text{(MeBO)}_3\ (\mathbf{13}) + t\text{-BuO}\cdot \longrightarrow t\text{-BuO—B(OBMe)}_2\text{O (cyclic)} + \text{Me}\cdot$$

$$\text{Me}\cdot + t\text{-BuOCl} \longrightarrow \text{MeCl} + t\text{-BuO}\cdot$$

autoxidation of 1-phenylethaneboronic acid (**14**) is a radical chain process.[23] The autoxidation is inhibited (temporarily) by efficient radical traps such as galvinoxyl or copper(II) *N,N*-dibutyldithiocarbamate, but not by diphenylpicrylhydrazyl, thiophenol, 2,6-di-*t*-butyl-4-methylphenol, or *N*-phenyl-β-naphthylamine. The product after hydrolysis is 1-phenylethyl hydroperoxide (**15**), which was characterized as the triphenylmethyl and 9-xanthenyl derivatives. In benzene solution saturated with oxygen, the reaction is first order in 1-phenylethaneboronic acid, with $k = 0.002\ \text{sec}^{-1}$ at 20°C. This rate law indicates that the chain-terminating step is combination of two alkylperoxy radicals (see previous kinetic analysis of iodination and a following example of autoxidation). Optically active 1-phenylethaneboronic acid yielded racemic 1-phenylethyl hydroperoxide. These observations are consistent with the radical chain propagation steps illustrated.

$$\underset{\mathbf{14}}{\text{Ph—CH(CH}_3\text{)—B(OH)}_2} + \text{Ph—CH(CH}_3\text{)—O—O}\cdot \longrightarrow \text{Ph—CH(CH}_3\text{)—O—O—B(OH)}_2 + \text{Ph—}\dot{\text{C}}\text{H—CH}_3$$

$$\text{Ph—}\dot{\text{C}}\text{H—CH}_3 + \text{O}_2 \longrightarrow \text{Ph—CH(CH}_3\text{)—O—O}\cdot$$

In a subsequent step,

$$\text{Ph—CH(CH}_3\text{)—O—O—B(OH)}_2 + \text{H}_2\text{O} \longrightarrow \underset{\mathbf{15}\ \text{(racemic)}}{\text{Ph—CH(CH}_3\text{)—O—OH}} + \text{B(OH)}_3$$

Allies and Brindley reported similar results with diisopinocampheylboranes.[24] Similar inhibitors were effective. Considerable racemization resulted on autoxidation of *B*-2-butyl-*B,B*-diisopinocampheylborane (in which the 2-butyl group is partially resolved by asymmetric induction in the synthesis from *cis*-2-butene and the optically active pinene-derived borane). However, racemization was not complete, being only about 60% when the autoxidation was carried out slowly, 87% when autoxidation was rapid. The optical rotations were measured on samples of 2-butanol derived from the initially formed 2-butyl hydroperoxide by reduction. Perhaps the 2-butanol formed with retention of configuration is derived from a competing polar process in which peroxide (R—O—O—H or R—O—O—BR_2) attacks the boron–carbon bond to form alkyl borate, $ROBR_2$, directly from the trialkylborane (see Chapter 4, Section II). The polar process would compete more effectively when the radical process is slow and the initially formed peroxide remains in contact with unreacted boronic ester for some time. Thus, there is no reason to doubt that the radical process involves free 2-butyl radicals. In related work, Mirviss has shown that trialkylboranes, BR_3, yield peroxides, R—O—O—BR_2, on autoxidation.[25] The probable chain-propagation steps for all these autoxidations probably resemble those for the boronic acid and are summarized in Eqs. (7-21) and (7-22).

$$\text{R—OO}\cdot + \text{BR}_3 \longrightarrow \text{R—OO—BR}_2 + \text{R}\cdot \qquad (7\text{-}21)$$

$$\text{R}\cdot + \text{O}_2 \longrightarrow \text{R—OO}\cdot \qquad (7\text{-}22)$$

Ingold has studied the kinetics of autoxidation of tri-2-butylboroxine (**16**) and has determined the rate of reaction of the 2-butylperoxy radical with the boroxine (**16**).[26] When the reaction is initiated by the thermolysis of di-*t*-butyl peroxyoxalate at 30°C, the rate is first order in boroxine **16** and half-order in the peroxide initiator and is independent of the partial pressure of oxygen in the 150–760 mm range, Eq. (7-23). The mechanism is outlined in Eqs. (7-24) through (7-28). The kinetic analysis is much like that of the iodination

$$\frac{-d[\text{O}_2]}{dt} = k[t\text{-Bu-OO-COCO-OO-}t\text{-Bu}]^{1/2}[(2\text{-BuBO})_3] \qquad (7\text{-}23)$$

$$t\text{-Bu—OO—}\overset{\text{O}}{\overset{\|}{\text{C}}}\text{—}\overset{\text{O}}{\overset{\|}{\text{C}}}\text{—OO—}t\text{-Bu} \xrightarrow{k_1} 2\text{CO}_2 + 2t\text{-BuO}\cdot \text{ (initiation)} \qquad (7\text{-}24)$$

$$t\text{-BuO}\cdot + \underset{\mathbf{16}}{(2\text{-BuBO})_3} \longrightarrow t\text{-BuO(BO)}_3(2\text{-Bu})_2 + 2\text{-Bu}\cdot \qquad (7\text{-}25)$$

$$2\text{-Bu}\cdot + \text{O}_2 \xrightarrow{k_2} 2\text{-Bu-OO}\cdot \text{ (propagation)} \qquad (7\text{-}26)$$

$$2\text{-Bu-OO}\cdot + \underset{\mathbf{16}}{(2\text{-BuBO})_3} \xrightarrow{k_3} 2\text{-Bu—OO—B(OB(2\text{-Bu}))_2\text{O [boroxine ring, two B bearing 2-Bu]}} + 2\text{-Bu}\cdot \text{ (propagation)} \qquad (7\text{-}27)$$

$$2\text{-Bu—OO}\cdot + 2\text{-Bu—OO}\cdot \xrightarrow{k_4} 2\text{-Bu—OO—2-Bu} + \text{O}_2 \text{ or other nonradical products (termination)} \qquad (7\text{-}28)$$

discussed earlier in this section. Both propagation steps must proceed at equal rates, therefore

$$k_2[\text{2-Bu}\cdot][\text{O}_2] = k_3[\text{2-Bu-OO}\cdot][(\text{2-BuBO})_3]$$

The rate of initiation [Eq. (7-24)] must equal the rate of termination [Eq. (7-28)], therefore

$$k_1[(t\text{-BuO-O}_2\text{C})_2] = k_4[\text{2-Bu-OO}\cdot]^2$$

Solving for [2-Bu-OO·] and substituting into the expression

$$\frac{-d[\text{O}_2]}{dt} = -\frac{d[(\text{2-BuBO})_3]}{dt} = k_3[\text{2-Bu-OO}\cdot][(\text{2-BuBO})_3]$$

yields the rate law

$$\frac{-d[\text{O}_2]}{dt} = k_3\left(\frac{k_1{}^{1/2}}{k_4}\right)[(t\text{-BuO-O}_2\text{C-})_2]^{1/2}[(\text{2-BuBO})_3]$$

which agrees with the experimental rate law in Eq. (7-23). The initiation rate was measured, and $k_3/(k_4)^{1/2}$ was found to be 39.6 (liters/mole second)$^{1/2}$. From previous work, k_4 was known to be 1.5×10^6 liters/mole second, which yields $k_3 = 4.8 \times 10^4$ liters/mole second, or 1.6×10^4 liters/mole second per carbon–boron bond. This is much larger than rate constants for attack of alkylperoxy radicals on carbon–hydrogen bonds of hydrocarbons.

Autoxidations of dimethylcadmium, dimethylzinc, trimethylaluminum dimer, triphenylaluminum dimer, 1-naphthylmagnesium bromide, alkylmagnesium bromide, and butyllithium all appear to follow similar mechanisms.[27] Galvinoxyl inhibited autoxidation of the cadmium and zinc compounds, slightly retarded autoxidation of the aluminum compounds, and failed to affect the magnesium and lithium compounds, apparently because these very reactive substances attack the inhibitor.

Iodine inhibits the initiation of autoxidation of trialkylboranes more effectively than does galvinoxyl. Brown and Midland found that the inhibition period is much longer with tri-2-butylborane than with tri-1-butylborane, suggesting that the initiation step is sterically hindered in the former case.[28] A possible mechanism for the initiation and inhibition is shown in Eqs. (7-29) through (7-31).[29] It should be noted that the temperature is much lower in

$$\text{O}_2 + \text{R}_3\text{B} \longrightarrow \text{R}_2\text{B—O—O}\cdot + \text{R}\cdot \tag{7-29}$$

$$\text{R}_2\text{B—O—O}\cdot + \text{I}_2 \longrightarrow \text{R}_2\text{BI} + \text{O}_2 + \text{I}\cdot \tag{7-30}$$

$$\text{R}\cdot + \text{I}_2 \longrightarrow \text{RI} + \text{I}\cdot \text{ (inert under these conditions)} \tag{7-31}$$

this case than in the cleavage of triethylborane by iodine mentioned earlier, and the iodine atoms are unable to attack the trialkylborane before they dimerize.

Chain transfer occurs when the autoxidation of trialkylboranes is carried out in the presence of allyl or benzyl iodide.[30] By adjusting the proportions of reactants the reaction can be made efficient either for the production of alkyl iodide or for the coupling of the allyl or benzyl radicals. Postulated steps in the reaction are shown in Eqs. (7-32) through (7-38). The chain-transfer step is, of course, fundamentally no different from any other chain propagation step, but is so marked to indicate clearly where the additional reactant, benzyl iodide, enters into the process.

$$Et_3B + O_2 \longrightarrow Et_2B{-}OO\cdot + Et\cdot \qquad \text{(initiation)} \qquad (7\text{-}32)$$

$$Et\cdot + O_2 \longrightarrow Et{-}OO\cdot \qquad \text{(propagation)} \qquad (7\text{-}33)$$

$$Et{-}OO\cdot + Et_3B \longrightarrow Et{-}OO{-}BEt_2 + Et\cdot \qquad \text{(propagation)} \qquad (7\text{-}34)$$

$$Et\cdot + PhCH_2I \longrightarrow EtI + PhCH_2\cdot \qquad \text{(chain transfer)} \qquad (7\text{-}35)$$

$$PhCH_2\cdot + O_2 \longrightarrow PhCH_2{-}OO\cdot \qquad \text{(propagation)} \qquad (7\text{-}36)$$

$$PhCH_2{-}OO\cdot + Et_3B \longrightarrow PhCH_2{-}OO{-}BEt_2 + Et\cdot \qquad \text{(propagation)} \qquad (7\text{-}37)$$

$$2PhCH_2\cdot \longrightarrow PhCH_2{-}CH_2Ph \qquad \text{(termination)} \qquad (7\text{-}38)$$

Dialkyl disulfides react with trialkylboranes by a free-radical mechanism analogous to autoxidation.[31] The reaction is initiated by oxygen or light and inhibited by iodine. Probable propagation steps are shown in Eqs. (7-39) and (7-40). The reaction continues until all three alkyl groups of the borane have been replaced by alkylthio groups.

$$MeS\cdot + BBu_3 \longrightarrow MeSBBu_2 + Bu\cdot \qquad (7\text{-}39)$$

$$Bu\cdot + MeS{-}SMe \longrightarrow Bu{-}SMe + MeS\cdot \qquad (7\text{-}40)$$

The 1,4-addition of trialkylboranes to acrolein or α,β-unsaturated ketones is a radical chain process and is inhibited by galvinoxyl.[32] Postulated chain-propagation steps are shown in Eqs. (7-41) and (7-42). The synthetically

$$R{-}CH_2{-}CH{=}C\begin{matrix}\diagup O\cdot \\ \diagdown H\end{matrix} + BR_3 \longrightarrow R{-}CH_2{-}CH{=}C\begin{matrix}\diagup O{-}BR_2 \\ \diagdown H\end{matrix} + R\cdot \qquad (7\text{-}41)$$

$$R\cdot + CH_2{=}CH{-}C\begin{matrix}\diagup\!\!\diagup O \\ \diagdown H\end{matrix} \longrightarrow R{-}CH_2{-}CH{=}C\begin{matrix}\diagup O\cdot \\ \diagdown H\end{matrix} \qquad (7\text{-}42)$$

useful product obtained on hydrolysis of $R{-}CH_2CH{=}CHO{-}BR_2$ is the aldehyde $R{-}CH_2CH_2CHO$. When this reaction was first discovered by Suzuki, Brown, and co-workers,[33–36] it was thought that the mechanism was a polar electrophilic displacement, and stereochemical evidence was obtained to show that the reaction proceeded with retention of configuration at the site where the boron is displaced.[37] However, the stereochemistry was only proved with tris(*trans*-2-methylcyclopentyl)borane, and the 2-methylcyclopentyl radical has such a strong steric preference for attack *trans* to the methyl group that apparent retention of configuration was inevitable. The practical synthetic importance of knowing something about the true mechanism becomes apparent when reactions of β-substituted α,β-unsaturated ketones and of ethynyl ketones are desired. These failed to react under the conditions used with acrolein, but when the need for a radical initiator such as ultraviolet light, peroxides,[38] or atmospheric oxygen[39,40] was provided for, these reactions were also carried out successfully.

Davies and co-workers have found that dimethylchloramine, Me_2NCl, cleaves trialkylboranes, R_3B, by competing radical and polar processes.[41] In the polar process, the nitrogen atom is the electrophilic site and the products are R_2BCl and Me_2NR (see Chapter 4, Section II,B). The radical process yields R_2BNMe_2 and RCl, and is inhibited by galvinoxyl. Generation of $Me_2N\cdot$ radicals from $Me_2N{-}N{=}N{-}NMe_2$ by photolysis in the presence of tributylborane yielded butyl radicals, which were detected by esr.[42] The chain-propagation steps of the Me_2NCl reaction therefore must be as shown in Eqs. (7-43) and (7-44).

$$Me_2N\cdot + BBu_3 \longrightarrow Me_2N{-}BBu_2 + Bu\cdot \qquad (7\text{-}43)$$

$$Bu\cdot + Me_2N{-}Cl \longrightarrow BuCl + Me_2N\cdot \qquad (7\text{-}44)$$

Radical reactions need not be chain processes if there is a transition metal present which can undergo a one-electron change of oxidation state. Lane has achieved the conversion of Bu_3B to BuBr, which cannot be carried out by radical bromination because of attack of $Br\cdot$ at the α-hydrogens, by using copper(II) bromide as the bromine source.[43] A reasonable nonchain pathway involving free butyl radicals can be written, Eqs. (7-45) and (7-46), though, of course, this is a suggestion, not a well-established mechanism.

$$Bu_3B + CuBr_2 + H_2O \longrightarrow Bu_2BOH + H^+ + Br^- + CuBr + Bu\cdot \qquad \text{(initiation)} \quad (7\text{-}45)$$

$$Bu\cdot + CuBr_2 \longrightarrow BuBr + CuBr \qquad \text{(termination)} \quad (7\text{-}46)$$

C. Groups IV and V

Boué *et al.* have found that the light-initiated reaction of tetraalkyltins with bromine in chlorobenzene can be made to proceed much faster than the polar reaction which occurs in the dark.[44] The products are alkyl bromide and trialkyltin bromide and the yields are about 95%. Probable chain-propagation steps are shown in Eqs. (7-47) and (7-48).

$$Br\cdot + R_4Sn \longrightarrow R_3SnBr + R\cdot \qquad (7\text{-}47)$$

$$R\cdot + Br_2 \longrightarrow RBr + Br\cdot \qquad (7\text{-}48)$$

Relative rates of cleavage of $R\cdot$ from Me_3SnR by $Br\cdot$ are allyl, 8.45; isopropyl, 6.95; *n*-butyl, 3.23; *n*-propyl, 3.23; ethyl, 3.7; benzyl, 2.15; methyl, 1.0.[44] These are very small differences in reactivity for the range of radical structures produced. The reaction of $Br\cdot$ with R_4Sn is probably highly exothermic. The low activation energy allows for little variation in the rates, and the transition state is probably early along the reaction coordinate, where there is little carbon–tin bond breaking and consequently little influence from the stability of the radical being formed.

Racemization of the alkyl group in conversion of $RSnMe_3$ to RBr provides evidence in favor of a free R· radical. Where R was optically active 1-methyl-2,2-diphenylcyclopropyl, Sisido and co-workers found 1.3% net retention of configuration in RBr.[45] Whether this slight net retention was the result of rapid reaction of R· before inversion of the cyclopropyl system, some sort of cage effect, or some other cause is impossible to decide.

Group IV metals in general appear to be among the least susceptible to radical attack. The tetrahedral configuration provides a good deal of steric shielding, and the α-hydrogens of the alkyl substituents are often more easily attacked than the metal atom itself (Section V). In contrast, Group V metalloids have an exposed unshared electron pair readily attacked by oxidizing radicals. Attack of RS· on PR'_3 yields $RS—\dot{P}R'_3$, which is readily detectable by esr, in contrast to the absence of any intermediate $RO—\dot{B}R'_3$ from boron.[46]

Trialkylantimony compounds react with benzenethiol in the presence of an initiator such as di-*t*-butyl hyponitrite, *t*-BuO—N═N—O—*t*-Bu, to form the alkane and the phenylthioantimony derivatives R_2SbSPh, $RSb(SPh)_2$, or $Sb(SPh)_3$.[47] Bismuth compounds behave similarly and are somewhat more reactive, and aryls of bismuth or antimony react much like the alkyls. The chain-propagation steps shown in Eqs. (7-49) and (7-50) are probable.

$$PhS\cdot + Sb(CH_3)_3 \longrightarrow PhS—\dot{S}b(CH_3)_3 \longrightarrow PhS—Sb(CH_3)_2 + CH_3\cdot \tag{7-49}$$

$$CH_3\cdot + PhSH \longrightarrow CH_4 + PhS\cdot \tag{7-50}$$

IV. Lithium and Sodium Compounds and CIDNP

Alkyllithiums are generally thought of as carbanion sources, but they are easily oxidized and thus also serve as electron sources. One-electron oxidation of RLi yields R· and Li^+. The details are no doubt complicated by the existence of alkyllithiums as tetramers, but such details are beyond present understanding of such mechanisms and will be ignored here. Until recently, it was not even possible to say unequivocally whether free radicals were involved in reactions of organolithium compounds, but a new nmr technique has changed that.

Ward found evidence for radical intermediates in the reaction of butyllithium with 6-bromo-1-phenyl-1-hexyne (**17**), which after hydrolysis yields 60% of benzylidenecyclopentane (**18**) and several hydrocarbon by-products.[48]

$$\underset{\mathbf{17}}{PhC{\equiv}C(CH_2)_4Br} \xrightarrow[2.\ H_2O]{1.\ BuLi} \underset{\mathbf{18}}{\underset{(60\%)}{\overset{Ph}{\underset{H}{>}}C{=}C_5H_8}} + CH_2{=}CHCH_2CH_3 +$$

$$\underset{\mathbf{19}}{\underset{(20\%)}{PhC{\equiv}C(CH_2)_7CH_3}} + \underset{(3\%)}{PhC{\equiv}C(CH_2)_3CH_3} + \underset{(1\%)}{PhC{\equiv}CCH_2CH_2CH{=}CH_2}$$

Carbanion intermediates had been postulated previously for related cyclizations.[49] The Wurtz coupling product, 1-phenyl-1-decyne (**19**), might be formed either from nucleophilic displacement of bromide from the alkyl bromide (**17**) by the butyl anion from butyllithium, or it might come from a radical reaction, RBr + BuLi yielding $R\cdot + Li^+Br^- + Bu\cdot$ followed by radical coupling to R—Bu. Since the yield is only 20% and both radicals are very reactive alkyl types, there is no need to invoke the usual prejudice against radical coupling in this case, and combination of the two radicals within the solvent cage in which they are formed or along the alkyllithium tetramer surface could easily be efficient enough to account for the observed yield. Positive evidence favoring a radical process is that when D_2O was used in place of H_2O in the work-up, the benzylidenecyclopentane (**18**) was only 25% deuterated. Most of the hydrogen must be incorporated into the product before the hydrolysis step, in accord with a radical mechanism and contrary to a carbanion mechanism, which would yield a substituted vinyllithium and consequently a deuterated product on treatment with D_2O. The presence of some deuterated product allows but does not require the possibility that some of the product is formed by a carbanionic path in competition with the radical path, and such competition is reasonable in view of the high reactivity and therefore perhaps low selectivity of the system. The operation of the radical path is further supported by the presence of 1-butene among the products, since this can arise readily from butyl radicals.

Dramatic proof of the existence of free-radical intermediates in reactions of this type was provided by the unusual nmr spectra observed by Ward and Lawler during the course of the reaction of butyllithium and butyl bromide in the presence of ether and diphenylacetylene.[50] The 1-butene freshly formed in this reaction in an nmr tube shows both emission and strongly enhanced absorption lines, illustrated in Fig. 7-5. Both the emission and the enhanced absorption lines in Fig. 7-5 correspond to part (but not all) of the vinyl proton spectrum of 1-butene. The enhanced absorption lines are far stronger than anything that could be achieved by concentration of the 1-butene, and of course the emission lines have to originate in some unprecedented effect.

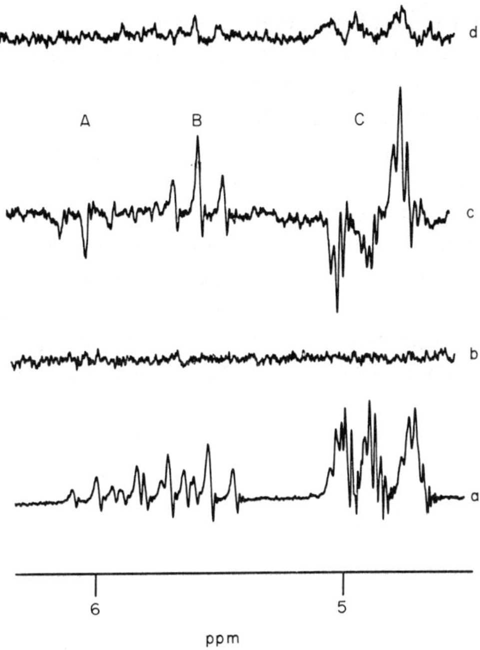

Fig. 7-5. Nmr spectra showing enhanced absorption and emission in 1-butene freshly formed from butyllithium and butyl bromide. Scan *a* is a concentrated solution of 1-butene; scan *b* is 1.6 *M* BuLi, 0.9 *M* BuBr, 0.8 *M* PhC≡CPh in hexane; scan *c* is the same 2 minutes after addition of 0.2 ml of ether to 0.5 ml of solution, which initiates the reaction; scan *d* is 6 minutes after addition of the ether. [From H. R. Ward and R. G. Lawler, *J. Amer. Chem. Soc.* **89**, 5518 (1967).]

This effect has been called "chemically induced dynamic nuclear polarization," abbreviated to "CIDNP."

In the particular experiment cited, if 2-pentyne is used in place of diphenylacetylene, the 1-butene emission lines of Fig. 7-5 are changed to absorptions and the absorptions to emissions. The role of the acetylene in the nuclear polarization is not understood. None is observed in its absence in this case, though most CIDNP observations have not required special additives.

Lawler pointed out that the emission and enhanced absorption lines observed require population differences between nuclear spin energy levels that are more than 10 times the Boltzmann population differences. To achieve this result, there must be coupling to a larger energy gap, which is provided by the unpaired electron spin of the radical.[51] Presumably the polarization occurs in the butyl radical, which is then converted to 1-butene which retains the imbalance of nuclear spin states. The imbalance, of course, decays within a few minutes by way of the normal spin relaxation processes.

Lawler first proposed that the polarization arose because of different rates of relaxation of different combinations of proton and electron spins.[51] However, this theory has been superseded by a related theory of rates of radical-pair recombination being different for different proton and electron spin states, as proposed by Closs[52] and by Kaptein and Oosterhoff.[53] According to the radical-pair model, some spin states are more prone to undergo radical recombination than others, so that diffusion out of the solvent cage is more probable for some radical pairs than for others. Thus, radicals with a slight excess number of protons polarized in one direction recombine within the solvent cage, and radicals with a slight excess number of protons polarized in the opposite direction escape from the solvent cage. In general, the products from recombination within the cage are different from those from diffusion apart. An excess of protons polarized in a lower energy spin state gives rise to enhanced absorption, and an excess in a higher energy spin state results in emission in the nmr. From the Boltzmann distribution law the excess of protons in the lower energy state at 60 MHz is only about 3 for every 10^7 in that state, and it does not take a very large change in the probabilities of radical recombination to produce emissions and enhanced absorptions that are hundreds of times stronger than normal absorptions.

As would be expected, the strengths of CIDNP emissions and absorptions are affected by the strength of the magnetic field in which the radicals are generated, but this is not a linear effect and CIDNP signals can be observed even when the radicals are generated in essentially a zero magnetic field and the products introduced into the magnetic field of the nmr instrument later. Calculations based on the radical-pair model are relatively easy when the radicals are generated in a strong magnetic field[52,53] and become much more complex for fields up to a few hundred gauss.[54] The computer calculations of Garst and co-workers[54] provide good verification of the theory, even though not all factors involved could be taken into account.

Continuing with experimental observations of CIDNP, the coupling reaction between ethyllithium and α,α-dichlorotoluene yields 1-chloro-1-phenylpropane with the protons polarized.[55] If the reaction is carried out in the magnetic field, enhanced absorption by the H–1 protons results, but if the reaction is carried out outside the magnetic field these protons yield an emission signal. The spectrum must be taken quickly, since the polarized state relaxes with $T_1 \cong 8$ seconds. Spectra obtained are shown in Fig. 7-6.

Numerous other examples of CIDNP involving alkyllithiums have been found. Reaction of *n*-butyllithium with 2-butyl iodide yields polarized protons in both the product 1-butyl iodide and in the unreacted 2-butyl iodide.[56] The polarized 2-butyl iodide presumably arises by way of a radical intermediate which can revert to this starting material. Lepley and Landau suggested that the CIDNP signals observed in recovered isopropyl iodide in the butyl-

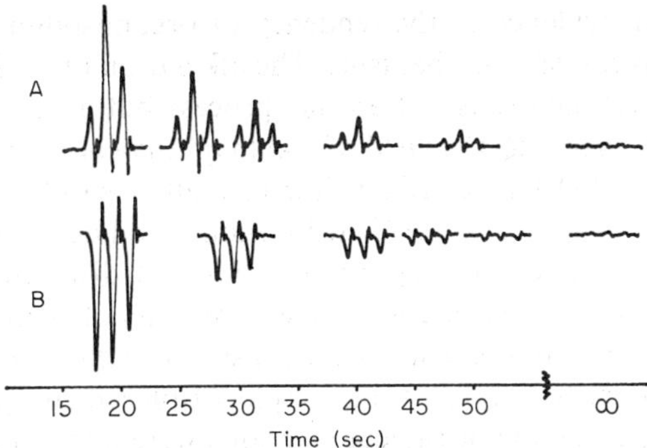

Fig. 7-6. The CIDNP spectra of 1-chloro-1-phenylpropane produced from ethyllithium and α,α-dichlorotoluene coupled inside and outside the magnetic field of the nmr instrument. [From H. R. Ward, R. G. Lawler, H. Y. Loken, and R. A. Cooper, *J. Amer. Chem. Soc.* **91**, 4929 (1969).]

lithium–isopropyl iodide system arise by transfer of an iodine atom from one radical to another, $R\cdot + RI \rightarrow RI + R\cdot$.[57] It was found that for the 1-butyl iodide–butyllithium reaction the polarized 1-butyl iodide concentration builds up to a maximum after about 5 minutes, the time required for consumption of half of the butyl iodide under the conditions used, and then decreases to zero in about 10 minutes.[58] An nmr emission spectrum has been obtained from $Et_2NCH{=}CH_2$ formed from the reaction of butyllithium with butyl bromide in the presence of triethylamine.[59] CIDNP signals with lifetimes of only a few seconds have been observed from 1,1-dimethylallene and 1-chloro-2,2-dimethylcyclopropane formed in the reaction of ethyllithium with 1,1-dichloro-2,2-dimethylcyclopropane.[60]

The widespread occurrence of radical intermediates in these reactions of alkyllithiums supports the suggestion made by Screttas and Eastham that many typical reactions of organolithium compounds involve electron transfer as the first step.[61]

Not too surprisingly, organosodium compounds also tend to react by radical mechanisms. Garst and Cox found an nmr emission signal from ethane generated by reaction of methyl iodide with a sodium mirror in dimethoxyethane.[62] They suggested that the mechanism is that shown in Eq. (7-51). In contrast, 1,4-diiodobutane gives unpolarized cyclobutane. This

$$CH_3Na + CH_3I \longrightarrow [CH_3\cdot + Na^+ + I^- + CH_3\cdot] \longrightarrow CH_3CH_3 + NaI \qquad (7\text{-}51)$$

does not necessarily mean that radicals are not involved, since diffusion apart of the two ends of the 1,4-tetramethylene diradical would not occur, and there may be no competing reactions to provide segregation of the different spin states.

There is other evidence of the tendency of organosodium compounds to react by electron-transfer mechanisms. The alkylation of sodium naphthalenide (**20**) with alkyl halides is not an S_N2 process but a radical reaction.[63,64] For a given alkyl group R, the yield of alkylnaphthalene is independent of the halide X in the alkyl halide, RX.[63] Tertiary alkyl halides give the highest yields of alkylnaphthalenes (61% with *t*-amyl iodide), consistent with a radical mechanism and inconsistent with an S_N2 displacement.[64] Bromobenzene yields mainly benzene, with lesser amounts of biphenyl, terphenyls, and α- and β-phenylnaphthalene, again consistent with a radical mechanism and inconsistent with any sort of simple displacement process.[65] Other by-products characteristic of radical reactions were also found.[63,64]

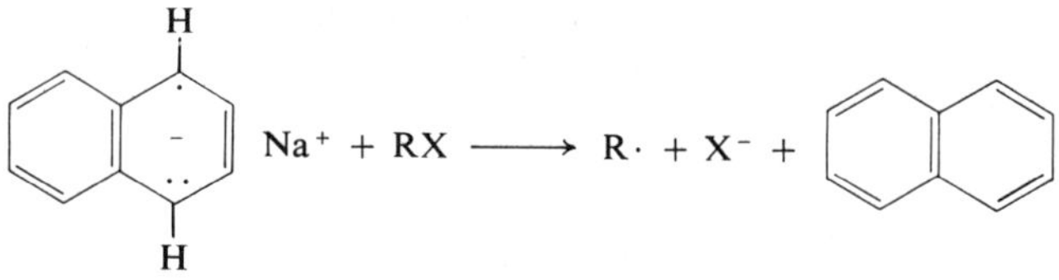

20 Na^+ + RX ⟶ R· + X^- + [naphthalene]

20 + R· ⟶ [H, R adduct] ⟶ Products

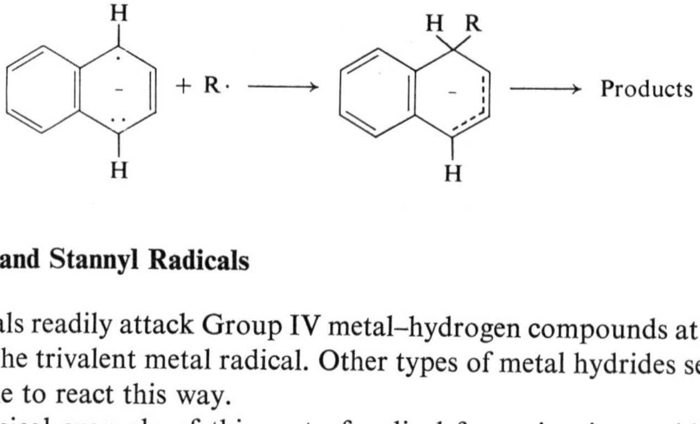

V. Silyl and Stannyl Radicals

Radicals readily attack Group IV metal–hydrogen compounds at hydrogen to yield the trivalent metal radical. Other types of metal hydrides seem much less prone to react this way.

A classical example of this sort of radical formation is provided by the radical addition of trichlorosilane to olefins, Eqs. (7-52) and (7-53).[66] Eaborn

$$Cl_3Si\cdot + CH_2{=}CHR \longrightarrow Cl_3Si{-}CH_2{-}\dot{C}HR \quad (7\text{-}52)$$

$$Cl_3Si{-}CH_2{-}\dot{C}HR + Cl_3SiH \longrightarrow Cl_3SiCH_2CH_2R + Cl_3Si\cdot \quad (7\text{-}53)$$

and co-workers have shown that the reaction of Cl_3SiH with 1-hexene initiated by *t*-butyl peroxide and ultraviolet light is not very temperature-sensitive,[67] implying low activation energies for both steps of the chain. The most efficient initiator system found for this reaction was ultraviolet light and $Me_3Si{-}Hg{-}SiMe_3$, which gave yields up to 97%.[68]

The silyl radicals $SiH_3\cdot$, $MeSiH_2\cdot$, $Me_2SiH\cdot$, and $Me_3Si\cdot$ have been generated from SiH_4, $MeSiH_3$, etc., and photochemically generated *t*-butoxy

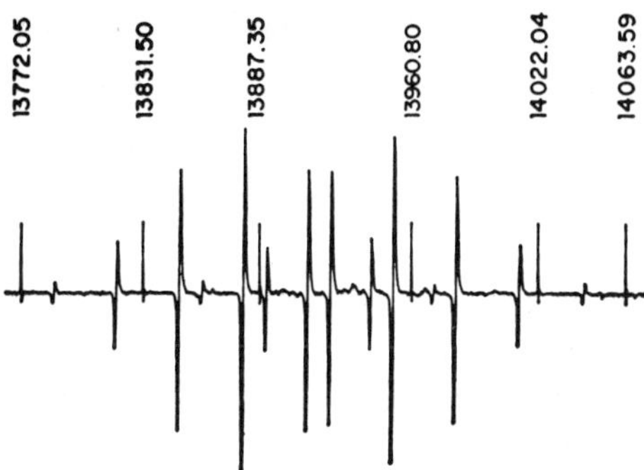

Fig. 7-7. The esr spectrum of the dimethylsilyl radical at $-123°C$. [From P. J. Krusic and J. K. Kochi, *J. Amer. Chem. Soc.* **91**, 3939 (1969).]

radicals in ethane at $-120°$ to $-150°C$ and their esr spectra have been observed.[69] A typical example, the dimethylsilyl radical, is shown in Fig. 7-7. The splittings of course follow the rules discussed in Section III,B.

Frangopol and Ingold have used the rotating sectored disk method to measure the rate of combination of 2 $Me_3Si\cdot$ to form $Me_3SiSiMe_3$, $k = 2.2 \times 10^9$ liters/mole second at 25°C.[70] The $Me_3Si\cdot$ radicals were generated photolytically from *t*-butyl peroxide and Me_3SiH. The rotating sectored disk method is described in texts on free-radical chemistry[7] and depends on the fact that radical chain reactions have rates proportional to the half-power of the initiator radical concentration. A rapidly rotating disk having a sector cut out to let flashes of light through effectively yields a low concentration of radicals continuously. A slowly rotating sectored disk yields intermittent higher radical concentrations. The total number of radicals formed in each case is the same, but the total amount of reaction is slower in the second case because the higher intermittent concentration of radicals is less effective at promoting chain reactions. There is an intermediate rate of disk rotation where the time between light flashes allows partial but not complete disappearance of the radicals by coupling, and with the aid of the proper mathematics this transition between the lower rate at slow rotation and the higher rate at very rapid rotation can be used to calculate the absolute rate of radical recombination (chain termination).

Silyl radicals are pyramidal and undergo inversion relatively slowly. Sakurai, Murakami, and Kumada observed 57–85% retention in the reaction of methylphenyl-α-naphthylsilane (**21**) with carbon tetrachloride, which was initiated with benzoyl peroxide.[71] Brook and Duff found somewhat lower degrees of retention when the same silyl radical was generated by photolysis of acetylmethylphenyl-α-naphthylsilane at $-40°C$.[72]

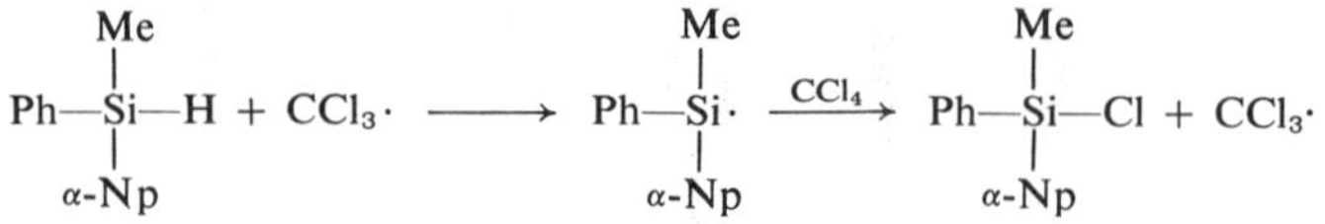

$$\text{Ph—Si(Me)(}\alpha\text{-Np)—H} + \text{CCl}_3\cdot \longrightarrow \text{Ph—Si(Me)(}\alpha\text{-Np)}\cdot \xrightarrow{\text{CCl}_4} \text{Ph—Si(Me)(}\alpha\text{-Np)—Cl} + \text{CCl}_3\cdot$$

21

For addition of $Me_3Si—SiMe_2\cdot$ to a variety of substituted benzenes, the Hammett ρ value is $+1.4$, indicating that this silyl radical is more nucleophilic than aryl or cyclohexyl radicals.[73]

The addition of organotin hydrides, R_3SnH, to terminal olefins, $CH_2{=}CHR'$, to form tetraalkyltins, $R_3SnCH_2CH_2R'$, was reported by van der Kerk, Luijten, and Noltes.[74] Kuivila and Sommer have shown that the addition of the trialkylstannyl radical to the olefin is reversible.[75] Treatment of either *trans*-1-deutero-1-hexene (**22**) or its *cis* isomer (**23**) with less than an equimolar amount of trimethyltin hydride or tributyltin hydride leads to recovered 1-deutero-1-hexene which is an equilibrium mixture of *cis* and

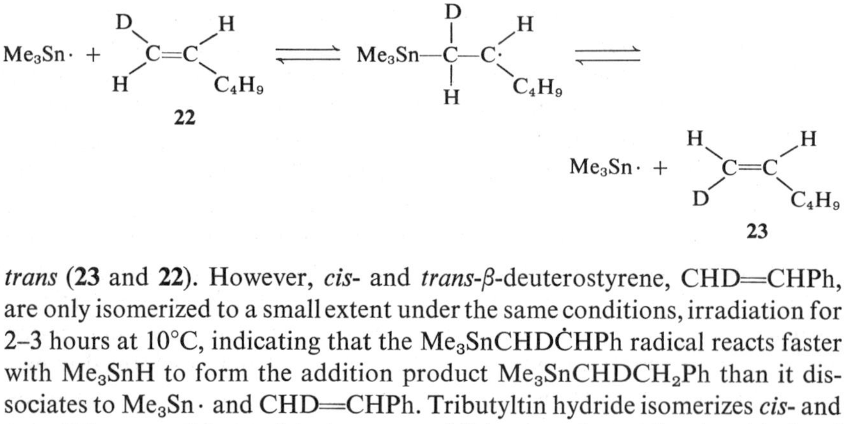

$$Me_3Sn\cdot + \text{(D)(H)C=C(H)(C}_4\text{H}_9\text{)} \; (\mathbf{22}) \rightleftharpoons Me_3Sn\text{—C(D)(H)—}\dot{\text{C}}\text{(H)(C}_4\text{H}_9\text{)} \rightleftharpoons Me_3Sn\cdot + \text{(H)(D)C=C(H)(C}_4\text{H}_9\text{)} \; (\mathbf{23})$$

trans (**23** and **22**). However, *cis*- and *trans*-β-deuterostyrene, $CHD{=}CHPh$, are only isomerized to a small extent under the same conditions, irradiation for 2–3 hours at 10°C, indicating that the $Me_3SnCHD\dot{C}HPh$ radical reacts faster with Me_3SnH to form the addition product $Me_3SnCHDCH_2Ph$ than it dissociates to $Me_3Sn\cdot$ and $CHD{=}CHPh$. Tributyltin hydride isomerizes *cis*- and *trans*-2-butene without forming any addition product. The less-hindered trimethyltin hydride does add to 2-butene to form $CH_3CH_2CH(SnMe_3)CH_3$.

The addition of $Bu_3Sn\cdot$, generated by photolysis of $(Bu_3Sn)_2$ or from the reaction of *t*-BuO· with Bu_3SnH, to butadiene gives the *trans* allylic radical **24**.[76] The esr splittings in gauss are indicated for each hydrogen atom in the

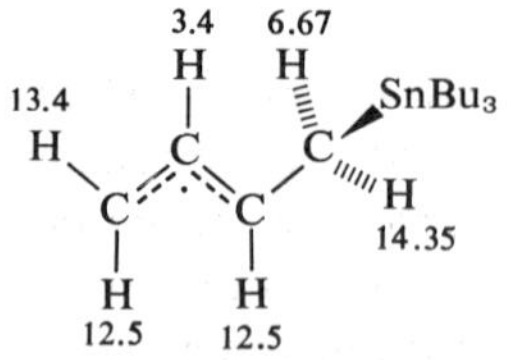

$$\text{H(13.4)(H 12.5)C}\cdots\text{C(H 3.4)}\cdots\text{C(H 12.5)—C(H 6.67)(H 14.35)SnBu}_3$$

24

(Numbers are esr splittings in gauss due to each H at −148°C.)

structure. The esr studies indicated that there is a 5 kcal/mole barrier to rotating the tin–carbon bond out of the most favorable position for hyperconjugation with the allylic radical π system.

Even though the *trans* allylic radical **24** is favored at low temperatures, the addition of trimethyltin hydride to butadiene in the presence of an azonitrile initiator yields more *cis*-1-trimethylstannyl-2-butene (**25**) than *trans* isomer (**26**).[77] There is no entirely satisfactory explanation of this discrepancy

$$Me_3SnH + CH_2{=}CH{-}CH{=}CH_2 \xrightarrow{Me_2C(CN)N{=}NC(CN)Me_2}$$

$$\underset{\substack{(55\%)\\ \mathbf{25}}}{(H)(Me_3SnCH_2)C{=}C(H)(CH_3)} + \underset{\substack{(38\%)\\ \mathbf{26}}}{(H)(Me_3SnCH_2)C{=}C(CH_3)(H)} + \underset{(7\%)}{Me_3SnCH_2CH_2CH{=}CH_2} \quad \text{(total yield 60\%)}$$

between the low-temperature esr data and the synthetic results at present. Fish, Kuivila, and Tyminski pointed out that on the basis of the data with styrene the addition of $Me_3Sn\cdot$ to butadiene should not be reversible within the time allowed before reaction with Me_3SnH, and they therefore favored intervention of a cyclic pentavalent tin radical intermediate.[77] However, this seems untenable in view of the esr results of Kawamura and Kochi.[76] Perhaps this addition is reversible, as could be the case if the allylic radical is relatively slow at abstracting hydrogen from trimethyltin hydride, and the product ratio could arise in the competing rate-determining steps where the equilibrating *cis*- and *trans*-allylic radicals abstract H from Me_3SnH.

The mechanism of reduction of alkyl halides by trialkyltin hydrides has been studied in considerable detail by Carlsson and Ingold.[78,79] Initiation was by photolysis of an azonitrile, and the rate of reduction was found to be half-order in initiator at constant light intensity or half-order in light intensity at constant initiator, corresponding to bimolecular chain termination. By the use of the rotating sectored disk method, rate constants for the termination steps

$$2R_3Sn \longrightarrow (R_3Sn)_2 \qquad 2R\cdot \longrightarrow R_2 \qquad R_3Sn\cdot + R\cdot \longrightarrow R_4Sn$$

were all found to be in the range 1.4×10^9 to 3.6×10^9 liters/mole second, which is below the theoretical limit of 8×10^9 for a diffusion-controlled rate in cyclohexane at 25°C. The chain-propagation steps, as usual, have considerably smaller rate constants. For example, k for the reaction of *t*-Bu· with Bu_3SnH to form *t*-BuH and $Bu_3Sn\cdot$ is 3×10^5 liters/mole second, and k for the reaction of $Bu_3Sn\cdot$ with *t*-BuCl to form Bu_3SnCl and *t*-Bu· is 1.5×10^4. The latter step is rate-controlling if the *t*-BuCl concentration is equal to or less than the Bu_3SnH, and the reaction is consequently first order in t-BuCl and zero order in Bu_3SnH, with the chain-termination step being mainly recombination of 2-$Bu_3Sn\cdot$ since these are present in the higher concentration and all the chain-termination rate constants are similar. At high enough t-BuCl concentrations, the other step becomes rate-determining, the rate law shifts to zero order in t-BuCl and first order in Bu_3SnH, and the principal chain-termination step becomes recombination or disproportionation of 2 t-Bu·. Several other cases of alkyl halides and trialkyltin or triphenyltin hydride were studied, and in some the first chain-propagation step had the higher k, in others the second step did. These mechanisms may be summarized by Eqs. (7-54) through (7-57).

$$R\cdot + R'_3SnH \longrightarrow R{-}H + R'_3Sn\cdot \qquad \text{(propagation)} \qquad (7\text{-}54)$$

$$R'_3Sn\cdot + R{-}Cl \longrightarrow R'_3SnCl + R\cdot \qquad \text{(propagation)} \qquad (7\text{-}55)$$

$$2R\cdot \longrightarrow R{-}R \text{ or (alkene + alkane)} \qquad \text{(termination)} \qquad (7\text{-}56)$$

$$2R'_3Sn\cdot \longrightarrow R'_3Sn{-}SnR'_3 \qquad \text{(termination)} \qquad (7\text{-}57)$$

In spite of the free-radical nature of the reductions with trialkyltin hydrides, reports have appeared that the reduction products from cyclopropyl halides are not entirely racemized, though no self-consistent pattern emerges. Triphenyltin hydride and optically active 1-bromo-1-methyl-2,2-diphenylcyclopropane apparently yielded 1% net inversion in the product, 1-methyl-2,2-diphenylcyclopropane.[80] Reduction of both diastereoisomers of 1-chloro-1-fluoronorcarane with tributyltin hydride has been reported to yield the corresponding 1-fluoronorcarane with retention of configuration.[81]

VI. Neighboring Metal Atoms in Carbon Radicals

A. α-Metalloalkyl Radicals

α-Metalloalkyl radicals can be formed either by the abstraction of a hydrogen atom or other labile group from the carbon atom adjacent to the metal or by the addition of a radical to the double bond of a vinylmetal compound.

Krusic and Kochi generated the radicals $(CH_3)_3MCH_2\cdot$ and $(CH_3CH_2)_3M\dot{C}HCH_3$, where M is Si, Ge, or Sn, by reaction of the tetraalkylmetal compound with t-butoxy radicals at $-77°$ to $-126°C$.[82] The neopentyl radical could not be generated by this method but was obtained from photolysis of di-t-butylacetyl peroxide for comparison with $(CH_3)_3MCH_2\cdot$. The esr coupling constants for the α-hydrogens of $(CH_3)_2MCH_2\cdot$ were 20.84–20.88 gauss, and for $(CH_3)_3CCH_2\cdot$, 21.81 gauss. The coupling constants for the CH_3 groups through the central metal atom M were 0.0–0.41 gauss, and through the central carbon atom of the neopentyl radical, 1.00 gauss. The greater ease of formation of $(CH_3)_3MCH_2\cdot$ compared to $(CH_3)_3CCH_2\cdot$ suggests stabilization of the former by $p\pi$–$d\pi$ bonding.

Rates of abstraction of chlorine atoms from α-chloro silicon compounds by sodium atoms in the gas phase at 520°K indicate stabilization of the adjacent radical site by silicon.[83] For Me_3CCH_2Cl, k is 1.06×10^8 liters/mole second; for Me_3SiCH_2Cl, k is 16.2×10^8 liters/mole second; for $(Me_3Si)_2CHCl$, k is 230×10^8 liters/mole second.

The triphenylsilylmethyl radical, $Ph_3SiCH_2\cdot$, has been generated from t-BuO· and Ph_3SiCH_2CHO at 150°C and been found to yield Ph_3SiCH_3 without rearrangement.[84] This contrasts with the complete rearrangement of the analogous carbon radical, $Ph_3CCH_2\cdot$, to $Ph_2\dot{C}CH_2Ph$ under similar conditions. Rearrangement of an α-silyl radical (**27**) has been observed in a case where the silicon migrates from silicon to sulfur, which, of course, oxidizes the migrating silicon.[85] The reaction is initiated by α,α'-azobis-

$$\underset{(\text{“}R_3SiSH\text{”})}{MeSi(SiMe_3)_2SH} \xrightarrow{R_3Si\cdot} \underset{\mathbf{27}}{MeSi(SiMe_3)_2S\cdot} \longrightarrow \underset{(\text{“}R_3Si\cdot\text{”})}{Me\dot{S}i(SiMe_3)S{-}SiMe_3} \xrightarrow{R_3SiSH} MeSiH(SiMe_3)S{-}SiMe_3$$

(isobutyronitrile) in cyclohexane at 150°C and is not affected by bases such as methyllithium or acids such as aluminum chloride.

Whenever it is possible to convert an M—CH_3 group to an M—$CH_2\cdot$ radical by reaction with an oxidizing radical, it is possible to convert M—$CH_2\cdot$ to M—CH_2X, where X is halogen, by suitable choice of halogenating agent. The following examples are all selected from boron chemistry and are not exhaustive.

Trimethylborane is chlorinated by chlorine in the gas phase at −95°C to form Me_2BCH_2Cl, probably by a free-radical process.[86] Lane and Brown found that bromination of trialkylboranes in methylene chloride solution leads to α-bromoalkylboron compounds by a free-radical mechanism, presumably with the chain-propagation steps shown in Eqs. (7-58) and (7-59). These α-bromoalkylboranes have not been isolated as such, but cleave with HBr as shown in Eq. (7-60) if no other reagent is added,[87] or rearrange as shown in Eq. (7-61) (see Chapter 4, Section III,B) if a Lewis base such as water is introduced.[88]

$$Br\cdot + (RCH_2)_3B \longrightarrow HBr + (RCH_2)_2B\dot{C}HR \tag{7-58}$$

$$(RCH_2)_2B\dot{C}HR + Br_2 \longrightarrow (RCH_2)_2BCH(Br)R + Br\cdot \tag{7-59}$$

$$(RCH_2)_2BCH(Br)R + HBr \longrightarrow (RCH_2)_2BBr + BrCH_2R \tag{7-60}$$

$$(RCH_2)_2BCH(Br)R + H_2O \longrightarrow RCH_2B(OH){-}CH(CH_2R)R + H^+ + Br^- \tag{7-61}$$

The gas-phase bromination of triethylborane also results in abstraction of the α-hydrogens by bromine atoms.[89] Minor amounts of cleavage to Et_2BBr and Et· occurred. Lissi *et al.* estimated the stabilization energy of the radical $Et_2B\dot{C}HCH_3$ to be about 14.5 kcal/mole compared to $CH_3\dot{C}HCH_3$ on the basis of relative rates of hydrogen atom abstraction by bromine.[89]

The attack of methyl radicals on triethylborane is closely balanced between hydrogen abstraction to form $Et_2B\dot{C}HCH_3$ and CH_4 and displacement of an ethyl group to form Et_2BMe and Et· at 27°C.[90] The rate constant for the hydrogen abstraction is 6.2×10^3 liters/mole second and that for the ethyl radical displacement is about the same at 27°C, but the activation energy of the former is 7.8 kcal/mole and for the latter is 5.2 kcal/mole, making the attack at hydrogen predominate at higher temperatures. Statistically corrected for the number of hydrogen atoms, hydrogen abstraction from Et_3B is 60 times faster than from toluene, suggesting that the α-boron atom greatly stabilizes the adjacent radical.

In spite of the evidence that an α-boron atom can stabilize a radical, *t*-BuO· abstracts a *B*-methyl hydrogen from $CH_3B(O\text{-}t\text{-}Bu)_2$ only with relative difficulty.[22] The estimated rate constant ratio for *B*-methyl/*C*-methyl attack is about 1.1–1.5 in this case. Chlorination with *t*-butyl hypochlorite gave only low yields of $ClCH_2B(OR)_2$, with much accompanying chlorination of the *t*-butyl groups.

The difference does not appear to be entirely between a trialkylborane and a boronic ester, though the —$B(OR)_2$ substituent is undoubtedly less stabilizing than —BR_2. Evidence for radical stabilization was found by Matteson when carbon tetrachloride was added to dibutyl vinylboronate (**28**).[91] The chain-transfer constant k_1/k_2 for the alternate reaction pathways of the

$$Cl_3C\cdot + \underset{\mathbf{28}}{CH_2{=}CH{-}B(OBu)_2} \longrightarrow$$

$$\underset{\mathbf{29}}{Cl_3CCH_2{-}\dot{C}H{-}B(OBu)_2} \xrightarrow[k_1]{CCl_4} Cl_3CCH_2{-}\underset{\underset{Cl}{|}}{CH}{-}B(OBu)_2 + Cl_3C\cdot$$

$$\underset{\mathbf{29}}{Cl_3CCH_2{-}\dot{C}H{-}B(OBu)_2} \xrightarrow[k_2]{CH_2{=}CH{-}B(OBu)_2} Cl_3CCH_2{-}\underset{\underset{CH_2{-}\dot{C}H{-}B(OBu)_2}{|}}{CH}{-}B(OBu)_2$$

intermediate radical **29** is about 3×10^{-3}, a low value characteristic of stabilized radicals in additions to double bonds. The stabilized radical **29** is reluctant to react with carbon tetrachloride and reacts preferentially with a second molecule of vinylboronic ester (**28**) to form another stabilized radical. The corresponding transfer constant for propylene is about 1 and for styrene, which yields a highly stabilized benzylic radical, it is 6×10^{-4}. It may be concluded that the vacant *p* orbital of the neighboring boron atom effectively delocalizes the radical, as represented by resonance structure **30**. Simple

Hückel calculations may also be used to describe this delocalization.[91]

$$\underset{H}{\overset{R}{>}}\dot{C}{-}B(OBu)_2 \longleftrightarrow \underset{H}{\overset{R}{>}}\overset{+}{C}{-}\overset{-}{\ddot{B}}(OBu)_2$$

30

Reasonable parameters for the boron atom suggest substantial delocalization energy and a substantial carbon–boron π-bond order, which is not evident on casual inspection of the resonance structures.

The failure of the boron atom to promote the attack of *t*-BuO· on the *B*-methyl hydrogens of $CH_3B(O\text{-}t\text{-}Bu)_2$ may be the result of little C—H bond breaking in the transition state for this exothermic process. Also, in order to delocalize the unpaired electron the boron atom must tend to withdraw it, contrary to the electron-seeking tendency of the attacking *t*-butoxy radical. Otherwise, all observations on α-boron-substituted radicals indicate substantial stabilization by carbon–boron π-bonding.

B. α-Metalloalkyl Radicals

Krusic and Kochi have generated Group IV β-metalloalkyl radicals, $R_3M{-}CH_2\dot{C}H_2$ or $R_3M{-}CH_2\dot{C}Me_2$, by the addition of $R_3M\cdot$ to the double bond of ethylene or isobutylene at $-100°C$.[92] The esr spectra indicate that rotation around the carbon–carbon bond is restricted, but the radical is definitely the open-chain structure **31** and not the cyclic structure **32**. The coupling constant of the electron to the β-protons is strongly temperature-

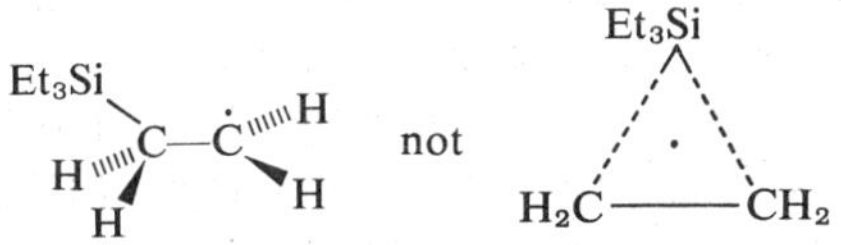

$Et_3Si{-}CH_2{-}\dot{C}H_2$ (**31**) not cyclic Et_3Si-bridged $H_2C{-}CH_2$ radical (**32**)

dependent. The behavior is accounted for by a twofold potential model in which the Group IV atom is aligned preferentially next to a lobe of the radical *p* orbital so that hyperconjugation is effective (see Chapter 5, Section II). The rotational barrier heights are only about 1–2 kcal/mole, but this seems enough rotational restriction to control stereochemistry in some rapid radical reactions.

β-Mercuri radicals are evidently unstable and (if formed at all) undergo elimination of the mercury to form olefins. (Compare the reversibility of the addition of $R_3Sn\cdot$ to olefins, Section V.) Jensen and Guard found a complex mixture of products from the reaction of dibutylmercury (**33**) with carbon tetrachloride in the presence of benzoyl peroxide.[93] The scheme outlined is a reasonable explanation of the results. The yields are given in mole percent based on Bu_2Hg.

$$Cl_3C\cdot + (C_4H_9)_2Hg \longrightarrow CHCl_3 + CH_3CH_2\dot{C}HCH_2HgC_4H_9 \longrightarrow$$

33 (64%) (lifetime short or nonexistent)

$$CH_3CH_2CH{=}CH_2 + C_4H_9Hg\cdot \xrightarrow{CCl_4} C_4H_9HgCl + Cl_3C\cdot \quad (25\%)$$

$$CH_3CH_2CH{=}CH_2 \xrightarrow{Cl_3C\cdot} CH_3CH_2\dot{C}HCH_2CCl_3 \xrightarrow{CCl_4} CH_3CH_2CH(Cl)CH_2CCl_3 + Cl_3C\cdot \quad (52\%)$$

$$C_4H_9Hg\cdot \longrightarrow Hg + C_4H_9\cdot \quad (63\%) \xrightarrow{CCl_4} \dot{C}_4H_9Cl + Cl_3C\cdot \quad (51\%)$$

These results of Jensen and Guard[93] correct earlier Russian claims for the isolation of $C_4H_9CCl_3$ from this reaction.[94] It is, of course, always a difficult problem to know when to challenge experimental data and when to believe strange results, but this was just a bit too strange. In order to account for the formation of $C_4H_9CCl_3$ one would have to postulate either wholesale coupling of $C_4H_9\cdot$ and $Cl_3C\cdot$, impossible because these reactive radicals are present at only very low concentrations and would do other things first, or displacement of RHg· from carbon by $Cl_3C\cdot$, which violates the rule that radical displacements of tetrahedral carbon do not take place. (See Section II for the rules of radical chemistry.) There is no reasonable way to arrive at $C_4H_9CCl_3$ as a major product. However, in the aromatic series the rules do not forbid displacement of PhHg· from phenyl carbon by $Cl_3C\cdot$, and Russian claims to have isolated $PhCCl_3$ (25% yield) as well as PhHgCl (31% yield) from the reaction of Ph_2Hg with CCl_4[94,95] were substantiated by Jensen and Guard.[93]

VII. Photochemistry

A. Photolysis to Radicals

Absorption of a quantum of ultraviolet light generally raises a molecule directly to an excited singlet state, that is, all the electrons remain paired. The unpairing of two electrons to form an excited triplet (biradical) state is a somewhat slower process involving changes in the positions of atomic nuclei and the loss of some energy by thermal collisions, the triplet state lying lower in energy than the singlet. Formation of the triplet state is rapid on the chemical time scale, and most photochemical reactions are reactions of triplet states. It is accordingly not surprising that light absorption often initiates cleavage of molecules to free radicals. Further details of the light absorption process may be found in Ramsey's earlier book in this series.[96]

Radicals may be generated either by direct cleavage of the excited triplet molecules or by secondary reactions of the excited triplet with another molecule. An example of the latter is the reaction of the triplet from irradiation of acetone (**34**) with trialkylboranes, which liberates alkyl radicals.[97,98]

$$CH_3\overset{O}{\overset{\|}{C}}CH_3 \xrightarrow{3130\ \text{Å}} CH_3-\overset{\dot{O}}{\overset{|}{\dot{C}}}-CH_3 \ (\mathbf{34}) \xrightarrow{R_3B} R\cdot + (CH_3)_2\dot{C}-O-BR_2$$

$$\mathbf{34} \xrightarrow{\text{no borane}} CH_3\cdot + CH_3-\dot{C}=O$$

Encina and Lissi estimated the rate constant for reaction of the acetone triplet (**34**) with triethylborane to be 1×10^8 liters/mole second at 20°C, 3.4×10^8 liters/mole second at 126°C.[97] The low temperature coefficient is consistent with the expectation of a low activation energy for such a rapid process. Davies and co-workers noted that the yields of Bu· and $Me_2\dot{C}$—O—

$[C_6H_5^*]_5Sb$ (**35**) $\xrightarrow{uv}$ $[C_6H_5^*]_4Sb\cdot$ + $C_6H_5^*\cdot$ $\longrightarrow$ $[C_6H_5^*]_3Sb$ + $C_6H_5^*\cdot$

C_6H_6 + $C_6H_5^*\cdot$ → phenylcyclohexadienyl radical (H) $\xrightarrow{\text{dimerizes}}$ dimer (H, H, H, H)

phenylcyclohexadienyl radical —transfers H→ biphenyl (*) —[O]→ C_6H_5—*—CO_2H + C_6H_5—C^*O_2H

dimer —$C_6H_5^*\cdot$→ quaterphenyl (*, *) + $C_6H_6^*$

BBu_2 radicals from BBu_3 were equal within the accuracy of the esr method.[98] Other ketones gave results similar to those from acetone.

Photolysis of organic compounds to pairs of free radicals (rather than biradicals) and subsequent reaction of these radicals with organometallic substrates has already been covered in Sections III and V.

Direct photolytic cleavage of carbon–metal bonds has been observed in many instances, a few of which were mentioned in Section I. An example which has been studied in some detail is the photolysis of pentaphenylantimony (**35**) in benzene.[99] Shen, McEwen, and Wolf found that the major products are triphenylantimony, biphenyl, and quaterphenyl. By the use of pentaphenylantimony (**35**) labeled with ^{14}C in the 1-position it was shown that the biphenyl incorporates solvent benzene as well as the labeled Ph_5Sb and that the label remains in the 1-position in the biphenyl.

In contrast to the photolytic mechanism, thermolysis of pentaphenylantimony at 220°C in benzene yields biphenyl and triphenylantimony by an intramolecular mechanism, keeping the label as concentrated in the product biphenyl as in the starting pentaphenylantimony.

Phenyl-*N*-*t*-butylnitrone (**36**) has been used by Janzen and Blackburn to trap free radicals generated by photolysis of organo lead, tin, and mercury compounds.[100] The resulting nitroxyl radicals can then be characterized by esr. For example, photolysis of tetraphenyllead yields phenyl radicals. However, the method failed to detect any $Ph_3Pb\cdot$ radicals. This

$$Ph_4Pb \xrightarrow{uv} \underset{\text{Not detected}}{Ph_3Pb\cdot} + Ph\cdot$$

$$Ph\cdot + \underset{\mathbf{36}}{PhCH{=}\overset{O^-}{\overset{|}{N}}{}^{+}{-}t\text{-}Bu} \longrightarrow Ph_2CH{-}\overset{O\cdot}{\overset{|}{N}}{-}t\text{-}Bu$$

study covered a variety of compounds, including, for example, Et_4Pb, Bu_4Pb, $Me_3PbOCOCH_3$, various R_2PbX_2 and $RPbX_3$, $Ph_3PbPbPh_3$, a similar range of organotin structures, and several organomercury compounds. Although $Ph_3PbPbPh_3$ irradiated in benzene yielded $Ph\cdot$ much faster than did Ph_4Pb, it was still not possible to detect any $Ph_3Pb\cdot$. Likewise, it was not possible to detect $R_3Sn\cdot$. Evidently these react with other substances in preference to the nitrone, or perhaps attack the nitrone at oxygen or nitrogen in preference to carbon and eliminate a radical so that no nitrone radical adduct results. Since $R_3Sn\cdot$ radicals are easily generated in other ways (see Section V and the following paragraph), it is hard to believe they are absent in photolyses of R_4Sn and R_3SnSnR_3.

Kuivila and co-workers have obtained good evidence for the formation of trimethylstannyl radicals, $Me_3Sn\cdot$, in the photolysis of 4-trimethylstannyl-2-butanone (**37**) and its homolog 5-trimethylstannyl-2-pentanone at 280 nm,

$$Me_3SnCH_2CH_2C(=O)CH_3 \ (\mathbf{37}) \xrightarrow{uv} \begin{cases} \xrightarrow{CCl_4} Me_3SnCl \\ \xrightarrow{\text{cyclohexane}} CH_4 + (Me_2Sn)_x \\ \xrightarrow{EtSH} Me_3SnSEt + CH_3CH_2C(=O)CH_3 \end{cases}$$

where the excitation is the $n \rightarrow \pi^*$ transition of the ketone group.[101] The solvents tested were cyclohexane, carbon tetrachloride, and ethyl mercaptan. The product Me_3SnCl in carbon tetrachloride probably arises from attack of $Me_3Sn\cdot$ on CCl_4, and the CH_4 and $(Me_2Sn)_x$ formed in cyclohexane presumably arise from decomposition of $Me_3Sn\cdot$ to $Me\cdot$ and Me_2Sn. The 2-butanone formed in ethyl mercaptan is evidence for the radical $CH_3COCH_2CH_2\cdot$, and Me_3SnSEt might arise either from $Me_3Sn\cdot$ or from attack of $EtS\cdot$ on $Me_3SnCH_2CH_2COCH_3$ (**37**). Tetramethyltin does not react under these photolysis conditions, and the analogous silicon compounds are not cleaved by irradiation at 280 nm.[102]

Photolytic boron–boron bond cleavage of 1,2-diphenyl-1,2-bis(dimethylamino)diborane(4) (**38**) has been observed by Hancock and Uriarte.[103] The photolysis was carried out in carbon tetrachloride with wavelengths of 300 nm and longer, where 99.9% of the light absorption is by the boron compound (**38**). Under conditions where carbon tetrachloride does not add to 1-octene, the boron compound (**38**) acts as a photosensitizer, and the $Cl_3C\cdot$ radicals attack the 1-octene in preference to the boron compound (**38**). Probable mechanistic steps in the photolysis of the boron–boron-bonded compound (**38**) in carbon tetrachloride are outlined in Eqs. (7-62) through (7-66). Since the ratio of products (**39**):(**40**):(C_2Cl_6) is about 4:2:1, a substantial amount of radical coupling must take place, this being the only route to C_2Cl_6. It may be concluded that the second chain-propagation step, Eq. (7-64), is relatively slow and might even be unimportant, but it was also shown that trichloromethyl radicals generated thermally from carbon tetrachloride and benzoyl peroxide lead to substantial amounts of the *B*-trichloromethyl compound (**40**). It therefore seems likely that both pathways [Eqs. (7-64) and (7-65)] contribute to the formation of the *B*-trichloromethyl compound (**40**).

$$(Me_2N)(Ph)B{-}B(Ph)(NMe_2) \ (\mathbf{38}) \xrightarrow{uv} 2\ (Me_2N)(Ph)B\cdot \qquad \text{(initiation)} \quad (7\text{-}62)$$

$$(Me_2N)(Ph)B\cdot + CCl_4 \longrightarrow (Me_2N)(Ph)B{-}Cl \ (\mathbf{39}) + Cl_3C\cdot \qquad \text{(propagation)} \quad (7\text{-}63)$$

$$Cl_3C\cdot + (Me_2N)(Ph)B{-}B(Ph)(NMe_2)\ (\mathbf{38}) \longrightarrow (Me_2N)(Ph)B{-}CCl_3\ (\mathbf{40}) + (Me_2N)(Ph)B\cdot \quad \text{(propagation)} \quad (7\text{-}64)$$

$$(Me_2N)(Ph)B\cdot + Cl_3C\cdot \longrightarrow (Me_2N)(Ph)B{-}CCl_3\ (\mathbf{40}) \quad \text{(termination)} \quad (7\text{-}65)$$

$$2Cl_3C\cdot \longrightarrow Cl_3C{-}CCl_3 \quad \text{(termination)} \quad (7\text{-}66)$$

Ethyl mercurybis(diazoacetate) (**41**) is set up for two kinds of photochemical cleavage. The reagent is prepared from ethyl diazoacetate and mercuric oxide. Photolysis with short-wavelength ultraviolet yields N_2, Hg, and carbethoxycarbyne, $EtO_2C{-}C$.[104] The esr spectrum of $EtO_2C{-}C$ was observed in Fluorolube glass at $-196°C$ as a single peak at $g = 2.001$, indicating one unpaired electron, and when the photolysis was carried out in cyclohexene, products of addition to the double bond and insertion in the allylic C—H bond were observed. With longer-wavelengths passed by a Pyrex filter (>280 nm), Strausz and co-workers found that the carbon–nitrogen bond was cleaved and the carbon–mercury bond left intact.[105]

$$Hg(C(=N{=}N){-}CO_2Et)_2\ (\mathbf{41}) \xrightarrow{\text{short-wavelength uv}} N_2 + Hg + :\dot{C}{-}CO_2Et$$

$$Hg(C(=N{=}N){-}CO_2Et)_2\ (\mathbf{41}) \xrightarrow{\text{longer-wavelength uv}} N_2 + (:C{-}CO_2Et){-}Hg{-}(C(=N{=}N){-}CO_2Et)$$

Flash photolysis of trimethylantimony produces several excited states of Sb atoms as well as Sb_2 and $CH_3\cdot$, which were detected spectroscopically.[106] Trimethylbismuth behaves in a similar manner.

Photolysis of arylthallium bis trifluoroacetates in benzene yields aryl radicals, which attack the solvent to yield 80–90% of substituted biphenyls.[107]

These examples certainly do not cover all the known photolytic cleavages of carbon–metal bonds to produce radicals, but should be enough to illustrate the types of results obtained. It is, of course, particularly difficult to prove detailed mechanisms in photochemical reactions, and the difficulties are reflected in the tentative nature of many of the conclusions cited.

B. Intramolecular Photochemical Rearrangements

Molecules in photoexcited states often rearrange if the excitation energy is insufficient to cause fragmentation. The singlet state initially formed on absorption of a quantum of light is generally thought to lose energy thermally to form a more stable, longer-lived triplet before rearrangement occurs. However, the gross electron distributions of the singlet and triplet are similar, and the differences are not revealed at all by simple molecular orbital approaches. It is also a difficult experimental problem to establish whether any given reaction involves a triplet or a singlet, and in the small number of organometallic examples to be cited this is an undecidable point. Organometallic chemists seriously interested in photochemistry should seek the theoretical details elsewhere.[96,108]

A simple example to begin with is the photochemical rearrangement of phenyl triphenylsilyl ketone (**42**) reported by Brook and Duff.[109] This rearrangement is initiated by the $n \rightarrow \pi^*$ transition of the carbonyl group, the characteristic low-intensity absorption at about 280 nm or longer wavelength. The $n \rightarrow \pi^*$ transition transfers electron density out of a molecular orbital centered mainly on the oxygen atom of the carbonyl group and into an antibonding π orbital centered mainly on both the carbon and oxygen atoms of the carbonyl π orbital centered mainly on both the carbon and oxygen atoms of the carbonyl group, with the highest electron density on the carbon. A simplified orbital picture of this $n \rightarrow \pi^*$ transition is shown in Fig. 7-8. (It is not correct to say that a particular electron leaves a particular atomic orbital to go to a particular antibonding orbital restricted to two atoms, but to a first approximation the total molecular wave function changes as if this were the case.) The net result of this $n \rightarrow \pi^*$ transition is to make the carbonyl oxygen atom electron-deficient, and the migration of the triphenylsilyl group in **42** is a typical migration to an electron-deficient atom (see Chapter 4).

$$\underset{\mathbf{42}}{Ph{-}C({=}O){-}SiPh_3} \xrightarrow{uv} Ph{-}\dot{C}^{-}({\vdots}\dot{O}^{+}){-}SiPh_3 \longrightarrow Ph{-}C{:}^{-}({=}O^{+}{-}SiPh_3)$$

$$\xrightarrow{MeOH,\ pyridine} Ph{-}CH(OSiPh_3)(OMe) \qquad \xrightarrow{uv,\ MeOH + H^+} PhCH(OMe)_2 + PhCHO + Ph_3SiOMe + Ph_3SiOH$$

Evidence that the migration of the silyl group is mostly intramolecular is provided by the rearrangement of optically active phenyl methylphenyl-α-naphthylsilyl ketone, which proceeds with 90% retention in the presence of pyridine. In the absence of pyridine, alcoholysis of the initial product analogous to that shown for the triphenylsilyl ketone **42** occurs and the net retention is only 78%. This alcoholysis is promoted by light and is slow in its absence. In view of the varying degrees of retention found under the different photolysis conditions and the secondary reactions which can occur, it seems likely

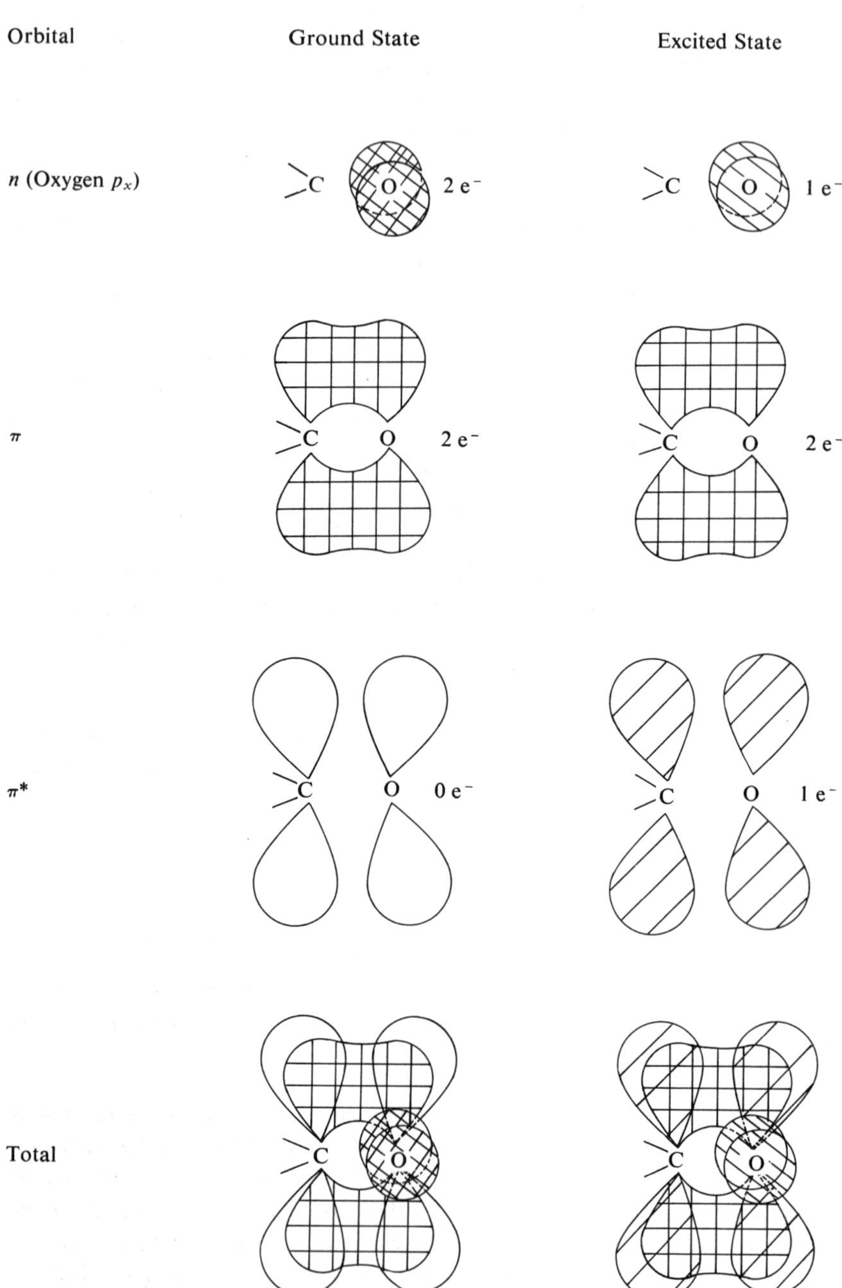

Fig. 7-8. Orbital picture of the $n \rightarrow \pi^*$ transition of the carbonyl group.

that the primary rearrangement step is entirely intramolecular with 100% retention, but this would be difficult to prove.

Several rearrangements of phenylmetal compounds have been reported. These probably involve the $\pi \rightarrow \pi^*$ transition of the benzene ring, which promotes an electron from one of the degenerate pair of bonding orbitals **43** or **44** to one of the degenerate pair of antibonding orbitals **45** or **46**. The + and − signs here are used to designate orbital symmetry (in effect, whether the atomic *p* orbital functions as if it were right side up, +, or upside down, −) and have nothing to do with electric charge.

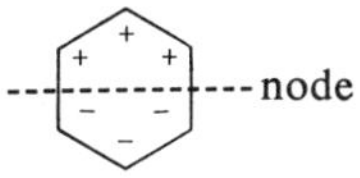

node node (bonding)

43 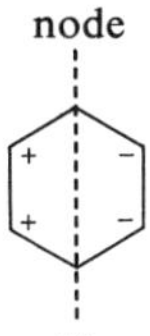 **44**

node node node node (antibonding)

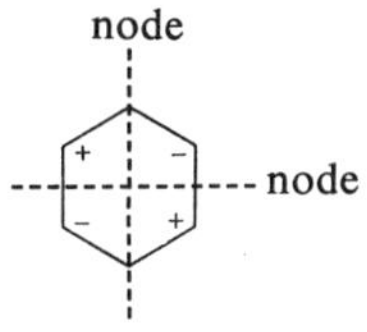

45 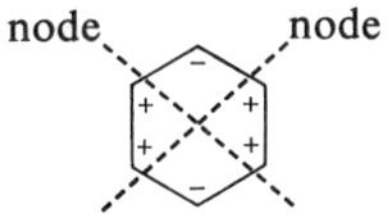 **46**

The excited state of the benzene ring does not have a closed-shell structure and may be regarded as an electron-deficient center. Migration of another group from a metal atom to the adjacent carbon of the benzene ring can be understood in these terms. However, there is not necessarily any significant change in electron density distribution between carbon atoms on photoexcitation. The Hückel electron densities are the same for bonding orbital **43** and antibonding orbital **46** and for orbitals **44** and **45**. Radicals are not normally prone to rearrange, but the photoexcitation energy will promote any reaction that can give relief.

Williams and co-workers have carried out extensive studies on the photochemical rearrangement of sodium tetraphenylboron (**47**) and related compounds.[110–112] Photolysis of aqueous solutions of $NaBPh_4$ (**47**) at 253.7 nm yields mostly 1-phenyl-1,4-cyclohexadiene (**48**) and the sodium salt of diphenylboronic acid (**49**) in the absence of oxygen. In the presence of oxygen, biphenyl was obtained instead of the cyclohexadiene (**48**).[110] A reasonable mechanism is outlined, with inclusion of some details that are not really known for purposes of clarity in presentation.

The main features of the mechanism outlined are supported by several additional kinds of evidence. When the reaction is carried out in D_2O under

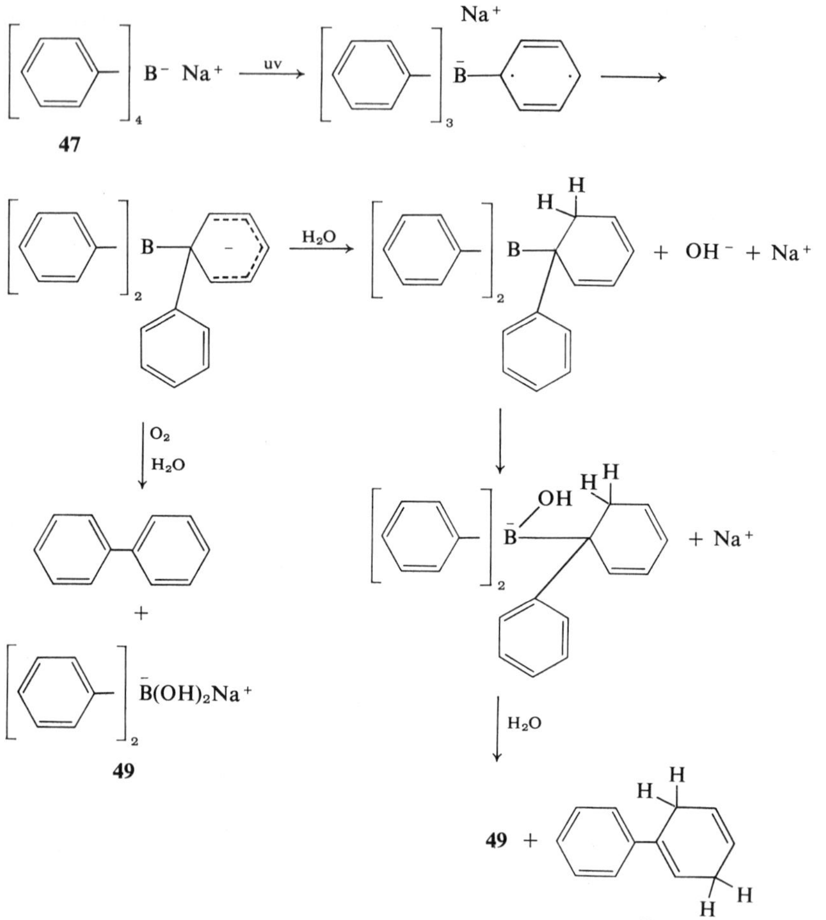

47

49

48
(and isomers)

nitrogen, the 1-phenyl-1,4-cyclohexadiene (**48**) has one deuterium each at the 3 and 6 positions (CHD groups in place of CH_2). Rearrangement of lithium tetra-*p*-tolylboron yields only bi-*p*-tolyl and its dihydro derivatives, no *meta*-isomers, which indicates a simple aryl group migration from boron to carbon, not attack of free aryl radicals on Ar_4B^- and not some more complex migration that would change the site of substitution on the aryl groups.[110] Similar results were obtained when the photolyses were carried out in various alcohol solvents.[111] Mixtures of tetraphenylboron and tetratolylboron anions yield only biphenyl and bitolyl derivatives, no tolylphenyl, showing that the reaction is intramolecular.

Irradiation of sodium tetraphenylboron in water containing oxygen and the dye Rose Bengal, which produces singlet oxygen, yields biphenyl.[112] In this case it was postulated that the singlet oxygen abstracts an electron from the tetraphenylboron anion to yield the radical cation **50**, which then undergoes a typical phenyl migration from boron to electron-deficient carbon.

$$Ph_4B^- + O_2^* \text{ (singlet)} \longrightarrow Ph_3\bar{B}\text{—}C_6H_5^{+} + O_2^{\dot{-}} \longrightarrow$$

50

$$Ph_2B(Ph)\text{—}C_6H_5^{\cdot} \longrightarrow Ph\text{—}Ph + Ph_2B\cdot$$

(?)

The photolysis of Al_2Ph_6 in benzene or toluene yields biphenyl, aluminum metal, and an organoaluminum hydride of undefined constitution.[113] In the presence of diphenylacetylene, the photolysis produces no aluminum metal, and the products after treatment with water were biphenyl, *cis*-stilbene, and 1,1,2-triphenylethylene in the ratio 2.7:1.0:2.5. Work-up with D_2O yielded a 60:40 mixture of dideutero and monodeutero *cis*-stilbene and about 36% of the biphenyl was monodeuterated. It is clear that this reaction is complex and any mechanisms written would be speculative. Eisch and Considine suggested the three-membered aluminum heterocycle **51** as the source of the *cis*-stilbene,[113] though in view of the presence of two aluminum atoms per Al_2Me_6 and the greater stability of an open-chain compound, it seems just as reasonable to suggest the open-chain alternative **52**. Perhaps the first step in the photolytic degradation is the formation of Ph—Ph and Ph_2Al—$AlPh_2$, which could add *cis* to an acetylene.

$$Al_2Ph_6 + Ph\text{—}C{\equiv}C\text{—}Ph \xrightarrow{uv} (Ph)C{=}C(Ph)\text{ bridged by }Al\text{—}Ph \text{ or } (Ph)(Ph_2Al)C{=}C(Ph)(AlPh_2) + Ph\text{—}Ph$$

51 (doubtful) **52** (possible)

More informative mechanistic data were obtained for the photolysis of lithium aryls in ether by van Tamelen *et al.*[114] The products are biaryls and lithium metal. 2-Naphthyllithium yields only bi-2-naphthyl, no 1-naphthyl coupling product, indicating that free radicals are not involved. Neither phenylsodium nor phenylmagnesium bromide yields biphenyl. In the presence of butyllithium, phenyllithium does not yield biphenyl or butylbenzene. These results indicate that it is phenyllithium dimer (**53**) which yields biphenyl. Phenylsodium and phenylmagnesium bromide are monomeric, and butyllithium will form mixed BuLi–PhLi tetramers with phenyllithium, thus removing the phenyllithium dimer (**53**) from the solution. Photolysis of the dimer (**53**) might yield an anion radical or might lead directly to biphenyl and solvated Li_2.

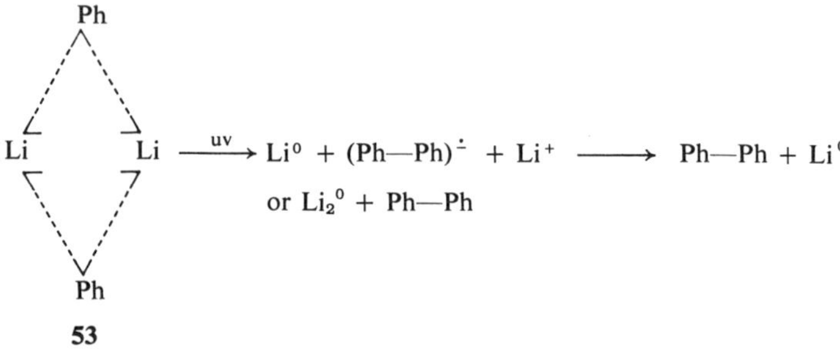

Glaze and co-workers have found that photolysis of triethylaluminum dimer, Al_2Et_6, at 253.7 nm in the presence of mercury metal as a photosensitizer yields ethane and $(Et_2Al)_2CHCH_3$, which was characterized by chlorination to CH_3CHCl_2.[115] This cleavage appears to be a direct reaction of the dimer, Al_2Et_6.

Photolysis of ethyllithium (tetramer or hexamer) in aliphatic hydrocarbon solvents yields mostly ethylene and lithium hydride (β elimination) as well as smaller amounts of ethane and lithium metal.[116] The mechanism for the production of ethane is evidently an intramolecular reaction of the ethyllithium aggregate and does not involve free ethyl radicals, as shown by the fact that photolysis in perdeuterocyclohexane yields no deuteroethane. When mercury is used as a photosensitizer, the major products are ethane and lithium amalgam, and if the solvent is fully deuterated the ethane is about 10–15% deuterated, indicating that production of ethyl radicals does occur under these conditions but is not the major reaction pathway.

Although the photolysis of dialkylmercuries generally produces free radicals, Ahlgren and co-workers have reported that photolysis of bis(cyanoethyl)mercury, $Hg(CH_2CH_2CN)_2$, yields adiponitrile, $NC(CH_2)_4CN$, apparently by a concerted process since radical coupling would be too inefficient.[117]

An example of photochemical ring closure has been found by Clark, Hancock and Zweifel, who irradiated the butadienylborane (**54**) obtained from isopropenylacetylene and dicyclohexylborane.[118]

To this miscellaneous collection of photochemical reactions might as well be added the observation that the chemiluminescence spectrum of phenylmagnesium bromide undergoing oxidation in benzene–triethylamine indicates that *p*-terphenyl is the light-emitting species.[119] Bardsley and Hercules

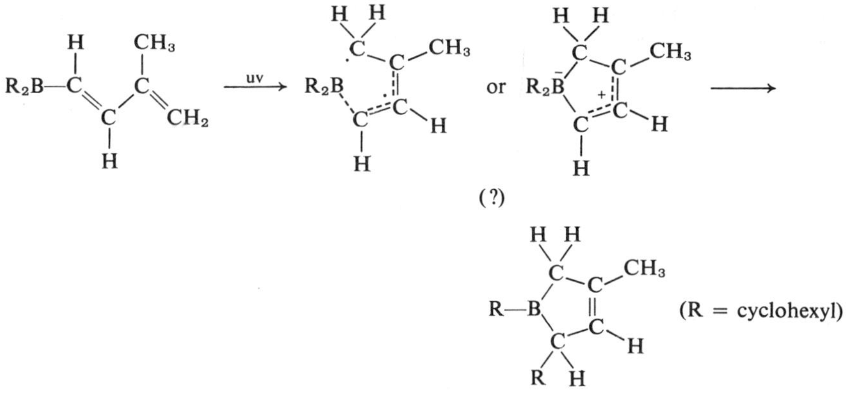

uv → or (?) →

(R = cyclohexyl)

suggested some ways in which the terphenyl might be produced in an excited state, but the evidence is confusing and any interpretation speculative.

From the examples presented in this section, it can be seen that development of the photochemistry of organometallic compounds is still in an exploratory phase, and there is not much firm evidence about mechanisms. Interpreting photochemical results is inherently more difficult than interpreting ordinary mechanistic studies, which is difficult enough. At present, mechanistic organometallic photochemistry may be described as a promising frontier particularly suitable for those who are eager to go out and get lost in the woods.

References

1. F. A. Paneth and W. Hofeditz, *Chem. Ber.* **62,** 1335 (1929).
2. G. A. Razuvaev, B. G. Zateev, and V. N. Myakov, *Proc. Akad. Sci. USSR* **154,** 37 (1964).
3. J. M. Blair and D. Bryce-Smith, *J. Chem. Soc.* 3174 (1959).
4. D. H. Hey, D. A. Shingleton, and G. H. Williams, *J. Chem. Soc.* 5612 (1963).
5. P. A. Leighton and R. A. Mortensen, *J. Amer. Chem. Soc.* **58,** 448 (1936).
6. C. Walling, "Free Radicals in Solution." Wiley, New York, 1957.
7. W. A. Pryor, "Free Radicals." McGraw-Hill, New York, 1966.
8. D. E. Applequist and R. Searle, *J. Amer. Chem. Soc.* **82,** 148 (1964).
9. K. Chakravorty, J. M. Pearson, and M. Szwarc, *J. Amer. Chem. Soc.* **90,** 283 (1968).
10. O. Dobis, J. M. Pearson, and M. Szwarc, *J. Amer. Chem. Soc.* **90,** 278 (1968).
11. F. R. Jensen, L. D. Whipple, D. K. Wedegaertner, and J. A. Landgrebe, *J. Amer. Chem. Soc.* **82,** 2466 (1960).
12. F. R. Jensen and L. H. Gale, *J. Amer. Chem. Soc.* **82,** 148 (1960).
13. F. R. Jensen, L. H. Gale, and J. E. Rodgers, *J. Amer. Chem. Soc.* **90,** 5793 (1968).
14. G. S. Hammond, *J. Amer. Chem. Soc.* **77,** 334 (1955).
15. D. J. Pasto and J. A. Gontarz, *J. Amer. Chem. Soc.* **91,** 719 (1969).
16. G. A. Gray and W. R. Jackson, *J. Amer. Chem. Soc.* **91,** 6205 (1969).

17. G. M. Whitesides and J. San Filippo, Jr., *J. Amer. Chem. Soc.* **92**, 6611 (1970).
18. F. G. Bordwell and M. L. Douglass, *J. Amer. Chem. Soc.* **88**, 993 (1966).
19. P. J. Krusic and J. K. Kochi, *J. Amer. Chem. Soc.* **91**, 3942 (1969).
20. A. G. Davies and B. P. Roberts, *J. Organometal. Chem.* **19**, P17 (1969).
21. E. A. Lissi and E. Sanhueza, *J. Organometal. Chem.* **32**, 285 (1971).
22. D. S. Matteson, *J. Org. Chem.* **29**, 3399 (1964).
23. A. G. Davies and B. P. Roberts, *J. Chem. Soc. B* 17 (1967).
24. P. G. Allies and P. B. Brindley, *Chem. Ind.* (*London*) 319 (1967).
25. S. B. Mirviss, *J. Org. Chem.* **32**, 1713 (1967).
26. K. U. Ingold, *J. Chem. Soc. D* 911 (1969).
27. A. G. Davies and B. P. Roberts, *J. Chem. Soc. B* 1074 (1968).
28. H. C. Brown and M. M. Midland, *J. Chem. Soc. D* 699 (1971).
29. M. M. Midland and H. C. Brown, *J. Amer. Chem. Soc.* **93**, 1506 (1971).
30. A. Suzuki, S. Nozawa, M. Harada, M. Itoh, H. C. Brown, and M. M. Midland, *J. Amer. Chem. Soc.* **93**, 1508 (1971).
31. H. C. Brown and M. M. Midland, *J. Amer. Chem. Soc.* **93**, 3291 (1971).
32. G. W. Kabalka, H. C. Brown, A. Suzuki, S. Honma, A. Arase, and M. Itoh, *J. Amer. Chem. Soc.* **92**, 710 (1970).
33. A. Suzuki, A. Arase, H. Matsumoto, M. Itoh, H. C. Brown, M. M. Rogić, and M. W. Rathke, *J. Amer. Chem. Soc.* **89**, 5708 (1967).
34. H. C. Brown, M. M. Rogić, M. W. Rathke, and G. W. Kabalka, *J. Amer. Chem. Soc.* **89**, 5709 (1967).
35. H. C. Brown, G. W. Kabalka, M. W. Rathke, and M. M. Rogić, *J. Amer. Chem. Soc.* **90**, 4165 (1968).
36. H. C. Brown, M. W. Rathke, G. W. Kabalka, and M. M. Rogić, *J. Amer. Chem. Soc.* **90**, 4166 (1968).
37. H. C. Brown, M. M. Rogić, M. W. Rathke, and G. W. Kabalka, *J. Amer. Chem. Soc.* **91**, 2150 (1969).
38. H. C. Brown and G. W. Kabalka, *J. Amer. Chem. Soc.* **92**, 712 (1970).
39. H. C. Brown and G. W. Kabalka, *J. Amer. Chem. Soc.* **92**, 714 (1970).
40. A. Suzuki, S. Nozawa, M. Itoh, H. C. Brown, G. W. Kabalka, and G. W. Holland, *J. Amer. Chem. Soc.* **92**, 3503 (1970).
41. A. G. Davies, S. C. W. Hook, and B. P. Roberts, *J. Organometal. Chem.* **22**, C11 (1970).
42. A. G. Davies, S. C. W. Hook, and B. P. Roberts, *J. Organometal. Chem.* **22**, C37 (1970).
43. C. F. Lane, *J. Organometal. Chem.* **31**, 421 (1971).
44. S. Boué, M. Gielen, and J. Nasielski, *J. Organometal. Chem.* **9**, 461 (1967).
45. K. Sisido, T. Miyanisi, T. Isida, and S. Kozima, *J. Organometal. Chem.* **23**, 117 (1970).
46. J. K. Kochi and P. J. Krusic, *J. Amer. Chem. Soc.* **91**, 3944 (1969).
47. A. G. Davies and S. C. W. Hook, *J. Chem. Soc. B* 735 (1970).
48. H. R. Ward, *J. Amer. Chem. Soc.* **89**, 5517 (1967).
49. S. A. Kandil and R. E. Dessy, *J. Org. Chem.* **30**, 3857 (1965); S. A. Kandil and R. E. Dessy, *J. Amer. Chem. Soc.* **88**, 3027 (1966).
50. H. R. Ward and R. G. Lawler, *J. Amer. Chem. Soc.* **89**, 5518 (1967).
51. R. G. Lawler, *J. Amer. Chem. Soc.* **89**, 5519 (1967).
52. G. L. Closs, *J. Amer. Chem. Soc.* **91**, 4552 (1969).
53. R. Kaptein and L. J. Oosterhoff, *Chem. Phys. Lett.* **4**, 195, 214 (1969).

54. J. I. Morris, R. C. Morrison, D. W. Smith, and J. F. Garst, *J. Amer. Chem. Soc.* **94**, 2406 (1972).
55. H. R. Ward, R. G. Lawler, H. Y. Loken, and R. A. Cooper, *J. Amer. Chem. Soc.* **91**, 4928 (1969).
56. H. R. Ward, R. G. Lawler, and R. A. Cooper, *J. Amer. Chem. Soc.* **91**, 746 (1969).
57. A. R. Lepley and R. L. Landau, *J. Amer. Chem. Soc.* **91**, 748 (1969).
58. A. R. Lepley, *J. Amer. Chem. Soc.* **91**, 749 (1969).
59. A. R. Lepley, *J. Amer. Chem. Soc.* **90**, 2710 (1968).
60. H. R. Ward, R. G. Lawler, and H. Y. Loken, *J. Amer. Chem. Soc.* **91**, 7359 (1968).
61. G. Screttas and J. F. Eastham, *J. Amer. Chem. Soc.* **88**, 5668 (1966).
62. J. F. Garst and R. H. Cox, *J. Amer. Chem. Soc.* **92**, 6389 (1970).
63. J. F. Garst, J. T. Barbas, and F. E. Barton, II, *J. Amer. Chem. Soc.* **90**, 7159 (1968).
64. G. D. Sargent and G. A. Lux, *J. Amer. Chem. Soc.* **90**, 7160 (1968).
65. T. C. Cheng, L. Headley, and A. F. Halasa, *J. Amer. Chem. Soc.* **93**, 1502 (1971).
66. L. H. Sommer, E. W. Pietrusza, and F. C. Whitmore, *J. Amer. Chem. Soc.* **69**, 188 (1947).
67. C. Eaborn, M. R. Harrison, and D. R. M. Walton, *J. Organometal. Chem.* **31**, 43 (1971).
68. S. W. Bennett, C. Eaborn, and R. A. Jackson, *J. Organometal. Chem.* **21**, 79 (1970).
69. P. J. Krusic and J. K. Kochi, *J. Amer. Chem. Soc.* **91**, 3938 (1969).
70. P. T. Frangopol and K. U. Ingold, *J. Organometal. Chem.* **25**, C9 (1970).
71. H. Sakurai, M. Murakami, and M. Kumada, *J. Amer. Chem. Soc.* **91**, 519 (1969).
72. A. G. Brook and J. M. Duff, *J. Amer. Chem. Soc.* **91**, 2118 (1969).
73. H. Sakurai and A. Hosomi, *J. Amer. Chem. Soc.* **93**, 1709 (1971).
74. G. J. M. van der Kerk, J. G. A. Luijten, and J. G. Noltes, *Chem. Ind.* (*London*) 352 (1956).
75. H. G. Kuivila and R. Sommer, *J. Amer. Chem. Soc.* **89**, 5616 (1967).
76. T. Kawamura and J. K. Kochi, *J. Organometal. Chem.* **30**, C8 (1971).
77. R. H. Fish, H. G. Kuivila, and I. J. Tyminski, *J. Amer. Chem. Soc.* **89**, 5861 (1967).
78. D. J. Carlsson and K. U. Ingold, *J. Amer. Chem. Soc.* **90**, 1055 (1968).
79. D. J. Carlsson and K. U. Ingold, *J. Amer. Chem. Soc.* **90**, 7047 (1968).
80. L. J. Altman and B. W. Nelson, *J. Amer. Chem. Soc.* **91**, 5163 (1969).
81. T. Ando, F. Namigata, H. Yamanaka, and W. Funasaka, *J. Amer. Chem. Soc.* **89**, 5719 (1967).
82. P. J. Krusic and J. K. Kochi, *J. Amer. Chem. Soc.* **91**, 6161 (1969).
83. D. S. Boak and B. G. Gowenlock, *J. Organometal. Chem.* **29**, 385 (1971).
84. J. W. Wilt, O. Kolewe, and J. F. Kraemer, *J. Amer. Chem. Soc.* **91**, 2624 (1969).
85. C. G. Pitt and M. S. Fowler, *J. Amer. Chem. Soc.* **90**, 1928 (1968).
86. R. Schaeffer and L. J. Todd, *J. Amer. Chem. Soc.* **87**, 488 (1965).
87. C. F. Lane and H. C. Brown, *J. Amer. Chem. Soc.* **92**, 7212 (1970).
88. C. F. Lane and H. C. Brown, *J. Amer. Chem. Soc.* **93**, 1025 (1971).
89. J. Grotewold, E. A. Lissi, and J. C. Scaiano, *J. Organometal. Chem.* **19**, 431 (1969).
90. J. C. Scaiano, J. Grotewold, and E. A. Lissi, *J. Chem. Soc. B* 1187 (1971).
91. D. S. Matteson, *J. Amer. Chem. Soc.* **82**, 4228 (1960).
92. P. J. Krusic and J. K. Kochi, *J. Amer. Chem. Soc.* **93**, 846 (1971).
93. F. R. Jensen and H. E. Guard, *J. Amer. Chem. Soc.* **90**, 3251 (1968).
94. A. N. Nesmeyanov, A. E. Borisov, E. T. Golubeva, and A. I. Kovredov, *Tetrahedron* **18**, 683 (1962).

95. A. E. Borisov, *Izv. Akad. Nauk SSSR, Otdel. Khim. Nauk* 524 (1951).
96. B. G. Ramsey, "Electronic Absorption of Organometalloids." Academic Press, New York, 1969.
97. M. V. Encina and E. A. Lissi, *J. Organometal. Chem.* **29**, 21 (1971).
98. A. G. Davies, B. P. Roberts, and J. C. Scaiano, *J. Chem. Soc. B* 2171 (1971).
99. K. Shen, W. E. McEwen, and A. P. Wolf, *J. Amer. Chem. Soc.* **91**, 1283 (1969).
100. E. G. Janzen and B. J. Blackburn, *J. Amer. Chem. Soc.* **91**, 4481 (1969).
101. H. G. Kuivila, K.-H. Tsai, and P. L. Maxfield, *J. Amer. Chem. Soc.* **92**, 6696 (1970).
102. H. G. Kuivila and P. L. Maxfield, *J. Organometal. Chem.* **10**, 41 (1967).
103. K. G. Hancock and A. K. Uriarte, *J. Amer. Chem. Soc.* **92**, 6374 (1970).
104. T. DoMinh, H. E. Gunning, and O. P. Strausz, *J. Amer. Chem. Soc.* **89**, 6785 (1967).
105. O. P. Strausz, T. DoMinh, and J. Font, *J. Amer. Chem. Soc.* **90**, 1930 (1968).
106. J. Connor, P. J. Young, and O. P. Strausz, *J. Amer. Chem. Soc.* **93**, 822 (1971).
107. E. C. Taylor, F. Kienzle, and A. McKillop, *J. Amer. Chem. Soc.* **92**, 6088 (1970).
108. J. G. Calvert and J. N. Pitts, Jr., "Photochemistry." Wiley, New York, 1966.
109. A. G. Brook and J. M. Duff, *J. Amer. Chem. Soc.* **89**, 454 (1967).
110. J. L. R. Williams, J. C. Doty, P. J. Grisdale, R. Searle, T. H. Regan, G. P. Happ, and D. P. Maier, *J. Amer. Chem. Soc.* **89**, 5153 (1967).
111. J. L. R. Williams, J. C. Doty, P. J. Grisdale, T. H. Regan, G. P. Happ, and D. P. Maier, *J. Amer. Chem. Soc.* **90**, 53 (1968).
112. J. C. Doty, P. J. Grisdale, T. R. Evans, and J. L. R. Williams, *J. Organometal. Chem.* **32**, C35 (1971).
113. J. J. Eisch and J. L. Considine, *J. Amer. Chem. Soc.* **90**, 6257 (1968).
114. E. E. van Tamelen, J. I. Brauman, and L. E. Ellis, *J. Amer. Chem. Soc.* **93**, 6141 (1971).
115. W. H. Glaze, T. L. Brewer, and A. C. Ranade, *J. Organometal. Chem.* **25**, C7 (1970).
116. W. H. Glaze and T. L. Brewer, *J. Amer. Chem. Soc.* **91**, 4490 (1969).
117. G. Ahlgren, B. Åkermark, and M. Nilsson, *J. Organometal. Chem.* **30**, 303 (1971).
118. G. M. Clark, K. G. Hancock, and G. Zweifel, *J. Amer. Chem. Soc.* **93**, 1308 (1971).
119. R. L. Beardsley and D. M. Hercules, *J. Amer. Chem. Soc.* **90**, 4545 (1968).

CHAPTER 8

Free Metals and Metal Anions

I. Introduction

The easy oxidation of most metals to cationic species dominates organometallic chemistry and underlies most of the chemistry in the first seven chapters, even the free-radical reactions often leading to oxidation products. However, anionic metal atoms can be obtained with some of the less electropositive elements, such as those of Group IV, and organometallic compounds of the more noble metals can decompose to yield the free metal. The reverse reaction, incorporation of a free metal and an organic halide into an organometallic compound, also falls in this chapter, except that little detailed mechanistic information is available on this topic. Organocopper compounds are particularly prone to decompose to the free metal, and more is known about this process than about the dissolution of more reactive metals. Accordingly, some organocopper chemistry is included for the insights it provides, even though it is on the edge of the transition series and formally outside the intended scope of this book.

II. Mercury Atoms and Mercury Surfaces

A. Solvolysis of Alkylmercury(II) Cations

Jensen and Ouellette found that alkylmercuric perchlorates, $RHg^+ClO_4^-$, undergo first-order solvolysis to the carbonium ion R^+, mercury metal, and perchlorate ion.[1] Alkylmercuric acetates, which ionize to form a small equilibrium concentration of RHg^+, solvolyze slowly. Alkylmercuric halides do not yield appreciable concentrations of RHg^+ and do not solvolyze. Nucleophilic solvents retard the solvolysis by coordinating with the RHg^+ cation and stabilizing it.[2] The strength of nucleophilic solvation of the carbonium ion center in the transition state is relatively small and the concentration of positive charge is very high, as shown by the extreme differences between the rates of solvolysis in acetic acid at 25°C: $MeHg^+$, 1.0 (relative rate); $EtHg^+$, 1.19×10^3; iso-$PrHg^+$, 4.1×10^7; *t*-$BuHg^+$, 6.6×10^{12}. However, the reaction does show considerable sensitivity to the nucleophilicity of the solvent, these relative rate differences becoming much smaller in ethanol.[3] Some net inversion was found in the solvolysis of optically active 2-butylmercuric ion in acetic acid, which provides further evidence for solvent participation. The high degree of positive charge on the developing carbonium ion in the transition state is perhaps related to the neutrality of the Hg^0 leaving group, which contrasts with the anionic leaving groups studied in most solvolyses. Jensen and Ouellette suggested that "rehybridization" of the mercury atom might also have something to do with the unusual characteristics of the reaction,[3] but orbital hybridization is a fiction in the first place (Chapter 1, Section II,C), and "rehybridization" is a meaningless term except insofar as it describes some sort of gross change in the bond angles around an atom, which hardly seems likely here. Nucleophilic participation by a neighboring phenyl group was also found to assist the reaction, $PhCH_2CH_2Hg^+$ solvolyzing 8.3 times faster than $HCH_2CH_2Hg^+$ in acetic acid at 75°C, 30 times faster in formic acid at 50°C.[4]

The first-order character of the decomposition implies that the rate-determining step involves generation of a solvated mercury atom, not deposition of a mercury atom onto a mercury surface, which would show an induction period and be accelerated by colloidal mercury.

Ouellette and co-workers have reported further applications of the solvolysis of RHg^+ to carbonium ion chemistry. For a series of ten polynuclear $ArCH_2Hg^+$ at 25°C in acetic acid, a plot of log k versus log k for solvolysis of the corresponding $ArCH_2Cl$ in 80% ethanol yields a linear correlation with a slope of 4.95, again illustrating the high degree of carbonium ion character in the transition state.[5] The solvolysis of $ArCH_2CH_2Hg^+$ correlates closely with that of $ArCH_2Hg^+$, a plot of log k for the latter versus log k for the former yielding a slope of 3.8 and a correlation coefficient of 0.996 with seven compounds.[6] This correlation provides strong evidence in favor of a phenonium ion or related intermediate (**1**) in the solvolysis of $ArCH_2CH_2Hg^+$. There is, of course, ample evidence for the phenonium ion from other studies not involving organometallic compounds.

The solvolysis of $Ar(CH_2)_4Hg^+$ proceeds by two competing paths, one involving internal assistance by the neighboring phenyl group to form a spiro

$C_6H_5CH_2CH_2Hg^+ \longrightarrow$ $H_2C—CH_2 + Hg^0$ (1) $\xrightarrow{HOAc}$ $C_6H_5CH_2CH_2OAc + H^+$

cation (**2**) and the other involving direct attack by solvent on carbon in the transition state (**3**).[7] The presence of two competing processes is supported by

$C_6H_5CH_2CH_2CH_2CH_2Hg^+ \longrightarrow$ (**2**) $+ Hg^0 \xrightarrow{HOAc}$ $C_6H_5CH_2CH_2CH_2CH_2OAc + H^+$

$C_6H_5CH_2CH_2CH_2CH_2Hg^+ \xrightarrow{HOAc}$ $Hg^{\delta+}$ ⋮ $C_6H_5CH_2CH_2CH_2CH_2{}^{\delta+}$ ⋮ HOAc (**3**) $\xrightarrow{-Hg^0}$ $C_6H_5CH_2CH_2CH_2CH_2OAc + H^+$

both the Arrhenius and the Hammett plots for the series of four compounds solvolyzed in acetic acid at 62°–112°C, and the solvent-assisted and internally assisted rate constants were obtained by computer analysis of the data. For $Ph(CH_2)_4Hg^+$, k for the reaction without phenyl assistance is 3.40×10^{-5} sec^{-1} at 100°C; ΔH^* is 35.7 kcal/mole, and ΔS^* is +16.8 eu. These values are similar to the results obtained with $C_4H_9Hg^+$, $k = 4.18 \times 10^{-5}$ sec^{-1}, $\Delta H^* = 35.5$ kcal/mole, and $\Delta S^* = +16.6$ eu. The phenyl-assisted reaction of $Ph(CH_2)_4Hg^+$ shows quite different parameters, $k = 1.47 \times 10^{-5}$ sec^{-1} at 100°C, $\Delta H^* = 28.5$ kcal/mole, and $\Delta S^* = -4.7$ eu. The other aryl-assisted reactions show similar ΔS^* values, and the ΔH^*-governed rate constants range from 1.0×10^{-5} sec^{-1} for *p*-chlorophenyl to 10.4×10^{-5} sec^{-1} for *p*-methoxyphenyl. These results are self-consistent and theoretically satisfying. Even so, one does not always win at referee roulette, and it was pointed out in a footnote, "Our original paper of six years ago was objected to so strenuously by one referee that we withdrew the manuscript."[7]

Winstein and co-workers have similarly separated other solvolyses into anchimerically assisted and solvent-assisted pathways, and Diaz and Winstein have used such interpretations in refuting arguments put forth by Brown in the well-known controversy over carbonium ion structures.[8]

B. Reaction of an Alkyl Iodide with Mercury

Mechanistic details are not known, but the reaction of methylene iodide with mercury metal is initiated by light and is evidently a free-radical process.[9] The reaction does not proceed with pure mercury and methylene iodide, but evidently requires mercurous iodide as a catalyst. A reasonable mechanism is outlined in Eqs. (8-1) through (8-4).

$$Hg_2I_2 \xrightarrow{\text{light}} 2HgI\cdot \qquad \text{(initiation)} \qquad (8\text{-}1)$$

$$HgI\cdot + CH_2I_2 \longrightarrow HgI_2 + ICH_2\cdot \qquad \text{(propagation)} \qquad (8\text{-}2)$$

$$ICH_2\cdot + Hg_2I_2 \longrightarrow ICH_2HgI + HgI\cdot \qquad \text{(propagation)} \qquad (8\text{-}3)$$

$$HgI_2 + Hg \longrightarrow Hg_2I_2 \qquad \text{(propagation)} \qquad (8\text{-}4)$$

It should be noted that this mechanism is peculiar to mercury and does not shed any light on mechanisms of reactions of other metals with alkyl halides.

C. Isotopic Exchange between Ar_2Hg and Hg

Reutov and Ostapchuk found that a variety of organomercury compounds will exchange mercury atoms with radioactively labeled mercury metal under mild conditions.[10] Arylmercury compounds generally work, and a few special alkyl groups also permit exchange. An optically active alkyl group retained its configuration in the exchange process.[11] Pollard and Westwood have extended these observations and shown that the rate is proportional to the amount of mercury surface present.[12,13] Working with surface reactions is inherently more difficult than running ordinary solution kinetics, and the data are consequently somewhat more scattered than typical solution data, but the results are convincing. The reaction is not diffusion-controlled.

Pollard and Westwood classified the reaction as "S_Ei" and wrote a four-center transition state (**4**). It is hard to see how mercury metal can be classified as an electrophile, since it has an unshared pair of 6*s* electrons that must become involved in the bonding, and if the four-center transition state were right it might be better be classified as "S_Ni," assuming one has to fit everything somehow into Ingold's preconceived categories. A symmetrical intermediate (**5**) has to be added to the Pollard and Westwood mechanism to satisfy requirements of microscopic reversibility. Both mercury atoms of **5** might be in the mercury surface or both might be out of it.

Ph—Hg—Ph + Hg (surface) ⇌ [Ph···Hg···Ph over surface Hg] **4** (doubtful) ⇌ [Hg, Ph, Ph, Hg ring] **5**

Kreevoy and Walters have provided a much more satisfying theoretical interpretation of this type of exchange reaction.[14] They verified the experimental findings of Pollard and Westwood, improved the measurements, and extended the measurements to new systems by the use of radioactively labeled organomercury compound, which permitted work at lower concentrations than did labeled mercury metal. They found the reactivity sequence ArHgI > ArHgBr > ArHgCl > ArHgOAc > Ar_2Hg, where Ar is *p*-$MeOC_6H_4$. The rate of exchange of *p*-$MeOC_6H_4HgBr$ with mercury in homogeneous solution in benzene (1.2×10^{-5} *M* is saturated at 25°C) yielded $k \cong 0.25$ liters/mole second, which at the concentrations used provides a much slower exchange than does the direct reaction of the organomercury compound with the mercury surface. The surface reaction is first order in ArHgBr and first order in the surface area.

Kreevoy and Walters concluded that the rate-determining step is some sort of chemisorption of the organomercurial onto the mercury surface. They suggested an electron transfer from ArHgBr to Hg, though it is hard to see why mere metallic mercury–mercury bonding would not suffice, with ArHgBr acting as an electron acceptor and being reduced to the mercurous state. Indeed, a mercurous compound (**6**) was postulated to be an intermediate in the exchange.[14]

Ar—Hg—Br + Hg (surface) ⇌ [Ar, Br on Hg···Hg surface] ⇌ Ar—Hg—Hg—Br **6**

The mercurous intermediate (**6**) probably remains adsorbed on the surface throughout the reaction, and the mercury atoms involved may remain part of the mercury surface. If the exchange is restricted to two mercury atoms at a time, it is necessary that both Ar and Br migrate from one mercury atom to the other in competing paths for the interconversion of ArBrHg—Hg and ArHg—HgBr in order to achieve isotopic exchange. Since $HgBr_2$ and an Hg surface react directly to produce BrHg—HgBr, which is thermodynamically stable in this case, it seems reasonable to postulate a thermodynamically unstable organomercurous intermediate in the exchange reactions involving ArHgBr to Ar_2Hg. It may also be noted that intramolecular migration of an alkyl group from one mercury atom to another is compatible with the observed retention of configuration,[11] and that this step is a type of electrophilic displacement.

The relative reactivities are also consistent with the organomercurous intermediate (**6**). The most stable organomercurous intermediate would be expected from ArHgI, which reacts fastest, and the least stable from Ar_2Hg, which reacts slowest. No correlation based entirely on electronegativities of the substituents is entirely self-consistent. Pollard and Westwood found that a Hammett correlation for various Ar_2Hg yielded a ρ of -0.5 per substituent (-1.0 for two aryl groups), which is a small effect insufficient to support their characterization of the reaction as an electrophilic displacement. The Hammett correlation is also contrary to the gross trend between ArHgX and Ar_2Hg, in which it has been estimated that the electron-withdrawing halogen increases the rate by three or four orders of magnitude.[14]

Kreevoy and Walters measured the activation parameters for the reaction of *p*-$MeOC_6H_4HgBr$ in toluene with a mercury surface. The value of ΔH^* is 8.1 ± 0.2 kcal/mole. To calculate ΔS^* it is necessary to assume a standard state for the transition state, which was arbitrarily taken to be 1 mole/cm^2. The value of ΔS^* then comes out to be -58.4 eu. This is not really as negative as it might seem at first glance, since there is a considerable contribution from the theoretical squeezing of 1 mole onto 1 square centimeter of surface to reach the hypothetical standard state. (A more realistic standard state for surface-bound materials would be 1 mole/km^2, which would be actually attainable in typical real cases and is very roughly equivalent to 1 mole/liter in terms of molecular space. By this criterion, ΔS^* would be -12.7 eu.) Kreevoy and Walters interpreted the ΔS^* value as being not particularly negative under the circumstances and suggested that considerable freedom of movement of the transition state along the surface was indicated. Such freedom could be attained by the electron transfer mechanism they suggested, or perhaps merely by metallic bonding of the mercury atom of ArHgBr to a cluster of mercury atoms at the mercury surface in the transition state.

In spite of the work of Kreevoy and Walters, Marshall and Pollard have persisted in calling the reaction "S_Ei" in a more recent publication.[15] (The rearrangement of an organomercurous intermediate would have electrophilic character, but this was not considered as a possibility.) Contrary to Kreevoy and Walters' conclusion that ArHgBr reacts faster than Ar_2Hg, Marshall and Pollard found that PhHgCl reacts only 1/7 as fast as Ph_2Hg in benzene. However, ΔH^* for the reaction of Ph_2Hg in benzene was reported to be 17.9 kcal/mole and ΔH^* for the reaction of PhHgCl only 14.3 kcal/mole, implying entropy control and not an electronic effect. These figures lose their credibility when it is noted that ΔH^* for Ph_2Hg in toluene is 9.2 kcal/mole. There is no reasonable explanation for such a large difference between similar solvents. A truly different solvent, pyridine, gave $\Delta H^* = 10.5$ kcal/mole. All the rates are within about two powers of 10 of each other. Evidently the rate constants observed were too inaccurate to yield meaningful activation parameters.

Marshall and Pollard have also observed that the reaction of Ph_2Hg in carbon tetrachloride with a mercury surface yields PhHgCl.[16] The rate is the same as that of incorporation of radioactive label from the mercury into Ph_2Hg and PhHgCl in solution. It was concluded that the CCl_4 solvates the "S_Ei" transition state for the exchange and diverts the reaction to produce PhHgCl. (It is wrong in principle to suggest that a transition state can be captured and diverted by another reactant, but this minor error can be corrected by replacing the words "transition state" by "intermediate.") Not considered was the likelihood that an organomercurous intermediate such as Ph—Hg—Hg—Ph would probably react rapidly with CCl_4 to produce PhHgCl and other fragments. It is hard to see why a transition state or intermediate requiring unusual partial bonding (the model of Pollard and co-workers) should be favored over a perfectly conventionally bonded organomercurous intermediate (the model of Kreevoy and Walters).

III. Thallium and Lead Compounds

Thallium and lead compounds resemble mercury compounds in being able to decompose to the free metal or a lower-valent compound of the metal, but mechanisms have not been well defined. Gilman and Haiduc obtained biphenyl and thallium metal from the thermal decomposition of triphenylthallium in xylene, or biphenyl and thallium benzoate if the decomposition was carried out in the presence of carbon dioxide.[17] A speculative radical chain mechanism would involve attack of Ph· on Ph_3Tl to form Ph—Ph and Ph_2Tl·, which could cleave to Ph· and PhTl, which could react with CO_2 or decompose to Ph· and Tl. McKillop *et al.* have found that biaryls result from the reaction of ArMgBr with TlBr, but $RTlBr_2$ and Tl are obtained from primary RMgBr and TlBr.[18] These results suggest an interesting area for mechanistic research.

Solvolytic cleavages of thallium(III) or lead(IV) compounds to thallium(I- or lead(II) and the carbonium ion evidently occur subsequent to cyclopropane ring openings (Chapter 4, Section VI,C) or oxythallations (Chapter 5, Section III,F). These are analogous to the solvolysis of RHg^+ to R^+ and Hg^0 (this chapter, Section II,A). Lead compounds can also yield the free metal by radical pathways (Chapter 7, Section I).

The acetolysis of the dilead compound Ph_3Pb—$PbPh_3$ and various ring-substituted derivatives has been studied by Kumar Das and Wells.[19] From thin-layer chromatography, the only compounds present during the reaction are Ph_3PbOAc, $Ph_2Pb(OAc)_2$, $Pb(OAc)_2$, and unchanged $Ph_3Pb—PbPh_3$. The stoichiometry is given by Eq. (8-5). A Hammett correlation with six

$$Ph_3Pb—PbPh_3 + 4\,HOAc \longrightarrow Ph_2Pb(OAc)_2 + Pb(OAc)_2 + 4\,PhH \quad (8\text{-}5)$$

compounds required the use of σ^+ for linearity and yielded $\rho = -2.3$. It appears that the rate-determining step is simple acid cleavage of a phenyl–lead bond, and that some acetoxydilead compound such as Ph_3Pb–$Pb(OAc)Ph_2$ or Ph_3Pb–$Pb(OAc)_3$ undergoes rapid disproportionation to lead(IV) and lead(II), probably by an internal migration of acetate ion. This disproportionation bears some analogy to the cleavage of RHg^+ to R^+ and Hg^0.

IV. Formation of Grignard and Lithium Reagents

Solid metal surfaces are complex, and reactions tend to take place at imperfections in the crystal lattice or sites where impurities are present. The interpretation of any data which might be gathered on such reactions is exceedingly difficult, to the point where a search of *Chemical Abstracts* for "Grignard reaction, mechanisms of" yields only (inappropriately classified) reactions of Grignard reagents with other compounds.

In spite of the difficulties, some features of the mechanism of reaction of alkyl halides with magnesium are known. Mechanistic studies prior to 1950 have been summarized by Kharasch and Reinmuth.[20] Electron transfer from the magnesium to the alkyl halide must be involved in some sense. The evidence appears to favor some sort of radical process, with the initial attack of the magnesium metal being on the halogen atom rather than the carbon of the alkyl halide. It is clear that the reaction does not involve any sort of S_N2 displacement of halide X^- from RX by Mg metal, since aryl halides react readily, and even the exceedingly hindered neophyl chloride, $PhCMe_2CH_2Cl$, yields a Grignard reagent.[21] Bridgehead halides such as 4-camphyl chloride react with lithium metal.[22]

Walborsky and Young have observed about 10–12% net retention of configuration in the reaction of optically active 1-bromo-1-methyl-2,2-diphenylcyclopropane (7) with magnesium in tetrahydrofuran or ether.[23] This result excludes free-radical intermediates, at least for the stereospecific portion of the reaction, but is compatible with a radical-pair mechanism. If the alkyl bromide RBr reacts with the magnesium surface to form a radical pair, $R\cdot$ $Br–Mg_n\cdot$, where electron loss from the magnesium is delocalized indefinitely among a number of metal atoms, then some of the $R\cdot$ may form an R—Mg linkage to the magnesium surface before it has time to racemize,

Br, CH_3, Ph, Ph → MgBr, CH_3, Ph, Ph

7 10–12% net retention

while most of the R· may have a long enough independent lifetime to racemize before forming the R—Mg link. A crucial stage in the reaction, not accessible to any experimental test so far, would then be the dissociation of RMgBr from the surface of the magnesium. This dissociation process would be actively aided by the solvent and might be the rate-limiting step in some or all reactions, hence the promotion of the reaction by ether or tetrahydrofuran (and the greater effectiveness of the latter).

Skell and Girard have condensed thermally generated magnesium atoms with alkyl halides at −196°C.[24] The initial product is a black matrix of unknown structure, which on warming rearranges to solvent-free alkylmagnesium halide. Perhaps the black intermediate is some sort of radical pair, R·Br–Mg· or R–Br·$^-$Mg·$^+$ or a resonance hybrid of structures of this sort. In any case, it is clear that the formation of RMgX is not a direct insertion of Mg into RX even with magnesium atoms, which must be much more reactive than a magnesium surface.

The coupling products R—R often obtained as by-products in the formation of RMgX are consistent with the idea of intermediate alkyl radicals bonded to the magnesium surface. The by-products are not consistent with wholesale formation of truly free radicals in the solution, which would lead to attack on the solvent and little formation of Grignard reagent.

The reaction of lithium metal with organic halides shows characteristics similar to the Grignard reaction. Letsinger showed that some optical activity is retained at low temperatures in hydrocarbon solvent.[25] The stereochemical preference is for retention, and this has found use in preparative chemistry with alkenyllithiums[26] and norbornyllithiums[27] (see Chapter 3, Section III,F).

Lithium is more reactive than magnesium toward alkyl halides, perhaps in part as a result of the higher oxidation potential, but other factors may contribute. The existence of alkyllithiums as tetramers or other aggregates (Chapter 1, Section III,F) probably facilitates separation from the lithium surface, reducing the need for assistance by an ethereal solvent. The higher stereoselectivity of the lithium reactions could result from more efficient capture of the alkyl radicals from the initially formed radical pairs. Lithium reagents formed in the presence of ethers are usually racemized, perhaps by pathways involving dissociation of the tetramers to dimers (Chapter 2, Section V,A) and further dissociation to carbanions.

There is, of course, much published information on the practical aspects of preparing both Grignard and lithium reagents, including factors such as activation of the metal surface, effects of impurities such as copper in magnesium or sodium in lithium, dilution of the alkyl halide, choice of solvent, and the like. Some of this has been summarized by Kharasch and Reinmuth.[20] The mechanistic insight provided by such information is limited, and therefore it will not be reviewed here. A full understanding of mechanisms of reactions at solid metal surfaces must await techniques that go beyond the methods customarily used by physical organic chemists.

V. Organocopper Chemistry

A. Reactions Yielding Copper Metal

Copper might be excluded from this book as a transition metal, and with that excuse the coverage here will be grossly incomplete, but some reactions of copper compounds are too interesting to ignore. Moreover, copper is a borderline transition metal, and organocopper(I) compounds show some relation to zinc(II) and mercury(II) compounds. A major point of difference is the easy oxidizability of copper(I) to copper(II), which provides some of the more distinctive and useful features of copper chemistry. The most compelling reason for including some copper chemistry here is that copper compounds provide some of the best understood examples of deposition of a free metal from organometallic compounds, and copper anions with strong nucleophilic properties are readily generated (Section V,B) and show behavior closely related to anions of nontransition metals.

Cairncross and Shepard have found that fluorinated arylcopper compounds such as C_6F_5Cu, *m*- and *p*-$CH_3C_6H_4Cu$, and *m*- and *p*-FC_6H_4Cu are stable, isolable solids which form biaryls and copper metal on heating.[28] These organocopper compounds are octameric, evidently with a polyhedron of copper atoms at the center.[29] Heating $(m\text{-}CF_3C_6H_4)_8Cu_8$ in benzene leads to decomposition according to the scheme outlined in Eqs. (8-6) and (8-7). The

$$Ar_8Cu_8 \longrightarrow Ar\text{—}Ar + Ar_6Cu_8 \tag{8-6}$$

$$Ar_6Cu_8 \longrightarrow Ar\text{—}Ar + Ar_4Cu_8 \longrightarrow \text{Cu metal} \tag{8-7}$$

first step [Eq. (8-6)] is an order of magnitude faster than the second [Eq. (8-7)]. Both are first-order decompositions. Free copper metal begins to appear only after the second pair of aryl groups has coupled. No cross-coupling with the solvent to form *m*-$CF_3C_6H_4$—C_6H_5 could be detected, which indicates that free radicals are not intermediates in the coupling reaction. The intermediate species Ar_6Cu_8 was found to be diamagnetic and to have a different ^{19}F nmr spectrum from Ar_8Cu_8.

The ease of generation of free copper from organocopper(I) compounds is evidently related to the existence of a preformed cluster of copper atoms, so that the free metal is not released as individual atoms but as stabilized metallic clusters. The reverse type of reaction, dissolving of a metal by an alkyl halide to form a Grignard or similar reagent, might also in some cases involve separation of clusters of alkylated metal atoms from the metal surface (as has already been suggested for organolithium reagents, Section IV). Copper is well known to promote reactions which dissolve more active metals, such as the reaction of zinc with methylene iodide (Chapter 6, Section II,A). Such promotion might be merely an electrochemical effect, but it also seems plausible that copper might enter into short-lived cluster compounds which aid in breaking away the organozinc reagent from the zinc surface, then regenerate the copper. This is, of course, pure speculation and would not be easy to test experimentally, but such possibilities should be kept in mind in studies of metal surface reactions.

Biaryl formation evidently involves the moving together of two aryl groups from separate bonding sites on the copper cluster. (Four-center bonding to triangular faces, as in alkyllithiums, is a likely structural feature of the aryl-copper octamers.) Such a mechanism would be expected to proceed with retention of configuration at carbon, not racemization, in systems other than arylcoppers. Whitesides and co-workers have observed retention in the formation of hexadienes from *cis*- and *trans*-1-propenylcopper(I) (**8** and **9**) prepared *in situ* from the corresponding propenyllithiums and cuprous iodide.[30] Silver compounds behave similarly. The stereospecificity excludes free radicals as intermediates.

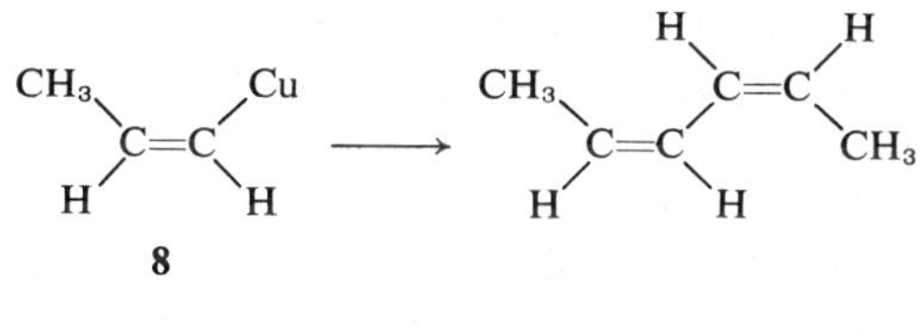

8

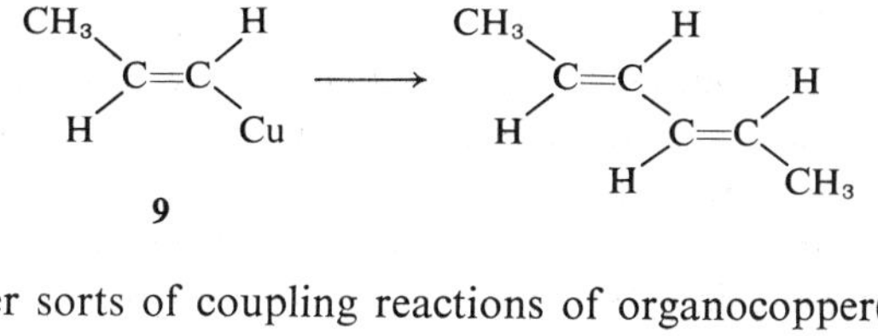

9

Several other sorts of coupling reactions of organocopper(I) compounds have been investigated, and none involve free radicals. Air oxidation of R_2CuLi yields R_2, and where R is $PhCMe_2CH_2$— no rearrangement occurs, indicating free radicals are not formed.[31] Reduction of $RCuPBu_3$ to RH, Cu, and PBu_3 by $HCuPBu_3$ proceeds with retention of configuration where R is

2-*endo*-norbornyl or *cis*- or *trans*-2-buten-2-yl, again excluding radical intermediates and implying that R and H merely migrate together on the surface of a cluster of copper atoms.[32] The thermal decomposition of $C_4H_9CuPR_3$ yields C_4H_{10}, $C_2H_5CH{=}CH_2$, Cu, and PR_3.[33] The mechanism was shown to involve elimination of $HCuPR_3$ from one molecule of $C_4H_9CuPR_3$ to form $C_2H_5CH{=}CH_2$, followed by reaction of the $HCuPR_3$ with a second molecule of $C_4H_9CuPR_3$ to form C_4H_{10}, Cu, and PR_3, with neither step involving free radicals. Tamura and Kochi have studied the copper(I)-catalyzed coupling of RMgBr and RBr and shown that radicals are not intermediates.[34] Again, the mechanism appears to involve formation and intramolecular rearrangement of $(RCu)_n$, where n is about 4 to 8. Silver(I) catalysis of the coupling of RMgBr and RBr behaves differently and does proceed by way of free radicals.[35]

B. Copper Anions

The anion Me_2Cu^-, if it exists as such, is isoelectronic with Me_2Zn. However, in view of the tendency of organocopper compounds to form clusters and the known formation of methyl-bridged complexes in such compounds as $MgAl_2Me_8$ (Chapter 1, Section III,C, Fig. 1-5) and $LiBMe_4$ (Chapter 1, Section III,E, it seems reasonable to formulate Me_2CuLi as a dimer, $Me_4Cu_2Li_2$, having a tetrahedron of copper and lithium atoms in the center with a methyl group bonded to each of the four triangular faces, analogous to the structure of methyllithium tetramer (Chapter 1, Section III,F).[36]

House and Umen have interpreted the β-methylation of α,β-unsaturated ketones and esters by Me_2CuLi on the basis of an electron-transfer mechanism.[36] If the α,β-unsaturated carbonyl compound will accept an electron at a potential less negative than -2.4 volts, Me_2CuLi will react with it. Otherwise, the carbonyl compound is recovered unchanged. If the reduction potential for adding the second electron is more negative than about -1.2 volts, the reaction is β-methylation, and if less negative (one example found), the result is reduction to the dilithium enolate instead. The postulated intermediate copper compound **10** is unlikely to be right in all its details, but the actual structure of the intermediate probably contains some of the essential structural features of **10**.

It should be pointed out that the proposed electron transfer preceding the formation of intermediate **10** cannot occur without being accompanied by some significant changes in the positions of atomic nuclei, that is, bond lengths and angles. In fact, it seems probable that the formation of an anion radical from $RCH{=}CHCO_2Et$ would be accompanied directly by transfer of Li^+ to oxygen from $(Me_2CuLi)_2$. The copper–lithium polyhedron might be

$(\text{Me}_2\text{CuLi})_2$ [Li, Li, Cu, Cu polyhedron with four Me] + $RCH{=}CHCO_2Et$ ⟶ [Li, Li, Cu, Cu polyhedron with four Me]$^{\dot{+}}$ $RCH{-}CH{-}CO_2Et^{\dot{-}}$ (?) ⟶

[Li^+, Li, Cu, Cu polyhedron with four Me] with Cu bonded to $CH(R){-}CH{=}C(O^-){-}OEt$

10

$\xrightarrow[\text{Li}^+\text{ transfer}]{\text{Me migration}}$ $(Me)(R)CH{-}CH{=}C(O^-Li^+){-}OEt + Me_2CuLi + MeCu$ (dimerize, etc.)

$\downarrow$ H^+, H_2O

$(Me)(R)CH{-}CH_2CO_2Et$

reduced to a Cu_2Li triangle in the intermediate **10**, or the polyhedron might be expanded by involvement of an additional molecule of $(Me_2CuLi)_2$. Whatever the complications, it seems probable that the copper–carbon bond and some sort of methylated polyhedral structure are present in the intermediate.

If the mechanism illustrated with intermediate **10** is reduced to its barest elements, the unit Me_2Cu^- adds to $RCH{=}CHCO_2Et$ to form Me_2Cu—CHR—$CHCO_2Et^-$, which then rearranges to MeCu and the enolate ion Me—CHR—$CHCO_2Et^-$. The Me_2Cu^- is postulated to be part of some larger aggregate at all times. Electron transfer (probably accompanied by gross structural change) prior to the formation of the intermediate Me_2Cu—CHR—$CHCO_2Et^-$ is postulated to explain the fact that the success of the reaction depends on reduction potential of the carbonyl compound and not on steric hindrance.

There are other reactions in which Ph_2Cu^- seems to function as a nucleophile, at least in a formal sense. Whitesides, House, and co-workers have found that optically active 2-BuBr and Ph_2CuLi yield 2-Bu—Ph with 84–92% net inversion at the 2-butyl carbon.[37] It is not known whether there is an intermediate 2-Bu—$CuPh_2$, in which the $CuPh_2$ group would probably be part of the usual sort of copper–lithium polyhedron, but it is certainly hard to imagine any other likely path. It is implausible that the copper reagent would just pop off a phenyl anion and ram it into place to cause a displacement of Br^- by Ph^-. It seems likely that the electron-transfer abilities of the organocopper reagent do have something to do with the result, or perhaps the ability of copper(I) to complex with bromide ion promotes the reaction.

In the coupling of RMgBr and R′Br by copper(I) salts, the relative reactivities of various R′ groups suggest that nucleophilic displacement of bromide (presumably by some form of R_2Cu^-) is involved in the rate-determining step.[32] However, nucleophilic displacement by R_2Cu^- cannot be the sole explanation of all couplings of alkyl halides with organocopper reagents in view of the observation by Corey and Posner that vinyl and aryl halides (R—X) as well as alkyl halides couple readily with Bu_2CuLi to form Bu—R in ether at −95°C to 0°C.[38] Again, electron transfer and probably complexing of the halide ion by copper(I) appear to be implicated.

This brief account of organocopper chemistry raises many more questions than it answers. Perhaps some of the anomalies or possibilities will catch the reader's fancy, and he may come up with a research idea that will open up more of this enticing frontier.

VI. Group IV Anions

A. Nucleophilic Displacements

Jensen and Davis have found that the anions Ph_3M^-, where M is Si, Ge, Sn, or Pb, displace bromide ion from 2-butyl bromide to form Ph_3Me-2-Bu with 50–90% net inversion of the 2-butyl carbon.[39] The stereochemistry was confirmed by correlations of optical rotations. Traylor and co-workers have found that bridgehead halides, which require retention, react readily with Me_3SnLi.[40] Kuivila and co-workers have reported that retention is favored in displacements on *syn*-7-bromonorbornene by Me_3SnNa or Me_3SnK unless the tight ion pairs are broken up by complexing the alkali metal cation with an ether such as tetraglyme, which results in inversion.[41] The inversion mechanism is probably an S_N2 process, and the retention mechanism probably involves some sort of electron transfer, most likely the reaction of R_3SnNa with R′Br to form R_3SnBr and R′Na, which then condense to R_3SnR'.

Pyridinebis(dimethylglyoximato)cobalt(I), which has all the *d* orbitals of the transition metal filled by back-bonding with the ligands, is better behaved as a nucleophile than the Group IV metal anions and inverts substituted cyclohexyl bromides or tosylates to form cyclohexylcobalt(III) derivatives.[42]

B. Reactions of Cl_3Si^-.

Benkeser and co-workers have discovered that $HSiCl_3$ reacts with tertiary amines under mild conditions to form Cl_3Si^-, which reacts readily with carbonyl compounds or alkyl halides to form silicon–carbon bonds. Synthetically useful examples include the conversion of PhCOPh to $PhCH(SiCl_3)Ph$ [with $(SiCl_2O)_n$ as the by-product], the conversion of $PhCH_2Cl$ to $PhCH_2SiCl_3$ and of $ArCHCl_2$ to $ArCH(SiCl_3)_2$, the conversion of EtCOCl to $EtCH(SiCl_3)_2$,[43] and the reduction of $ArCO_2H$ to $PhCH_2SiCl_3$.[44]

Acetonitrile appears to be the best solvent for most of these reactions. Solutions of $HSiCl_3$ and tripropylamine in acetonitrile appear from the proton nmr spectrum to consist of Pr_3NH^+ and Cl_3Si^- ions.[45] Bernstein found little or no deuterium isotope effect in the reaction of Bu_3N with $HSiCl_3$ in methylene chloride, suggesting that the rate-determining step is formation of a complex $Bu_3NSiHCl_3$ rather than the Si—H bond breaking itself.[46] Further evidence for the existence of Cl_3Si^- is provided by the preparation of solid $Me_3NH^+Cl_3Si^-$ by Ring and co-workers.[47] Benkeser has reviewed the work of his research group on Cl_3Si^-.[48]

VII. Notes on some Topics Not Covered

It is impossible in a book of finite length to cover every topic related in some way to organometallic reaction mechanisms. Some peripheral topics that have been omitted but might interest some readers are mentioned here.

Radical anions, often well characterized by esr spectra, have been generated by reduction of a variety of organometallic compounds, including mercury,[49,50] boron,[51–53] and silicon[54–58] compounds.

Carbon–metal π-bonding can strongly influence radical reactions (Chapter 7, Section VI,A) as well as the dienophilic properties of unsaturated compounds of boron,[59] silicon,[60] or tin.[61] Boron participates especially strongly in π-bonding, sufficiently to stabilize aromatic systems including borazaro compounds,[62] *B*-phenylborabenzene anion,[63] and a benzoborepin,[64] and to destabilize the antiaromatic pentaphenylborole.[65] There is also considerable evidence for carbanions stabilized by neighboring boron.[66,67] The author has reviewed carbon–boron π-bonding in detail elsewhere.[68]

Carborane chemistry has provided a variety of mechanistic studies. Most of these have been reviewed recently by Grimes.[69] Points of special interest include the mechanism of cage formation from decaborane and acetylene, which has been studied only indirectly by Hawthorne and co-workers,[70] and evidently involves dissociation of a decaborane–ligand complex in the rate-determining step. Rearrangement of icosahedral 1,2-carborane to 1,7-carborane proceeds by way of a cuboctahedral intermediate, which can also undergo a triangle rotation at higher temperatures to yield 1,12-carborane.[71] The author has scrutinized this work of Lipscomb and co-workers very carefully and has been totally unable to think of any other interpretation which would be self-consistent and in agreement with the complex experimental data. The reader who likes puzzles might like to try for himself. Finally, the useless but interesting rearrangement of 1,12-carborane to 1,7-carborane on reduction to the dianion followed by reoxidation[72] evidently involves a species in which the icosahedron has been opened to expose a distorted pentagonal face. Hawthorne and co-workers have captured the analogous dianion from 1,2-carborane as its cyclopentadienylcobalt(III) complex, which contains a larger-than-icosahedral 13-atom cage.[73]

VIII. Prognosis for Future Research

It should be readily apparent to the reader who has come this far that a wide variety of mechanisms of reactions of organometallic compounds have been discovered and are well understood. Areas of uncharted territory may be readily found in photochemistry, reactions at metal surfaces, the chemistry of highly reactive organometallic compounds, neighboring-group effects, and transition metal compounds. There could still be some surprises in worked-over areas such as electrophilic displacement, elimination, or free-radical chemistry, but many of the results should by now be qualitatively predictable.

The tools of the future are sophisticated instruments, such as nmr, esr, mass spectrometry, photoelectron spectroscopy, X-ray crystallography, and that universal assistant, the computer. The theory of the future is quantum theory. Experimental approaches are the only way to find out the gross mechanisms of complex reactions, but our only reasonable hope for obtaining a detailed description of a (necessarily simple) potential energy surface is to carry out accurate calculations with a computer.

Finally, it may be pointed out that as synthetic organometallic chemists learn how to make new types of structures, we can expect new types of reactions and new mechanistic insights. In turn, the mechanistic understanding we have already gained will be of considerable aid in designing new syntheses.

References

1. F. R. Jensen and R. J. Ouellette, *J. Amer. Chem. Soc.* **83**, 4477 (1961).
2. F. R. Jensen and R. J. Ouellette, *J. Amer. Chem. Soc.* **83**, 4478 (1961).
3. F. R. Jensen and R. J. Ouellette, *J. Amer. Chem. Soc.* **85**, 363 (1963).
4. F. R. Jensen and R. J. Ouellette, *J. Amer. Chem. Soc.* **85**, 367 (1963).
5. B. G. van Leuwen and R. J. Ouellette, *J. Amer. Chem. Soc.* **90**, 7056 (1968).
6. R. J. Ouellette and B. G. van Leuwen, *J. Amer. Chem. Soc.* **91**, 7061 (1968).
7. R. J. Ouellette, R. Papa, M. Attea, and C. Levin, *J. Amer. Chem. Soc.* **92**, 4893 (1970).
8. A. F. Diaz and S. Winstein, *J. Amer. Chem. Soc.* **91**, 4300 (1969).
9. E. P. Blanchard, Jr., D. C. Blomstrom, and H. E. Simmons, *J. Organometal. Chem.* **3**, 97 (1965).
10. O. A. Reutov and G. M. Ostapchuk, *Dokl. Akad. Nauk SSSR* **117**, 826 (1957).

11. O. A. Reutov, *Angew. Chem.* **72**, 198 (1960).
12. D. R. Pollard and J. V. Westwood, *J. Amer. Chem. Soc.* **87**, 2809 (1965).
13. D. R. Pollard and J. V. Westwood, *J. Amer. Chem. Soc.* **88**, 1404 (1966).
14. M. M. Kreevoy and E. A. Walters, *J. Amer. Chem. Soc.* **89**, 2986 (1967).
15. R. A. G. Marshall and D. R. Pollard, *J. Amer. Chem. Soc.* **92**, 6723 (1970).
16. R. A. G. Marshall and D. R. Pollard, *J. Organometal. Chem.* **25**, 287 (1970).
17. H. Gilman and I. Haiduc, *J. Amer. Chem. Soc.* **90**, 5912 (1968).
18. A. McKillop, L. F. Elsom, and E. C. Taylor, *J. Amer. Chem. Soc.* **90**, 2423 (1968).
19. V. G. Kumar Das and P. R. Wells, *J. Organometal. Chem.* **23**, 143 (1970).
20. M. S. Kharasch and O. Reinmuth, "Grignard Reactions of Nonmetallic Substances," pp. 20–22, 56–66. Prentice-Hall, Englewood Cliffs, New Jersey, 1954.
21. S. Winstein and T. G. Traylor, *J. Amer. Chem. Soc.* **77**, 3747 (1955).
22. S. Winstein and T. G. Traylor, *J. Amer. Chem. Soc.* **78**, 2597 (1956).
23. H. M. Walborsky and A. E. Young, *J. Amer. Chem. Soc.* **86**, 3288 (1964).
24. P. S. Skell and J. E. Girard, *J. Amer. Chem. Soc.* **94**, 5518 (1972).
25. R. L. Letsinger, *J. Amer. Chem. Soc.* **72**, 4842 (1950).
26. A. S. Dreiding and R. J. Pratt, *J. Amer. Chem. Soc.* **76**, 1902 (1954).
27. D. E. Applequist and G. N. Chmurny, *J. Amer. Chem. Soc.* **89**, 875 (1967).
28. A. Cairncross and W. A. Sheppard, *J. Amer. Chem. Soc.* **90**, 2187 (1968).
29. A. Cairncross and W. A. Sheppard, *J. Amer. Chem. Soc.* **93**, 247 (1971).
30. G. M. Whitesides, C. P. Casey, and J. K. Krieger, *J. Amer. Chem. Soc.* **93**, 1379 (1971).
31. G. M. Whitesides, J. San Filippo, Jr., C. P. Casey, and E. J. Panek, *J. Amer. Chem. Soc.* **89**, 5302 (1967).
32. G. M. Whitesides, J. San Filippo, Jr., E. R. Stedronsky, and C. P. Casey, *J. Amer. Chem. Soc.* **91**, 6542 (1969).
33. G. M. Whitesides, E. R. Stedronsky, C. P. Casey, and J. San Filippo, Jr., *J. Amer. Chem. Soc.* **92**, 1426 (1970).
34. M. Tamura and J. Kochi, *J. Amer. Chem. Soc.* **93**, 1485 (1971).
35. M. Tamura and J. Kochi, *J. Amer. Chem. Soc.* **93**, 1483 (1971).
36. H. O. House and M. J. Umen, *J. Amer. Chem. Soc.* **94**, 5495 (1972).
37. G. M. Whitesides, W. F. Fischer, Jr., J. San Filippo, Jr., R. W. Bashe, and H. O. House, *J. Amer. Chem. Soc.* **91**, 4871 (1969).
38. E. J. Corey and G. H. Posner, *J. Amer. Chem. Soc.* **90**, 5615 (1968).
39. F. R. Jensen and D. D. Davis, *J. Amer. Chem. Soc.* **93**, 4047 (1971).
40. G. S. Koermer, M. L. Hall, and T. G. Traylor, *J. Amer. Chem. Soc.* **94**, 7205 (1972).
41. H. G. Kuivila, J. L. Considine, and J. D. Kennedy, *J. Amer. Chem. Soc.* **94**, 7206 (1972).
42. F. R. Jensen, V. Madan, and D. H. Buchanan, *J. Amer. Chem. Soc.* **92**, 1414 (1970).
43. (a) R. A. Benkeser and W. E. Smith, *J. Amer. Chem. Soc.* **91**, 1556 (1969); (b) R. A. Benkeser, J. M. Gaul, and W. E. Smith, *ibid.* **91**, 3666 (1969); (c) R. A. Benkeser, K. M. Foley, J. M. Gaul, G. S. Li, and W. E. Smith, *ibid.* **91**, 4578 (1969).
44. R. A. Benkeser and R. M. Gaul, *J. Amer. Chem. Soc.* **92**, 720 (1970).
45. R. A. Benkeser, K. M. Foley, J. B. Grutzner, and W. E. Smith, *J. Amer. Chem. Soc.* **92**, 697 (1970).
46. S. C. Bernstein, *J. Amer. Chem. Soc.* **92**, 699 (1970).
47. M. A. Ring, R. L. Jenkins, R. Zanganeh, and Henry C. Brown, *J. Amer. Chem. Soc.* **93**, 265 (1971).
48. R. A. Benkeser, *Accounts Chem. Res.* **4**, 94 (1971).
49. R. E. Dessy, M. Kleiner, and S. C. Cohen, *J. Amer. Chem. Soc.* **91**, 6800 (1969).

50. J. L. Webb, C. K. Mann, and H. M. Walborsky, *J. Amer. Chem. Soc.* **92**, 2042 (1970).
51. H. J. Shine, L. D. Hughes, and P. Gesting, *J. Organometal. Chem.* **24**, 53 (1970).
52. S. D. Darling, O. N. Devgen, and R. E. Cosgrove, *J. Amer. Chem. Soc.* **92**, 696 (1970).
53. J. E. Leffler, G. B. Watts, T. Tanigaki, E. Dolan, and D. S. Miller, *J. Amer. Chem. Soc.* **92**, 6825 (1970).
54. E. Carberry, R. West, and G. E. Glass, *J. Amer. Chem. Soc.* **91**, 5446 (1969).
55. E. G. Janzen, J. B. Pickett, and W. H. Atwell, *J. Amer. Chem. Soc.* **90**, 2719 (1968).
56. E. G. Janzen, W. B. Harrison, and J. B. Pickett, *J. Organometal. Chem.* **20**, P13 (1969).
57. J. Eisch and G. Gupta, *J. Organometal. Chem.* **20**, P9 (1969).
58. A. L. Allred and L. W. Bush, *J. Amer. Chem. Soc.* **90**, 3352 (1968).
59. (a) D. S. Matteson and J. O. Waldbillig, *J. Org. Chem.* **28**, 366 (1963); (b) D. S. Matteson, *ibid.* **27**, 4293 (1962).
60. (a) D. Seyferth, D. R. Blank, and A. B. Evnin, *J. Amer. Chem. Soc.* **89**, 4793 (1968); (b) D. Seyferth and D. L. White, *ibid.* **94**, 3132 (1972); (c) D. Seyferth and D. L. White, *J. Organometal. Chem.* **34**, 119 (1972).
61. (a) A. B. Evnin and D. Seyferth, *J. Amer. Chem. Soc.* **89**, 952 (1967); (b) D. Seyferth and A. B. Evnin, *ibid.* **89**, 1468 (1967).
62. M. J. S. Dewar, *Progr. Boron Chem.* **1**, 235 (1964).
63. A. J. Ashe, III, and P. Shu, *J. Amer. Chem. Soc.* **93**, 1804 (1971).
64. A. J. Leusink, W. Drenth, J. G. Noltes, and G. J. M. van der Kerk, *Tetrahedron Lett.* 1263 (1967).
65. J. J. Eisch, N. K. Hota, and S. Kozima, *J. Amer. Chem. Soc.* **91**, 4575 (1969).
66. G. Zweifel and H. Arzoumanian, *J. Amer. Chem. Soc.* **89**, 291 (1967).
67. (a) R. B. Castle and D. S. Matteson, *J. Organometal. Chem.* **20**, 19 (1969); (b) D. S. Matteson and P. B. Tripathy, *ibid.* **21**, P6 (1970); (c) D. S. Matteson and J. R. Thomas, *ibid.* **24**, 263 (1970); (d) D. S. Matteson and G. L. Larson, *J. Amer. Chem. Soc.* **91**, 6541 (1969); (e) D. S. Matteson, L. A. Hagelee, and R. J. Wilcsek, *ibid.* **95**, 5096 (1973).
68. D. S. Matteson, *Progr. Boron Chem.* **3**, 117 (1970).
69. R. N. Grimes, "Carboranes." Academic Press, New York, 1970.
70. M. F. Hawthorne, R. L. Pilling, and R. C. Vasavada, *J. Amer. Chem. Soc.* **89**, 1075 (1967).
71. (a) H. D. Kaesz, P. Bau, H. A. Beall, and W. N. Lipscomb, *J. Amer. Chem. Soc.* **89**, 4218 (1967); (b) H. V. Hart and W. N. Lipscomb, *ibid.* **91**, 771 (1969).
72. L. I. Zakharkin and V. N. Kalinin, *Izv. Akad. Nauk SSSR, Ser. Khim.* 194 (1969).
73. G. B. Dunks, M. M. McKown, and M. F. Hawthorne, *J. Amer. Chem. Soc.* **93**, 2541 (1971).

Author Index

Numbers in parentheses are reference numbers and indicate that an author's work is referred to although his name is not cited in the text. Numbers in italics show the page on which the complete reference is listed.

H

I

J

K

L

S

W

Y

Z

Subject Index

A

B